Lecture Notes in Computer Science   12924

More information about this subseries at http://www.springer.com/series/7409

Christine Strauss · Gabriele Kotsis ·
A Min Tjoa · Ismail Khalil (Eds.)

Database and Expert Systems Applications

32nd International Conference, DEXA 2021
Virtual Event, September 27–30, 2021
Proceedings, Part II

 Springer

Editors
Christine Strauss
University of Vienna
Vienna, Austria

Gabriele Kotsis
Johannes Kepler University of Linz
Linz, Oberösterreich, Austria

A Min Tjoa
Vienna University of Technology
Vienna, Austria

Ismail Khalil
Johannes Kepler University of Linz
Linz, Austria

ISSN 0302-9743   ISSN 1611-3349   (electronic)
Lecture Notes in Computer Science
ISBN 978-3-030-86474-3   ISBN 978-3-030-86475-0   (eBook)
https://doi.org/10.1007/978-3-030-86475-0

LNCS Sublibrary: SL3 – Information Systems and Applications, incl. Internet/Web, and HCI

This Springer imprint is published by the registered company Springer Nature Switzerland AG
The registered company address is: Gewerbestrasse 11, 6330 Cham, Switzerland

Preface

The volume at hand represents the result of joint efforts of contributing researchers, reviewers, and organizers, and contains the papers presented at the 32nd International Conference on Database and Expert Systems Applications (DEXA 2021). This year, DEXA was held for the second time as a virtual conference during September 27–30, 2021, instead of in Linz, Austria, as originally planned. The decision to organize another virtual version of DEXA was driven by the intention to provide stable conditions for all DEXA participants and set a good example in temporarily suspending on-site meetings. We put our trust in the loyalty of DEXA community and look forward to personal DEXA meetings in 2022.

We are proud to report that authors from 43 different countries submitted papers to DEXA this year. The number of submissions was similar to those of the past few years. Our Program Committee conducted more than 500 reviews. We would like to sincerely thank our Program Committee members for their rigorous and critical, and at the same time motivating, reviews of these submissions. Based on the total number of accepted papers, we can report that the acceptance rate this year was 27%, a rate comparable to DEXA conferences of the last few years.

The conference program this year covered a wide range of important topics such as data management and analytics; consistency; integrity; quality of data; data analysis and data modeling; data mining; databases and data management; information retrieval; prediction and decision support; authenticity, privacy, security, and trust; cloud databases and workflows; data and information processing; knowledge discovery; machine learning; semantic web and ontologies; stream data processing; and temporal, spatial, and high dimensional databases.

We tried to follow our on-site face-to-face format. Thus, the authors of the accepted papers presented their research online using video conference software over four days. Presentations were performed live in 12 different thematic clusters structured as 15 sessions, each one with an assigned session chair. The scientific presentations, discussions, and question-and-answer time were all live and part of each session. As we were aware of time difference issues, for example, for participants from Australia or South American countries having to present or participate during unusual times of the day, we tried to minimize this inconvenience.

We would like to express our gratitude to the distinguished keynote speakers for illuminating us on their leading-edge topics: Elisa Bertino (Purdue University, USA) for her talk on "Privacy in the Era of Big Data, Machine Learning, IoT, and 5G", Amit Sheth (University of South Carolina, USA) for his talk on the third wave of AI, and Torben Bach Pedersen (Aalborg University, Denmark) for his talk on "Extreme-Scale Model-Based Time Series Management with ModelarDB".

In addition, we had a panel discussion on "Big Minds Sharing their Vision on the Future of AI" led by Bernhard Moser (SCCH, Austria), with Battista Biggio (University of Cagliari, Italy), Claudia Diaz (Katholieke Universiteit Leuven,

Belgium), Heiko Paulheim (University of Mannheim, Germany), and Olga Saukh (Complexity Science Hub, Austria).

As is the tradition of DEXA, all accepted papers were published in "Lecture Notes in Computer Science" (LNCS) and made available by Springer. Authors of selected papers presented at the conference will be invited to submit substantially extended versions of their conference papers for publication in special issues of international journals. The submitted extended versions will undergo a further review process.

The 32nd edition of DEXA featured six international workshops – three established ones and three brand-new ones – covering a variety of specific topics:

- The 12th International Workshop on Biological Knowledge Discovery from Data (BIOKDD 2021)
- The 5th International Workshop on Cyber-Security and Functional Safety in Cyber-Physical Systems (IWCFS 2021)
- The 3rd International Workshop on Machine Learning and Knowledge Graphs (MLKgraphs 2021)
- The 1st International Workshop on Artificial Intelligence for Clean, Affordable, and Reliable Energy Supply (AI-CARES 2021)
- The 1st International Workshop on Time Ordered Data (ProTime2021)
- The 1st International Workshop on AI System Engineering: Math, Modelling, and Software (AISys2021)

The success of the conference is due to the continuous and generous support of its participants and their relentless efforts. Our sincere thanks go to the dedicated authors, renowned Program Committee members, session chairs, organizing and steering committee members, and student volunteers who worked tirelessly to ensure the continuity and high quality of DEXA 2021.

We would also like to express our thanks to all institutions actively supporting this event, namely:

- Institute of Telekooperation, Johannes Kepler University Linz (JKU), Austria
- Software Competence Center Hagenberg (SCCH), Austria
- Web Applications Society (@WAS)

We hope you have enjoyed the conference! We are looking forward to seeing you again next year.

September 2021   Christine Strauss

Organization

Program Committee Chair

Christine Strauss   University of Vienna, Austria

Steering Committee

Gabriele Kotsis   Johannes Kepler University Linz, Austria
A Min Tjoa   Vienna University of Technology, Austria
Robert Wille   Software Competence Center Hagenberg, Austria
Bernhard Moser   Software Competence Center Hagenberg, Austria
Ismail Khalil   Johannes Kepler University Linz, Austria

Program Committee

Susan Ariel Aaronson	George Washington University, USA
Javier Nieves Acedo	Azterlan, Spain
Sonali Agarwal	IIIT, India
Hamid Aghajan	Ghent University, Belgium
Hans Akkermans	Vrije Universiteit Amsterdam, The Netherlands
Riccardo Albertoni	CNR-IMATI, Italy
Idir Amine Amarouche	USTHB, Algeria
Rachid Anane	Coventry University, UK
Mustafa Atay	Winston-Salem State University, USA
Sören Auer	Leibniz Universität Hannover, Germany
Juan Carlos Augusto	Middlessex University London, UK
Monica Barratt	RMIT University, Australia
Ladjel Bellatreche	LIAS, ENSMA, France
Nadia Bennani	LIRIS, INSA de Lyon, France
Karim Benouaret	Université Claude Bernard Lyon 1, France
Djamal Benslimane	Université de Lyon, France
Morad Benyoucef	University of Ottawa, Canada
Mikael Berndtsson	University of Skövde, Sweden
Catherine Berrut	LIG, Université Joseph Fourier, France
Vasudha Bhatnagar	University of Delhi, India
Didier Bigo	King's College London, UK
Steven Bird	Charles Darwin University, Australia
Ankur Singh Bist	KIET Ghaziabad, India
Joseph Bonneau	New York University, USA
Johan Bos	University of Groningen, The Netherlands
Athman Bouguettaya	University of Sydney, Australia
Olivier Bousquet	Google Brain, Zurich, Switzerland

Noura Faci   Université de Lyon, France
Hany Farid   Berkeley School of Information, USA
Bettina Fazzinga   ICAR-CNR, Rende, Italy
Stefano Ferilli   Universita' di Bari, Italy
Miriam Fernandez   Open University, UK
Flavio Ferrarotti   Software Competence Centre Hagenberg, Austria
Mariel Finucane   Mathematica Policy Research, Cambridge, USA
Seth Flaxman   Imperial College London, UK
Luciano Floridi   University of Oxford, UK
Vladimir Fomichov   National Research University Higher School
                             of Economics, Russia

Flavius Frasincar   Erasmus University Rotterdam, The Netherlands
Bernhard Freudenthaler   Software Competence Center Hagenberg, Austria
Andrea Fumagalli   Universita di Pavia, Italy
Steven Furnell   Plymouth University, UK
Aryya Gangopadhyay   University of Maryland Baltimore County, USA
David Garcia   Complexity Science Hub Vienna, Austria
Jorge Lloret Gazo   University of Zaragoza, Spain
David Geary   University of Missouri, USA
Claudio Gennaro   ISTI-CNR Pisa, Italy
George Gerard   Singapore Management University, Singapore
Manolis Gergatsoulis   Ionian University, Greece
Carlo Ghezzi   Politecnico di Milano, Italy
Javad Ghofrani   HTW Dresden University of Applied Sciences,
                             Germany
Dmitry Goldgof   University of South Florida, USA
Don Gotterbarn   Access East Tennessee State University, USA
Vikram Goyal   IIIT-Delhi, India
Carmine Gravino   University of Salerno, Italy
Sven Groppe   University of Lübeck, Germany
William Grosky   University of Michigan, USA
Francesco Guerra   Università di Modena e Reggio Emilia, Italy
Giovanna Guerrini   University of Genova, Italy
Allel Hadjali   LIAS, ENSMA, France
Abdelkader Hameurlain   IRIT, Paul Sabatier University, France
Ibrahim Hamidah   Universiti Putra Malaysia, Malaysia
Takahiro Hara   Osaka University, Japan
Lynda Hardman   The Centrum Wiskunde and Informatica,
                             The Netherlands
Eszter Hargittai   University of Zurich, Switzerland
Sven Hartmann   Clausthal University of Technology, Germany
Manfred Hauswirth   The Fraunhofer Institute for Open Communication
                             Systems FOKUS, Germany
Eva Heiskanen   University of Helsinki, Finland
Julio Hernandez-Castro   University of Kent, UK
Antonio Hidalgo   Universidad Politécnica de Madrid, Spain

Magdalena Hurtado	Arizona State University, USA
Ionut Iacob	Georgia Southern University, USA
Sergio Ilarri	University of Zaragoza, Spain
Abdessamad Imine	Loria, France
Yasunori Ishihara	Nanzan University, Japan
Ivan Izonin	Lviv Polytechnic National University, Ukraine
Peiquan Jin	University of Science and Technology of China, China
Deborah Johnson	University of Virginia, USA
Anne Kao	Boeing, USA
Dimitris Karagiannis	University of Vienna, Austria
Stefan Katzenbeisser	TU Darmstand, Germany
Anne Kayem	Hasso Plattner Institute, University of Potsdam, Germany
Deanna Kemp	University of Queensland, Australia
Faisal Khan	University of Calgary, Canada
Ewan Klein	University of Edinburgh, UK
Carsten Kleiner	University of Applied Science and Arts Hannover, Germany
Peter Knees	Vienna University of Technology, Austria
Henning Koehler	Massey University, New Zealand
Michal Kratky	VSB-Technical University of Ostrava, Czech Republic
Petr Kremen	Czech Technical University in Prague, Czech Republic
David Kreps	Stanford University, USA
Agnes Kukulska-Hulme	Open University, UK
Tahu Kukutai	University of Waikato, New Zealand
Josef Küng	Johannes Kepler University Linz, Austria
Nhien-An Le Khac	University College Dublin, Ireland
Lenka Lhotska	Czech Technical University in Prague, Czech Republic
Wenxin Liang	Chongqing University of Posts and Telecommunications, China
Chuan-Ming Liu	National Taipei University of Technology, Taiwan
Oscar Pastor Lopez	Universitat Politècnica de València, Spain
Hui Ma	Victoria University of Wellington, New Zealand
Qiang Ma	Kyoto University, Japan
Zakaria Maamar	Zayed University, UAE
Sanjay Madria	Missouri University of Science and Technology, USA
Elio Masciari	Federico II University, Italy
Brahim Medjahed	University of Michigan, USA
Jun Miyazaki	Tokyo Institute of Technology, Japan
Lars Moench	University of Hagen, Germany
Riad Mokadem	Paul Sabatier University, France
Anirban Mondal	University of Tokyo, Japan
Yang-Sae Moon	Kangwon National University, South Korea
Franck Morvan	IRIT, Paul Sabatier University, France
Cedric du Mouza	CNAM, France
Francesc Munoz-Escoi	Universitat Politècnica de València, Spain

External Reviewers

Tooba Aamir
Amani Abusafia
Abdulwahab Aljubairy
Mohammed Bahutair
Andrea Baraldi
Nabila Berkani
Francesco Del Buono
Loredana Caruccio
Olivier De Casanove
Dipankar Chaki
Rachid Chelouah
Stefano Cirillo
Labbe Cyril
Matthew Damigos
Jonathan Debure
Abir Farouzi
Sheik Mohammad Mostakim Fattah
Angelo Ferrando
Lukas Fischer
Arnaud Flori
Jorge Galicia
María del Carmen Rodríguez Hernández
Akm Tauhidul Islam
Eleftherios Kalogeros
Julius Köpke
Bogdan Kostov
Cyril Labbe
Chuan-Chi Lai
Hieu Hanh Le
Xuhong Li

Ji Liu
Jin Lu
Qiuhao Lu
Jia-Ning Luo
Jorge Martinez-Gil
Ahcene Menasria
Niccolo Meneghetti
Quoc Hung Ngo
Daria Novoseltseva
Matteo Paganelli
Louise Parkin
Gang Qian
Subhash Sagar
Nadouri Sana
Chayma Sellami
Mohamed Sellami
Vladimir A. Shekhovtsov
Tao Shi
Hannes Sochor
Manel Souibgui
Sofia Stamou
Carlos Telleria-Orriols
Daniele Traversaro
Oscar Urra
Francesco Visalli
Shuang Wang
Yi-Hung Wu
Fa Yao Yin
Feng Yu
Eric Zhang

Organizers

Abstracts of Keynote Talks

Privacy in the Era of Big Data, Machine Learning, IoT, and 5G

Elisa Bertino

Samuel Conte Professor of Computer Science, Cyber2SLab, Director,
CS Department, Purdue University, West Lafayette, Indiana, USA

Abstract. Technological advances, such as IoT devices, cyber-physical systems, smart mobile devices, data analytics, social networks, and increased communication capabilities are making possible to capture and to quickly process and analyze huge amounts of data from which to extract information critical for many critical tasks, such as healthcare and cyber security. In the area of cyber security, such tasks include user authentication, access control, anomaly detection, user monitoring, and protection from insider threat. By analyzing and integrating data collected on the Internet and the Web one can identify connections and relationships among individuals that may in turn help with homeland protection. By collecting and mining data concerning user travels, contacts and disease outbreaks one can predict disease spreading across geographical areas. And those are just a few examples. The use of data for those tasks raises however major privacy concerns. Collected data, even if anonymized by removing identifiers such as names or social security numbers, when linked with other data may lead to re-identify the individuals to which specific data items are related to. Also, as organizations, such as governmental agencies, often need to collaborate on security tasks, data sets are exchanged across different organizations, resulting in these data sets being available to many different parties. Privacy breaches may occur at different layers and components in our interconnected systems. In this talk, I first present an interesting privacy attack that exploits paging occasion in 5G cellular networks and possible defenses. Such attack shows that achieving privacy is challenging and there is no unique technique that one can use; rather one must combine different techniques depending also on the intended use of data. Examples of these techniques and their applications are presented. Finally, I discuss the notion of data transparency – critical when dealing with user sensitive data, and elaborate on the different dimensions of data transparency.

Don't Handicap AI without Explicit Knowledge

Amit Sheth

University of South Carolina, USA

Abstract. Knowledge representation as expert system rules or using frames and variety of logics, played a key role in capturing explicit knowledge during the hay days of AI in the past century. Such knowledge, aligned with planning and reasoning are part of what we refer to as Symbolic AI. The resurgent AI of this century in the form of Statistical AI has benefitted from massive data and computing. On some tasks, deep learning methods have even exceeded human performance levels. This gave the false sense that data alone is enough, and explicit knowledge is not needed. But as we start chasing machine intelligence that is comparable with human intelligence, there is an increasing realization that we cannot do without explicit knowledge. Neuroscience (role of long-term memory, strong interactions between different specialized regions of data on tasks such as multimodal sensing), cognitive science (bottom brain versus top brain, perception versus cognition), brain-inspired computing, behavioral economics (system 1 versus system 2), and other disciplines point to need for furthering AI to neuro-symbolic AI (i.e., hybrid of Statistical AI and Symbolic AI, also referred to as the third wave of AI). As we make this progress, the role of explicit knowledge becomes more evident. I will specifically look at our endeavor to support human-like intelligence, our desire for AI systems to interact with humans naturally, and our need to explain the path and reasons for AI systems' workings. Nevertheless, the variety of knowledge needed to support understanding and intelligence is varied and complex. Using the example of progressing from NLP to NLU, I will demonstrate the dimensions of explicit knowledge, which may include, linguistic, language syntax, common sense, general (world model), specialized (e.g., geographic), and domain-specific (e.g., mental health) knowledge. I will also argue that despite this complexity, such knowledge can be scalability created and maintained (even dynamically or continually). Finally, I will describe our work on knowledge-infused learning as an example strategy for fusing statistical and symbolic AI in a variety of ways.

Extreme-Scale Model-Based Time Series Management with ModelarDB

Torben Bach Pedersen

Aalborg University, Denmark

Abstract. To monitor critical industrial devices such as wind turbines, high quality sensors sampled at a high frequency are increasingly used. Current technology does not handle these extreme-scale time series well, so only simple aggregates are traditionally stored, removing outliers and fluctuations that could indicate problems. As a remedy, we present a model-based approach for managing extreme-scale time series that approximates the time series values using mathematical functions (models) and stores only model coefficients rather than data values. Compression is done both for individual time series and for correlated groups of time series. The keynote will present concepts, techniques, and algorithms from model-based time series management and our implementation of these in the open source Time Series Management System (TSMS) ModelarDB. Furthermore, it will present our experimental evaluation of ModelarDB on extreme-scale real-world time series, which shows that that compared to widely used Big Data formats, ModelarDB provides up to 14x faster ingestion due to high compression, 113x better compression due to its adaptability, 573x faster aggregation by using models, and close to linear scale-out scalability.

Big Minds Sharing their Vision on the Future of AI (Panel)

Panelists

Battista Biggio, University of Cagliari, Italy
Claudia Diaz, Katholieke Universiteit Leuven, Belgium
Heiko Paulheim, University Mannheim, Germany
Olga Saukh, Complexity Science Hub, Austria

Moderator

Bernhard Moser, Software Competence Center Hagenberg and Austrian Society
for Artificial Intelligence, Austria

Abstract. While we are currently mainly talking about narrow AI systems, in the future, neural networks will increasingly be combined with graph-based and symbolic-logical approaches (3rd wave of AI).

How will this technological trend affect the key issues of security such as integrity protection or privacy protection, and environmental impact? In this context, in this interactive panel discussion, technology experts will discuss current and envisioned challenges to AI from the research perspective of their respective fields.

Contents – Part II

Knowledge Discovery

Machine Learning

Contents – Part I

Data Mining

Databases and Data Management

Information Retrieval

Prediction and Decision Support

Authenticity, Privacy, Security and Trust

Less is More: Feature Choosing under Privacy-Preservation for Efficient Web Spam Detection

Jia-Qing Wang[1], Yan Zhu[1]([⊠]), Huan He[1], and Chun-Ping Li[2]

[1] Southwest Jiaotong University, Chengdu 611756, China
yzhu@swjtu.edu.cn
[2] School of Software, Tsinghua University, Beijing 100084, China

Abstract. Researches on detecting Web spam are in full swing. However, very high feature dimension and sensitive information leakage restrict the mining. In this paper, a cascade feature selection for mining spam is proposed, which bases on Privacy Preservation (PP) method and Genetic Algorithm (GA). Two criteria, privacy protection degree and maximum classification reliability, are used to pick the representative features to form an optimal minimum feature subset. Discretization, data balancing, feature selection, and ensemble learning method are integrated to detect Web spam. The approach not only greatly reduces the data dimension but also protects the sensitive features from detection. Good spam detection performance is achieved by using only 22 features.

Keywords: Web spam detection · Privacy Preservation · Feature selection · Genetic Algorithm

1 Introduction

Many researchers are working on detecting Web spam with data mining techniques. For example, Zhuang et al. [1] introduced deep belief network for Web spam demotion. Wang et al. [2] considered both time and network effect to detect collusive spammers and their activities. Dou et al. [3] proposed an algorithm using reinforcement learning for shallow graphs and behavior-based spam detectors. Wei et al. [4] proposed a cascade detection mechanism based on entropy-based outlier mining (EOM) approach, where 280 features including Web quality features and Web semantic features are used for improving performance. Nevertheless, performance improvement is often at the price of the high dimensionality issue and huge sensitive information disclosure risk. USCS [5] considered feature reduction in the Web spam detection system based on decision tree, clone selection and undersampling. However, no single optimal feature subset is generated for the detector and the feature number is still high to some extent.

In spam detection, sensitive information such as phone number or home address is at risk of being leaked. To minimize the risk, feature selection is crucial. Pattuk et al. [6] selected a set of specific features to generate view. However, choosing features during the

© Springer Nature Switzerland AG 2021
C. Strauss et al. (Eds.): DEXA 2021, LNCS 12924, pp. 3–8, 2021.
https://doi.org/10.1007/978-3-030-86475-0_1

view updating totally depends on the previous selected ones. Their algorithm is strongly dependent on the application domain. Zhang et al. [7] added differential privacy noise to the features, but the feature may not be effectively protected when the noise interference is small. Sheikhalishahi et al. [8] balanced privacy and utility by removing irrelevant features safely.

This paper proposes a cascade feature selection mechanism with sensitive feature protection (PPGAFS) for spam detection. Our major work is summarized as follows:

a) Hiding essential sensitive features is generally in conflict with improving detection performance. PPGAFS integrates the techniques such as entropy theory and genetic algorithm for solving the problems in feature selection.
b) Rough set is applied for discretization, which can obtain better detection quality than other traditional methods. Two uneven datasets are balanced by SMOTE.
c) A Web spam detection mechanism is constructed, which utilizes the ensemble learning method Random Forest (RF). The performance is superior to that of related algorithms in terms of fewer features and better privacy preservation.

2 The PPGAFS Approach

2.1 Preselecting Privacy-Preserving Features

The proposed algorithm (PPGAFS) consists of two cascade steps, preserving data privacy and selecting the optimal minimum feature subset. Its key idea is to hide the most sensitive features and at the same time disclose the necessary features (less sensitive or insensitive ones) for spam detection.

Conditional entropy is suitable to represent for the variable privacy degree under the increasing disclosed *feature set $Fsub = \{f_a, f_b, \ldots\}$*. Based on the definition of [6], the *privacy protection degree (privacy)* under F_{sub} is calculated with Eqs. (1–2).

$$privacy = -\sum_{i=1}^{n} P(C_i|Fsub) \log_2 P(C_i|Fsub) \tag{1}$$

$$P(C_i|Fsub) = NUM\,(drec,\, C_i|Fsub)\big/S \tag{2}$$

where C_i denotes the class label (normal or spammed), $n = 2$. $P(C_i|Fsub)$ means the probability of each class C_i under F_{sub}. NUM denotes the number of records of C_i under a certain F_{sub}, S is the number of all records under the same F_{sub}, and *drec* is data record.

As to feature f_j, the *expected confidence* is calculated by Eq. (3) in terms of its each value v_j. If the value v_{max} which produces maximal *expected confidence* is found and the values of f_j exactly contain v_{max}, f_j is picked into F_{sub}. F_{sub} should be updated with those features which enable the classifier to own the *maximum classification reliability (maximum confidence*, ref. Eq. (4)). Eqs. (3) and (4) are inspired by the relevant definitions of [6].

$$expected\ confidence = P_{v_j} \max P(C_i|F_{sub}) \tag{3}$$

$$confidence = \max P(C_i|F_{sub} \leftarrow f_j) \tag{4}$$

The *confidence threshold* ***tsc*** and the *privacy degree threshold* ***tsp*** are determined in advance through Random Forest (RF). In our PPFS-Algorithm, a feature subset F_{sub} named *PPFS* is generated, where features are filtered from the original feature set *FS* one by one. f is put into F_{sub} at end of each of iterations only when its values contain v_{max} and the criteria *privacy* and *confidence* meet *tsp* and *tsc*, respectively.

2.2 Generating Minimum Feature Subset Based on the Improved GA

The feature number of PPFS should be further reduced. Two points are studied to improve Genetic Algorithm (GA) for gaining the best feature subset in generations.

Firstly, a new encoding method is developed. Chro-Algorithm is developed to produce chromosome (feature set) *chro*. In order to encode all features in the original set *FS*, $(|Log_2 N|+1)$ bits are used to represent for one gene (feature), N is the feature number of *FS*. Chromosome denotes a feature set, where each feature is a gene. For example, if 6 out of 200 features ($N = 200$) are chosen, the binary string length of a single feature is dramatically reduced from 200 to 8 (as in Fig. 1). PPFS-Algorithm and Chro-Algorithm are left out due to the space limitation.

f130	f45	f128	f1	f11	f77
10000010	00101101	10000000	00000001	00001011	01001101

Fig. 1. Instance of new binary encoding for genes and the chromosome

Secondly, selection operator is also improved in ImpSel-Algorithm to optimize the convergence speed and efficiency of GA. Roulette wheel method and the best individual preservation are combined to generate the best chromosome in one generation, which can be inherited by the next generation. The commonly used measure in classification, auc, is as the fitness function.

ImpSel-Algorithm: The improved selection operator

Input: Population P; the best fitness value *best-fit*;
  the chromosome *best-chro* which has the best fitness value.
Output: the population after selection *PAS*; the updated *best-fit* and *best-chro*.
Begin
  For each chro in P// P is the chromosome from Chro-Algorithm{
    Decode *chro*. Calculate *auc* in terms of corresponding feature subset with RF classifier.
    If *auc* > *best-fit*   Then { *best-fit* = *auc*, *best-chro* = *chro*. }}
  For each chromosome *chro'* except the *best-chro* {
    Calculate the selection probability *sp* and cumulative probability *cp* based on *auc*.
    Generate random number *r* with 0-1 uniform distribution as roulette parameter.
    If *cp* > *r*   Then {add *chro'* into *PAS* }}
  Put *best-chro* into *PAS*.
  Return *PAS*.
End

PPGAFS Algorithm produces an optimal minimal feature set based on the improved GA.

PPGAFS Algorithm:

Input: feature number in FS *num*; number of selected features *gn*; crossover rate *cross rate*; mutation rate *mutation rate;* number of evolutional generation *P (count)*; The maximum evolution generation (*mp*).

Output: the minimum feature subset *PPOMFS*

Begin

    Use Random Forest classifier to adjust *tsp* and *tsc*.

    Call PPFS-Algorithm to preselect and obtain PPFS.

    Set *FS = PPFS*.

    Define *best-fit* as the largest AUC in each generation.

    Define *best-chro* as the chromosome in terms of the maximal AUC.

    Call Chro-Algorithm generate *ip* chromosomes for initial population *P(count)* on *gn*.

    Use *PPFS* Do { Call ImpSel-Algorithm to implement the selection operation.

        Implement Crossover and Mutation.

        count←count +1. }

    Until *count = mp*.

    Decode the *best-chro* to get the feature subset PPOMFS.

    Return PPOMFS.

End

---

3   Spam Detection and Verification Experiment Analysis

3.1   Web Spam Detection Procedure

A novel mechanism for filtering Web spam is developed. At Stage 1, Z-score is used for normalization, and equal frequency binning based on rough set method is applied to the discretized data. Since the number of spam samples is much smaller than that of the normal samples, SMOTE is used to balance the training set at Stage 2. Selecting feature for privacy preservation and dimension reduction (Sect. 2) is carried out at Stage 3. Web spam is detected with Random Forest approach at Stage 4.

3.2   Dataset and Evaluation Measures

WEBSPAM-UK2007 [9] has 137 features is applied to verify the proposed approaches. The data set is preprocessed at Stage 1–2. Apart from Accuracy (Acc), TPR, TNR, AUC and F1 measure, Compared Decision Ability (CDA) is defined to observe the performance change based on the above criteria, where $m_{i(dis)}$ means the ith measure after processing and $m_{i(orig)}$ means the ith original measure. H is indicator function. The higher the CDA is, the better the classification results are. *Privacy, confidence* and the computational time are additionally used to evaluate the PPOMFS generation procedure.

$$CDA = \sum_{i}^{n} H \times (m_{i(dis)} - m_{i(orig)})^2$$

$$H = \begin{cases} -1, \text{if } m_{i(dis)} < m_{i(orig)} \\ 1, \text{if } m_{i(dis)} \geq m_{i(orig)} \end{cases} \tag{5}$$

3.3 Experiment Design and Result Analysis

Experiment 1. K-Means, equal-frequency binning (EFB), and entropy-based rough set model (EBR) are applied to discretize the original data set. EBR used in our spam detection produces the highest CDA value.

Experiment 2. In PPFS-Algorithm the *privacy threshold tsp* and the *classification capability (confidence) threshold tsc* are key for preselecting a group of features. A smaller *tsc* can significantly reduce the number of disclosed features and can terminate the feature selection procedure early than a larger one. However, such a *tsc* will seriously degrade the detection performance. Through experiments, two maximal values of the *confidence*, 1 and 0.957, are obtained when *tsc* is fixed as 0.96. CDA is the highest when *privacy* value 0.017 is smaller than *tsp* 0.2, while *confidence* value 1 is bigger than *tsc* 0.96. Therefore, 96 features are pre-selected and fed to next step of Stage 3.

Experiment 3. PPGAFS Algorithm is verified. The maximum evolution generation (*mp*) and the initial population size (*ip*) are assigned as 100 and 10, according to the experience. The number of the selected feature set *gn* varies from 5 to 30. Results are shown in Fig. 2.

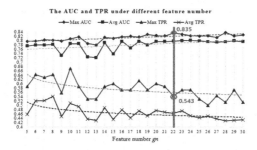

Fig. 2. The distributions of AUC and TPR for different feature numbers

The first maximum AUC is 0.835 and the corresponding TPR is relatively high (0.543) when *gn* is 22. Therefore 22 features are chosen in the privacy-preserving optimal minimum feature subset, which contains both link- and content-based features but no private features.

Experiment 4. Our approach achieves good spam detection performance by using only 22 features and meanwhile protecting sensitive data (Table 1). In comparison, EOM classifier performs the best AUC with 280 features, but has very high computa-tion cost and privacy disclosure risk.

Table 1. The comparison of different web spam detection methods

Methods	Feature number	Accuracy	TPR	AUC
Original-balanced data	137	0.929	0.400	0.735
USCS ensemble [5]	32 +	0.773	/	0.843
EOM [4]	280	0.886	0.752	0.867
PPOMFS-RF (this paper)	**22**	**0.868**	0.543	**0.835**

4   Conclusion

Less is more. This stand point is supported by our achievement on spam detection and sensitive information protection with fewer features. The experimental results show that effectively mining and privacy protection can be a good combination. In the future, the approach will be further developed by exactly capturing and dealing with the influence of sensitive features. The effectiveness of private feature protection and availability in data mining should be studied comprehensively.

Acknowledgments. This work is supported by the Sichuan Science and Technology Program (No 2019YFSY0032).

References

1. Zhuang, X., Zhu, Y., Peng, Q., Khurshid, F.: Using deep belief network to demote web spam. Fut. Gener. Comput. Syst. **118**, 94–106 (2021)
2. Wang, Z., Hu, R., Chen, Q., Gao, P., Xu, X.: ColluEagle: collusive review spammer detection using Markov random fields. Data Min. Knowl. Disc. **34**(6), 1621–1641 (2020). https://doi.org/10.1007/s10618-020-00693-w
3. Dou, Y., Ma, G., Yu, P., et al. Robust spammer detection by nash reinforcement learning. In: Proceedings of the 26th ACM SIGKDD Conference, pp. 924–933, San Diego (2020)
4. Wei, S., Zhu, Y.: Cleaning out web spam by entropy-based cascade outlier detection. In: Benslimane, D., Damiani, E., Grosky, W.I., Hameurlain, A., Sheth, A., Wagner, R.R. (eds.) DEXA 2017. LNCS, vol. 10439, pp. 232–246. Springer, Cham (2017). https://doi.org/10.1007/978-3-319-64471-4_19
5. Lu, X.-Y., Chen, M.-S., Wu, J.-L., Chang, P.-C., Chen, M.-H.: A novel ensemble decision tree based on under-sampling and clonal selection for web spam detection. Pattern Anal. Appl. **21**(3), 741–754 (2017). https://doi.org/10.1007/s10044-017-0602-2
6. Pattuk, E., Kantarcioglu, M., Ulusoy, H., Malin, B.: Privacy-aware dynamic feature selection. In: Proceedings of ICDE Conference, pp. 78–88, Seoul (2015)
7. Zhang, T., Zhu, T., Xiong, P., et al.: Correlated differential privacy: feature selection in machine learning. IEEE Trans. Ind. Inform. **16**(3), 2115–2124 (2020)
8. Sheikhalishahi, M., Martinelli, F.: Privacy-utility feature selection as a tool in private data classification. In: Proceedings of International Symposium on Distributed Computing and Artificial Intelligence, pp. 254–261, Porto (2017)
9. Web Spam Challenge: Results. http://webspam.lip6.fr/wiki/pmwiki.php?n=Main.PhaseIII

Construction of Differentially Private Summaries Over Fully Homomorphic Encryption

Shojiro Ushiyama[1](✉), Tsubasa Takahashi[2](✉), Masashi Kudo[1](✉),
and Hayato Yamana[1](✉)

[1] Waseda University, Tokyo, Japan
{s-ushiyama,kudoma34,yamana}@yama.info.waseda.ac.jp
[2] LINE Corporation, Tokyo, Japan
tsubasa.takahashi@linecorp.com

Abstract. Cloud computing has garnered attention as a platform of query processing systems. However, data privacy leakage is a critical problem. Chowdhury et al. proposed Cryptε, which executes differential privacy (DP) over encrypted data on two non-colluding semi-honest servers. Further, the DP index proposed by these authors summarizes a dataset to prevent information leakage while improving the performance. However, two problems persist: 1) the original data are decrypted to apply sorting via a garbled circuit, and 2) the added noise becomes large because the sorted data are partitioned with equal width, regardless of the data distribution. To solve these problems, we propose a new method called DP-summary that summarizes a dataset into differentially private data over a homomorphic encryption without decryption, thereby enhancing data security. Furthermore, our scheme adopts Li et al.'s data-aware and workload-aware (DAWA) algorithm for the encrypted data, thereby minimizing the noise caused by DP and reducing the errors of query responses. An experimental evaluation using torus fully homomorphic encryption (TFHE), a bit-wise fully homomorphic encryption library, confirms the applicability of the proposed method, which summarized eight 16-bit data in 12.5 h. We also confirmed that there was no accuracy degradation even after adopting TFHE along with the DAWA algorithm.

Keywords: Differential privacy · Differentially private summary · Fully Homomorphic encryption · TFHE

1 Introduction

In recent years, cloud computing has garnered significant attention as a system that facilitates query processing. However, data leakage is considered a serious problem, especially when processing sensitive data on cloud servers. Contextually, we assume three entities, namely data owners that provide original data to be analyzed, a cloud server that processes the original data, and data analysts that perform arbitrary analysis through the query responses of the cloud server. In this setting, the original data could

© Springer Nature Switzerland AG 2021
C. Strauss et al. (Eds.): DEXA 2021, LNCS 12924, pp. 9–21, 2021.
https://doi.org/10.1007/978-3-030-86475-0_2

be leaked to either or both the cloud server and data analysts. This paper aims to solve the aforenoted problems by adopting fully homomorphic encryption [1] and differential privacy [2].

Fully homomorphic encryption (FHE) [1] evaluates arbitrary functions in addition and multiplication operations over encrypted data without decryption. By adopting FHE to handle the original data on the cloud server, we can preserve the privacy of this data. Note that homomorphic encryption (HE) is a limited version of FHE that enables additions or multiplication, or an arbitrary number of additions and a limited (or few) number of multiplications over encrypted data.

Differential privacy (DP) [2] is a promising privacy-preserving technique that hinders the estimation of input data by adding noise to output data. In query processing systems, we can adopt DP to preserve the privacy of individual data from query responses by adding noise. DP gives an information theoretic privacy guarantee. However, we must trust the cloud server because the cloud server handles the original data, whose privacy must be maintained, to respond queries; this means that both the original data and the query response data are revealed to the cloud server.

Research on combining HE and DP to take advantage of both techniques to preserve both the privacy of the original data provided by data owners and the privacy of output data has gained increasing research focus since around 2015 [3]. In 2020, Chowdhury et al. [4] proposed a query processing system called Cryptε that protects original data against both two cloud servers and data analysts by combining HE and DP. The DP index that they proposed summarizes a dataset via DP to successfully bound the information leakage. However, two problems persist: 1) the original data are decrypted to apply sorting via a garbled circuit, and 2) the added noise becomes large because the sorted data are partitioned with equal width, regardless of the data distribution.

To tackle these problems, we combine FHE and DP to protect the privacy of original data against both two cloud servers and data analysts by summarizing the original data without decrypting the original data, which we call DP-summary. We construct the DP-summaries in advance from the original data over FHE, followed by decrypting them to handle query processing. Then, the cloud server processes data analysts' queries with the DP-summaries, which are plaintext, to speed up the query response time such that it is the same as that of Cryptε. Since all queries are processed on the DP-summaries whose privacy is guaranteed by DP, data analysts cannot make statistical guesses about the data owners' original data even if they query many times. Moreover, we adopt a part of the data-aware and workload-aware (DAWA) algorithm proposed by Li et al. [5] over FHE to reduce the query response errors caused by DP, which is another characteristic issue of Cryptε.

Our contribution is stated below:

– We combine FHE and DP to protect the privacy of original data owned by data owners against both a cloud server and data analysts. During the process, we never decrypt the original data until DP is adopted to enhance the security. Moreover, we adopt the DAWA algorithm [5] over FHE to reduce the errors of query responses while Cryptε exhibits substantial errors.

This paper is organized as follows. Section 2 summarizes preliminary information regarding HE and DP. Related work is discussed in Sect. 3. The details of the proposed method are presented in Sect. 4, followed by the experimental evaluation in Sect. 5. Finally, we provide conclusions and discuss future work in Sect. 6.

2 Preliminaries

2.1 Homomorphic Encryption

FHE [1], proposed by Gentry, enables an arbitrary number of multiplications and additions over encrypted data without decryption, whereas HE enables multiplications or additions over encrypted data, or an arbitrary number of additions and a limited (or few) number of multiplications over encrypted data. Generally, HE enables faster execution than FHE. Although we do not need to decrypt the encrypted data even during the calculation by adopting HE, we cannot execute any branch operations because we cannot know the Boolean conditions. Thus, complicated functions such as square root and trigonometric functions are difficult to implement. By contrast, the adoption of bit-wise FHE such as torus fully homomorphic encryption (TFHE) [6] enables the implementation of arbitrary functions by constructing circuits. In this study, we adopt TFHE.

2.2 Differential Privacy

DP [2] protects the privacy of data by adding noise. A trade off exists between the strength of privacy preservation and the usefulness of differentially private data. Specifically, while adding more noise improves the privacy preservation strength, the differentially private data has a larger deviation from the original value, and the usefulness of the differentially private data decreases.

DP has a thorough mathematical basis. It is said that a randomized mechanism m satisfies ϵ-DP if and only if it satisfies Definition 1, given below. Then, ϵ is called a privacy parameter and takes a real value larger than 0. The size of ϵ can be used to adjust the privacy strength. Specifically, the smaller the value of ϵ, the stronger the guaranteed privacy.

Definition 1 (ϵ-differential privacy [2]). A randomized mechanism m satisfies differential privacy if and only if the following holds:

$$\frac{\Pr(m(D) \in S)}{\Pr(m(D') \in S)} \leq exp(\epsilon), \tag{1}$$

where D and D' are any pair of databases with $d(D, D') = 1$[1], S is any subset of the output of the randomized mechanism, and Pr() means the probability that the event in () occurs.

---

[1] $d(D, D') = 1$ means that the two databases D and D' are exactly the same except for one record, and the rest of the records are the same.

Here, a randomized mechanism, e.g., Laplace mechanism [2], is a function that adds a random value to its input value to satisfy DP. The Laplace mechanism samples noise according to the Laplace distribution with a zero mean as follow: $m_{LAP}(D) = q(D) + r$, where q is a query, r is sampled from $Lap\left(\frac{\Delta q}{\epsilon}\right)$, and Δ_q is the sensitivity of query q.

Since the noise added by DP is sampled according to a probability distribution having a mean of zero, collecting numerous differentially private output data can reconstruct its statistical properties, i.e., attackers who collect numerous differentially private output data can guess the original data probabilistically. Thus, the upper limit on the number of times that add the randomized noise is controlled by setting a privacy budget to prevent such attacks.

3   Related Work

3.1   Combination of Homomorphic Encryption and Differential Privacy

In 2020, Chowdhury et al. [4] proposed a privacy-preserving query processing system called Cryptε by combining labeled homomorphic encryption (labHE) [7], an extension of linearly homomorphic encryption, and DP. It consists of two cloud servers, one computation server and one decryption server, as shown in Fig. 1. The system protects the privacy of original data against both the two cloud servers and data analysts by applying DP under labHE. The original Cryptε has the disadvantage of a slow query response time because it performs homomorphic operations to apply DP after receiving a query. As a countermeasure to the above disadvantage, Cryptε was amended with a differentially private index, called a DP index, to accelerate range query responses. Once the DP index is constructed, all the queries are executed with the DP index. However, building the DP index requires the decryption of original data to apply sorting via a garbled circuit, which means it is possible for the sorting result to be leaked to the computation server[2]. Besides, the added noise becomes large because the sorted data are partitioned with equal widths, regardless of the data distribution.

3.2   Range Queries Under Differential Privacy

The range query algorithm under DP proposed by Li et al. [5] in 2014 achieves low errors for range queries corresponding to one-dimensional and two-dimensional data represented by histograms. Low errors obtained via DP improve the accuracy of query responses. The algorithm consists of two steps: 1) partitioning input data represented by histograms into clusters, each of which consists of close values, and 2) optimizing the response results for a set of range queries (i.e., workload) to reduce the error. The DAWA [5] algorithm includes these two steps. Here, we call step 1) the data-aware algorithm, and step 2) the workload-aware algorithm. In the data-aware algorithm, noise is added to the total sum value corresponding to each cluster; then, the average over the cluster is calculated. Hence, this step generates two types of errors, aggregation errors

---

[2] Details of Cryptε's possible privacy leakage are unknown because of no detailed implementation described in the paper [4]; thereby, some other information might be leaked.

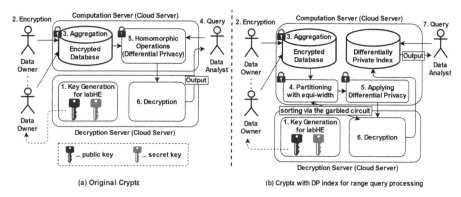

Fig. 1. Cryptε system [4]

by averaging and perturbation errors by adding noise. To reduce the query response errors, the data-aware algorithm seeks the best partitioning that minimizes both types of errors. Therefore, the data-aware partitioning reduces the total amount of noise rather than adding noise to the raw pieces of data. Recently, several workload-aware approaches that have outperformed the DAWA algorithm have been proposed [8]. However, our focus is constructing a differentially private summary without given workloads. Among data-aware approaches without given workloads, the DAWA algorithm is the state-of-the-art in terms of average errors [9].

4 Proposed Method

4.1 Overview

In this section, we propose a range query processing system that responds to data analysts' queries quickly, with no limit on the number of query responses, by constructing a differentially private summary (DP-summary) over FHE in advance. Specifically, our proposed method improves the query response time by responding to data analysts' queries quickly using a pre-constructed DP-summary in plaintext. Moreover, it also overcomes the limitation on the number of query responses, caused by DP, by responding to all queries from the pre-constructed DP-summary instead of applying DP to the response for each query. These advantages are the same as those of Cryptε with a DP index.

Besides, our proposed method solves two problems of Cryptε with DP index: 1) decryption of the original data before adopting DP and 2) the large amount of added noise caused by partitioning the sorted data with equi-width regardless of the data distribution in the DP index. Our proposed method solves the problem 1) by applying DP over FHE before decryption, which does not decrypt any original data until DP is adopted to enhance data security. To tackle the problem 2), we reduce the amount of added noise by adopting the data-aware algorithm, which optimizes the partitioning depending on the data distribution over FHE.

Figure 2 shows an overview of our proposed method. We assume four entities: data owners (DOs), a computation server (CS), a decryption server (DS), and data analysts

(DAs). The DOs, CS, and DS are assumed to be semi-honest; that is, they follow the protocol of our proposed system but attempt to steal the original data owned by the DOs. The DAs are assumed to be untrustworthy. It is also assumed that the CS and the DS collude neither with each other nor with other entities. Descriptions of each entity are presented below. In the following descriptions, N and M are arbitrary positive integers larger than or equal to 1.

- **Data Owner ($DO\_j(\forall j, 1 \le j \le N)$)**
  The number of DOs is N. DOs encrypt their own data using symmetric keys received from the DS and then send the data to the CS. After sending the data, the DOs are not involved in the system.
- **Computation Server (CS)**
  The CS aggregates the received encrypted data and applies DP to the encrypted data over TFHE. It also cooperates with the DS to construct the DP-summary for the received encrypted data. Any data stored on the CS are always protected by either or both TFHE and DP.
- **Decryption Server (DS)**
  The DS has two roles: key generation for TFHE and decryption of the DP-enabled encrypted data received from the CS. Any data decrypted by the DS is always differentially private; thus, the original data owned by the DOs is protected.
- **Data Analyst ($DA\_i(\forall i, 1 \le i \le M)$)**
  The number of DAs is M. The DAs query the CS and obtain responses to the queries from the CS. DAs can attempt to make statistical guesses regarding the DOs' original data by receiving many query responses.

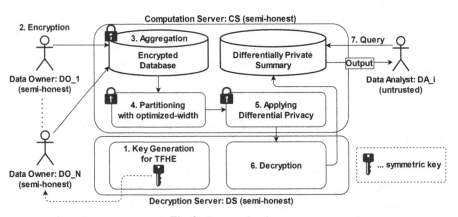

Fig. 2. Proposed system

Our proposed system protects the privacy of original data owned by the DOs against the CS, DS, and DAs. Further, the CS and DS are allowed to store the data secured via DP. By guaranteeing DP for the partitioning result, the CS and DS are allowed to store the partitioning result in plaintext. The symmetric keys are held only by the DOs and

DS. Encryption is performed only by the DOs, and decryption is performed only by the DS.

The procedure of our proposed method is presented below based on Fig. 2. Among the following steps, the preprocessing steps from "1. **Key Generation**" to "6. **Decryption**" are completed before the DAs' queries are input.

1. **Key Generation**: The DS generates a symmetric key via TFHE. The generated symmetric key is sent securely to the DOs.
2. **Encryption**: The DOs encrypt their data using the symmetric key received from the DS and send the encrypted data to the CS.
3. **Aggregation**: After receiving the encrypted data from the DOs, the CS aggregates multiple encrypted data points whose domain is the same over homomorphic operations to construct a histogram for each pre-defined domain.
4. **Partitioning**: The CS partitions the aggregated encrypted data using homomorphic operations without decryption.
5. **Applying Differential Privacy**: The CS adds noise to each partitioned cluster using homomorphic operations and the Laplace mechanism without decryption followed by sending the differentially private encrypted data (DP-summary) to the DS.
6. **Decryption**: The DS decrypts the encrypted DP-summary received from the CS. The decrypted DP-summary is sent to the CS.
7. **Query**: The DAs query the CS and obtain the query response result.

4.2 Adoption of Differential Privacy over Fully Homomorphic Encryption

In this section, we explain how to adopt DP to a range query processing system based on FHE. Here, we only target range queries for histogram data; thus, we adopt the data-aware algorithm, a part of the DAWA algorithm proposed by Li et al. [5]. We consider only one-dimensional data in our present implementation. The data-aware algorithm performs the partitioning such that the sum of the deviations for each cluster is minimized. To perform the partitioning over ciphertexts, we need to calculate absolute values and minimum values over ciphertexts. Thus, we adopt TFHE [6] as an FHE scheme, specifically, as a bit-wise fully homomorphic encryption scheme. By adopting TFHE, arbitrary logic circuits consisting of binary gates can be constructed for the encrypted data to calculate absolute values and minimum values over ciphertexts.

In the data-aware algorithm, the aggregated histogram data, which is the aggregation of input data sent from the DOs to the CS, is partitioned into a set of clusters, each of which has close values such that the deviation in the cluster is small; in other words, there are approximately uniform histogram data in each cluster. Then, the noise is added to the total sum value of each cluster. Rather than adding noise to each data point, adding noise to each partitioned cluster reduces the total amount of noise added. Since partitioning the data to minimize the deviation within each cluster would result in a privacy violation, the data-aware algorithm consumes privacy budget ϵ_1 to perform differentially private partitioning. Assuming that privacy budget ϵ_2 is entailed when adding noise to the total sum value of the data in each partitioned cluster, the data-aware algorithm satisfies ϵ-differential privacy, where $\epsilon = \epsilon_1 + \epsilon_2$.

Here, we call each histogram data a *domain*, while a cluster of domains classified by partitioning is called a *bucket*. A set of buckets is called a *partition*. We assume a set of histogram data represented by $x = (x_1, x_2, \ldots, x_i, \ldots, x_n)$, where $1 \leq i \leq n$, n is the number of domains, and x_i represents the data of the i-th domain. A set of buckets is defined as $B = (b_1, b_2, \ldots, b_j, \ldots, b_k)$, where $1 \leq j \leq k \leq n$ and b_j represents the j-th bucket. For example, if b_j is a set of domains from the third to the sixth domains, it is expressed as $b_j = \{3, 4, 5, 6\}$. B is calculated from x and ϵ_1 in the way that minimizes the total cost of buckets; specifically, the cost of each bucket is the deviation of the data in the bucket, same as algorithm 1 of DAWA [6]. Noise generated by the Laplace mechanism by consuming ϵ_1 is added to each bucket's deviation to calculate its cost; then, the final B is chosen to minimize the total cost, making B differentially private. Further, we define a set of the total sum values of each bucket as $S = (s_1, s_2, \ldots, s_j, \ldots, s_k)$, where s_j is the sum value over the bucket b_j and $1 \leq j \leq k$. To make S differentially private, the CS adds noise generated from the Laplace mechanism by consuming ϵ_2 to S. The differentially private total sum values of a given partition are defined as $S' = \left(s'_1, s'_2, \ldots, s'_j, \ldots, s'_k\right)$, where $1 \leq j \leq k$ and s'_j represents the differentially private total sum value of the histogram data in the j-th bucket.

To deploy a DP-summary, the CS sends S' to the DS to decrypt S' using the symmetric key of TFHE, and then the DS sends it back to the CS as plaintext. The CS applies uniform expansion to the decrypted S', i.e., dividing s'_j by the number of elements in b_j. For example, if $s'_j = 10$ and $b_j = \{3, 4, 5, 6\}$, the uniform histogram value becomes 2.5 so that $\left(x'_3, x'_4, x'_5, x'_6\right) = (2.5, 2.5, 2.5, 2.5)$. Finally, the CS obtains the uniformly expanded data $x' = \left(x'_1, x'_2, \ldots, x'_i, \ldots, x'_n\right)$. We use x' as the DP-summary to respond to DAs' queries. Note that x and S are represented in ciphertext and B and x' are represented in plaintext. S' is represented in ciphertext until it is decrypted by the DS.

Figure 3 shows an example with $x = (E(3), E(2), E(6), E(5), E(6), E(3), E(4))$ and $B = \{\{1, 2\}, \{3, 4, 5\}, \{6, 7\}\}$ as a calculated partition, where E is the encryption algorithm, i.e., TFHE. In this example, $S = (E(5), E(17), E(7))$. The differentially private total sum value per bucket is assumed to be $S' = (E(4.6), E(16.2), E(7.8))$; then, the values of the uniformly expanded data are $x' = (2.3, 2.3, 5.4, 5.4, 5.4, 3.9, 3.9)$.

4.3 Security Analysis

The security assumption of our proposed method is presented below.

- The DOs, CS, and DS are assumed to be semi-honest; that is, they follow the protocol of our proposed system but attempt to steal the original data owned by the DOs.
- The DAs are assumed to be untrusted.
- The CS and the DS collude neither with each other nor with other entities.
- The symmetric keys are held only by the DOs and DS. (Encryption is performed only by the DOs, and decryption is performed only by the DS.)

Our proposed system protects the privacy of original data owned by the DOs against the CS, DS, and DAs. The original data are encrypted at the DO to be sent to the CS,

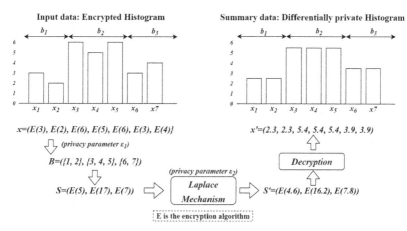

Fig. 3. Construction of differentially private histogram over fully homomorphic encryption

so that the CS cannot see the original data. Then, the CS executes partitioning over the encrypted original data followed by applying differential privacy without decryption, which guarantees the CS cannot see any information related to the original data. After the DS receives the differentially private encrypted partitioned data consisting of B and S', the DS can decrypt them, which also guarantees that the DS only know differentially private data. Thus, any information related to the original data does not reveal to any parties under the condition where CS and DS never collude each other.

5 Experimental Evaluation

In the experimental evaluation, we examined the execution time to construct the DP-summary and its accuracy.

5.1 Experimental Setup

The programs used in the evaluation were written in C++ with TFHE [6] version 1.1 and run with single-threaded execution in the environment presented in Table 1. We adopted fixed-point number representation and two's complement arithmetic. In the implementation using fixed-point arithmetic, the value of the fractional part that cannot be expressed was truncated. Although approximate arithmetic (CKKS) [10] enables arbitrary polynomial functions over encrypted complex-number vectors for handling real numbers, we cannot execute branch operations such as greater-than without decryption, resulting in no partitioning. Thus, we adopt TFHE.

The privacy parameter, ϵ, used in the evaluation experiment was 1.00. The ratio of ϵ_1 and ϵ_2 was 1:3, same as that used by Li et al. [5], i.e., $\epsilon_1 = 0.25$ and $\epsilon_2 = 0.75$. The histogram data x_i, $1 \leq i \leq n$ used in the experiment was generated randomly between 0 and 10. The upper limit of the data was determined to ensure that no overflow occurs during the computation process. Since negative numbers are not assumed as the numerical data, if the differentially private numerical data becomes negative, the numerical data is replaced with 0.

Table 1. Experimental environment

Name	Value
CPU model	Intel(R) Xeon(R) Platinum 8280
Socket	2
Core	56
Memory size	1.5 TB
OS	CentOS Linux release 7.6.1810(Core)
Linux version	3.10.0-957.21.3
g++ version	7.3.1

5.2 DP-Summary Construction Time

To validate our proposed method's applicability, we examined the DP-summary construction time, which is the execution time taken to construct a differentially private database from the encrypted database with TFHE. The construction time depends on the domain size and the number of bits representing ciphertexts in TFHE. Thus, we changed the domain size and the bit size representing the ciphertexts to measure the construction time.

The construction time was measured from the beginning of partitioning to the end of the uniform expansion 10 times to determine the average by using the *chrono* function, which is included in C++ standard library. We changed the domain size from 2 to 8 and the bit sizes as 10(2) bits, 12(4) bits, and 16(8) bits, representing the total bit size (the bit size in the fractional part). For example, 10(2) shows 10 bits in total, in which 2 bits are used for the fractional part, 7 bits are used for the integer part, and the remaining 1 bit is used for the code part.

Figure 4 shows the construction time based on different domain sizes; this confirms that the DP-summary construction time increases exponentially with the domain size because the number of domain combinations to merge increases exponentially to identify the best one. However, the proposed method is still feasible when the domain size is less than or equal to 8 because the DP-summary construction requires only one execution.

Figure 5 shows the construction time based on different bit sizes; this confirms that the DP-summary construction time increases linearly with the bit size representing ciphertexts. Thus, we confirm that our proposed method is feasible with a small domain size regardless of the bit size representing ciphertexts.

The DP-summary construction time is slow for two reasons. One is because of homomorphic operations. The computation cost over ciphertexts using homomorphic operations is large compared to that over plaintexts. In particular, TFHE is the bit-wise FHE, and the computation cost using TFHE is likely to be large compared to that using the integer-based FHE. The other is because any optimizations are impossible to be adapted. Original data-aware algorithm in DAWA [5] adopts some optimizations to speed up the partitioning. On the other hand, in our proposed method, any handled data is ciphertexts, which results in unavailability of any optimizations because we cannot see the values, i.e., plain texts, when encrypted. When partitioning, we need to seek the optimal partition from all possible partitions, whose computational complexity order is $O(2^n)$ (n is domain size). Including these two reasons, it was reported that the computation using TFHE is approximately 10^9 times slower than that over plaintexts [11].

During the measurement of the construction time, we also measured the maximum consumed memory size with the DP-command "bin/time --format = %M." The maximum consumed memory size varied from 513 MB to 530 MB depending on the domain size and the bit size.

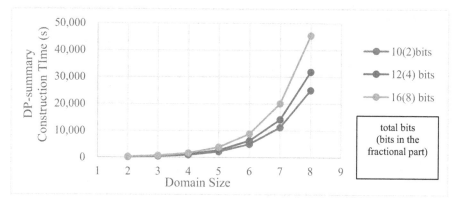

Fig. 4. DP-summary construction time v.s. domain size

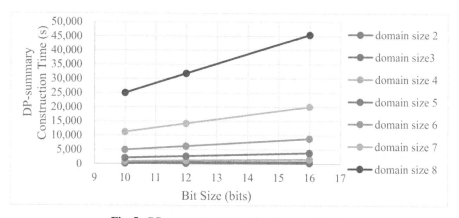

Fig. 5. DP-summary construction time v.s. bit size

5.3 Accuracy of DP-Summary

In this experiment, we evaluated the accuracy of the DP-summary. In the implementation, we adopted a fixed-point number representation to truncate the fractional value according to the bit size representing the fractional part, which may affect the accuracy of the DP-summary. When the fractional part's bit size is small, the truncated fractional value becomes large, i.e., the accuracy of the DP-summary is expected to change depending on the fractional part's bit size. Thus, we measured the changes in the accuracy of the constructed DP-summary with different bit sizes of the fractional part.

We implemented the plaintext program that performs the same processing as our proposed method and the baseline using a floating-point number representation. This number representation is expressed in 64 bits:1 bit for the code part, 52 bits for the mantissa part, and 11 bits for the exponent part.

We measured the error between the constructed DP-summary and the aggregated data on which differential privacy was not applied. We examined the error 100 times to determine the average for three different bit sizes—10(2) bits, 12(4) bits, and 16(8) bits—in a histogram with domain sizes from 2 to 10.

Figure 6 shows the results of accuracy. We cannot verify the difference in accuracy according to the bit size. The reason is that, in our proposed method, the size of the truncated fractional part is negligibly small compared to the size of noise added by differential privacy; this implies that the effect on accuracy caused by changing the bit size of the fractional part is not significant.

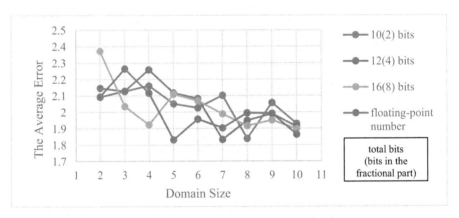

Fig. 6. Comparison of accuracy v.s. domain size

6 Conclusion

In this study, we propose a privacy-aware query processing system that constructs differentially private summary (DP-summary) in advance over fully homomorphic encryption to respond to the queries using the DP-summary in plaintext. We enhance the data security in comparison with DP index proposed by Chowdhury et al. [4] in Cryptε system by applying differential privacy over the encrypted original data with FHE. Furthermore, we solve the problem of large added noise caused by partitioning the sorted data with equal -width regardless of the data distribution, by adopting the data-aware algorithm [5], which optimizes the partitioning depending on the data distribution.

Although our proposed method responds quickly to the queries, it requires high computational cost for DP-summary construction over fully homomorphic encryption. Our experimental evaluation with TFHE, which is a bit-wise fully homomorphic encryption library, shows that the proposed method requires 12.5 h to construct DP-summary for eight 16-bits data; this is still feasible because the DP-summary requires only one construction. We also confirm that accuracy is not degraded even after adopting TFHE to the data-aware algorithm.

Our future work includes increasing the speed to prepare DP-summary for a larger dataset.

Acknowledgment. The research was supported by NII CRIS collaborative research program operated by NII CRIS and LINE Corporation.

References

1. Gentry, C.: Fully homomorphic encryption using ideal lattices. In: Proceedings of the 41st Annual ACM Symposium on Theory of Computing (STOC 2009), pp.169–178 (2009)
2. Dwork, C., McSherry, F., Nissim, K., Smith, A.: Calibrating noise to sensitivity in private data analysis. In: Halevi, S., Rabin, T. (eds.) TCC 2006. LNCS, vol. 3876, pp. 265–284. Springer, Heidelberg (2006). https://doi.org/10.1007/11681878_14
3. Ushiyama, S., Masashi, K., Takahashi, T., Inoue, K., Suzuki, T., Yamana, H.: Survey on the combination of differential privacy and homomorphic encryption. In: Proceedings of Computer Security Symposium 2020, pp. 207–214 (2020). (in Japanese)
4. Chowdhury, A.R., Wang, C., He, X., Machanavajjhala, A., Jha, S.: Cryptε: crypto-assisted differential privacy on untrusted servers. In: Proceedings of the 2020 SIGMOD International Conference on Management of Data (SIGMOD 2020), pp. 603–619 (2020)
5. Li, C., Hay, M., Miklau, G., Wang, Y.: A data- and workload-aware algorithm for range queries under differential privacy. In: Proceedings of VLDB 2014, pp. 341–352 (2014)
6. Chillotti, I., Gama, N., Georgieva, M., Izabachène, M.: Faster fully homomorphic encryption: bootstrapping in less than 0.1 seconds. In: Cheon, J.H., Takagi, T. (eds.) ASIACRYPT 2016. LNCS, vol. 10031, pp. 3–33. Springer, Heidelberg (2016). https://doi.org/10.1007/978-3-662-53887-6_1
7. Barbosa, M., Catalano, D., Fiore, D.: Labeled homomorphic encryption. In: Foley, S.N., Gollmann, D., Snekkenes, E. (eds.) ESORICS 2017. LNCS, vol. 10492, pp. 146–166. Springer, Cham (2017). https://doi.org/10.1007/978-3-319-66402-6_10
8. McKenna, R., Miklau, G., Hay, M., Machanavajjhala, A.: Optimizing error of high-imensional statistical queries under differential privacy. Proc. VLDB Endow. **11**(10), 1206–1219 (2018)
9. Hay, M., Machanavajjhala, A., Miklau, G., Chen, Y., Zhang, D.: Principled evaluation of differentially private algorithms using DPBench. In: Proceedings of the 2016 SIGMOD International Conference on Management of Data (SIGMOD 2016), pp. 139–154 (2016)
10. Cheon, J.H., Kim, A., Kim, M., Song, Y.: Homomorphic encryption for arithmetic of approximate numbers. In: Takagi, T., Peyrin, T. (eds.) ASIACRYPT 2017. LNCS, vol. 10624, pp. 409–437. Springer, Cham (2017). https://doi.org/10.1007/978-3-319-70694-8_15
11. Matsuoka, K., Banno, R., Matsumoto, N., Sato, T., Bian, S.: Virtual secure platform: a five-stage pipeline processor over TFHE. In: Proceedings of the 30th USENIX Security Symposium, pp. 1–18 (2021, in press)

SafecareOnto: A Cyber-Physical Security Ontology for Healthcare Systems

Fatma-Zohra Hannou[✉], Faten Atigui, Nadira Lammari,
and Samira Si-said Cherfi

CEDRIC Lab, CNAM - Conservatoire National des Arts et Métiers Paris,
Paris, France
{fatma-zohra.hannou,faten.atigui,samira.cherfi}@lecnam.net,
lammari@cnam.fr

Abstract. Vital to society, healthcare infrastructures are frequently subject to many threats that exploit their vulnerabilities. Many cyber and physical attacks are triggered, leading to many high-impact incidents. There is a growing need for innovative solutions that combine cyber and physical security features. To improve the response to incidents caused by attacks combining cyber and physical threats, we have produced within the H2020 project "Safecare", an ontology-based solution. The Safecare ontology is designed to support an impact propagation model application, integrating cyber-physical interactions. In this paper, we present the different steps carried out to develop this ontology and two use cases on asset management and incident propagation.

1 Introduction

Like all critical infrastructures, hospitals are of increasing complexity, particularly following the expanding integration of cyber-physical systems and connected objects. Hospital's critical assets are exposed to growing risks exploiting their vulnerabilities. The ISO/TS 11633-1:2019 define an asset as anything that has value to the organization, which includes technical data (credentials, passwords), non-health data(financial data), IT services, hardware, software, communications facilities, media, IT facilities, and medical devices that record or report data. It is considered critical if its malfunction induces a high impact on the overall system's operation and the patients [5].

Hospital risks are generally associated with various threats, whether natural (floods), unintentional as human errors and technical device failures, or deliberate threats like malicious or criminal acts. This paper considers only deliberate threats (malicious man-made hazards) that exploit one or many vulnerabilities to trigger cyber and/or physical attacks. Authors in [4] distinguish a cyber-enabled physical attack, identified as a physical attack involving cyber activities, from a physical-enabled cyber-attack that is a cyber-attack where the attacker gains physical access to a site location before launching the cyber attack.

When a cyber or physical attack (incident) occurs, it may provoke far-reaching cascading effects throughout the entire critical infrastructure, which

C. Strauss et al. (Eds.): DEXA 2021, LNCS 12924, pp. 22–34, 2021.
https://doi.org/10.1007/978-3-030-86475-0_3

need to be identified and estimated to ensure precise risk management [22]. Many critical assets could be compromised, and measures put in place to protect them could fail. This situation might cause harm, perceived as "injuries or damages to the health of people, or damage to property or the environment"[1]. To increase situational awareness and ensure a quick and effective response to incidents and their cascading effects, a healthcare organization must have a comprehensive inventory of its critical assets and record their interconnections.

The work presented in this paper is developed within the European project Safecare[2]. The project aims to cover the integration of both cyber and physical security for healthcare infrastructures in a combined solution. To fulfill the project objective, we propose SafecareOnto, an ontology for a uniform representation of critical healthcare assets and their dependencies. This ontology is used to simulate impact propagation and potential cascading effects of a cyber or a physical incident within healthcare infrastructures. The global solution supports the prevention and mitigation of incident propagation by evaluating risks and alerting stakeholders.

This paper is organized as follows. Section 2 introduces a global view of SafecareOnto, including its development approach. Section 3 describes the knowledge acquisition process. In Sect. 4, the formalization and implementation of the ontological framework are explained. A cyber-physical attack scenario allows in Sect. 5 to illustrate some project real use cases before discussing related works in Sect. 6.

2   Safecare Ontology

Several definitions of ontology exist, but only one predominates in the information system field. In [10] Gruber defines an ontology as an explicit specification of a conceptualization (an abstract representation of the world intended to represent). The definition of the ontological framework covers three main components: concepts, relations, and axioms. Different link types relate concepts: equivalence links (synonyms), hierarchical links (generalization/specialization), and associative links such as "cause/effect link"[3]. Each Concept represents a set of different Individuals, also called instances. Axioms allow defining the semantics of concepts, relations and express some restrictions on their values or cardinalities. The use of axioms enables representing specific capabilities or features of a concept and avoids adding new concepts that would not be reused [24].

To meet the project goals, including the propagation of incidents and to provide the right balance between expressiveness and complexity, we chose to design a modular ontology, named "SafecareOnto", organized in three sub-ontologies (modules in the ontology engineering terms): a central ontology named `Asset ontology` and two related sub-ontologies `protection` and `impact` (Fig. 1). **The asset sub-ontology** captures the static knowledge about healthcare critical

---

[1] ISO 14971:2019.

[2] https://www.safecare-project.eu/.

[3] ANSI/NISO Z39.19-2005.

assets and their structural relationships. These assets are organized in concepts. **The impact subontology** defines the concepts that are essential to the computation of impact propagation and provide indicators to help decide about the suitable countermeasures to face attacks. It relies on `Incident` and `Impact` concepts. An incident that result from an attack may negatively impact the healthcare infrastructure. The risk of its spread or the spread of its effects on related assets is not zero. The impact is the result of such propagation. This propagation needs to be precisely qualified and/or quantified to help decide the mitigation plans efficiently. **The protection sub-ontology** manages the protection of assets against attacks. The latter usually exploit one or many vulnerabilities that make an asset sensitive to a threat. A `vulnerability` is any identified weakness. It can be known and in this case, referenced in an existing knowledge base such as that of the MITRE[4]. It can also emerge after an attack. A `protection` is a countermeasure that protects an asset from `threats`. For example, a <u>camera</u> is a protection for an office against <u>unauthorized access</u>.

The choice of the most suitable ontology development approach relies on defining the starting point, whether the ontology will be built by integrating existing ontologies or created from scratch [20]. Our literature study shows that existing security ontologies are either exclusively cyber or physical and do not (or rarely and in a minimal way) integrate cyber-physical

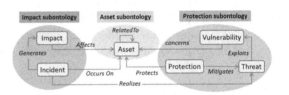

Fig. 1: The conceptual view of SafecareOnto

interactions. These interactions are complex, highly required for the impact propagation study, and their formalization can not be limited to putting the two separate sides together. Furthermore, healthcare ontologies [9] focus generally on the medical process terminologies and do not cover security aspects.

The "from scratch" method [21] include top-down, buttom-up and middle-out approaches. `Top-down` approaches identify a core of abstract generic concepts, and expand it by specialization to more domain-specific concepts [8,18]. `Bottom-up` approaches capture the task specifications, gather terms, and define new concepts at a low level [11,19]. Concepts can be later generalized to a higher level. `Middle-out` approaches define core concepts before deriving new concepts through specialization and generalization using experts' data [28].

Considering our project specifications, we adopt a "from scratch" approach with a bottom-up fashion. This allows formalizing experts' knowledge and fitting the propagation task. We drew on 1, 2, and 7 of the Neon methodology [25] and followed an incremental and iterative process. We considered existing semantic resources and standards to feed some ontology modules for better reusability

---

[4] https://cve.mitre.org/.

and genericity. In the following sections, we focus on Knowledge acquisition, Conceptualization, Formalization, and Implementation phases.

3   Knowledge Acquisition

The knowledge acquisition phase is a critical step in the "from scratch" ontology building approach, ensuring collecting the necessary data for the ontology design and population. The SAFECARE project gave the ground for a direct acquisition process from cyber-physical security experts employed by the hospitals(project end-users) or their stakeholders(security systems suppliers).

During the process, we had to manage several issues including the heterogeneity of terminologies. Indeed, the interviewed experts came from several hospitals and countries. Consequently, for better genericity, we had to integrate an alignment step to homogenize the vocabularies based on literature taxonomies and security standards. Moreover, due to the time-consuming task of collecting business experts' knowledge we had some difficulty to get engagement. Therefore, to maximize the amount of data and increase its quality, we mixed a passive collection process where experts autonomously fill pre-formatted files and active collection phases where ontology designers discuss with experts (online meetings) to check, complete, and validate the acquired data.

With our project partners, we defined twelve complex cyber-security attack scenarios. Each attack scenario depicts a set of actions an attacker performs to accomplish his malicious aims. For each scenario, we carried out the following:

1. **Phase 1:** identify the list of involved assets, the related risks, and the protections in place;
2. **Phase 2:** identify the assets inter-dependencies and the surrounding infrastructures (cyber and physical). Asset relations represent potential vectors for incident propagation;
3. **Phase 3:** for each asset, identify how its risks might propagate to create new impacts on its connected asset graph.

4   Formalization and Implementation

The purpose of this step is to provide a formal model with a fine-grained description of hospitals' security management concepts raised by the knowledge acquisition task. The ontological framework includes concepts, relations, and underlying axioms.

4.1   Concepts Identification

The following concepts constitute the asset subontology. The most important are illustrated in Fig. 2.

Asset concept ($Asset \sqsubseteq \top$). The asset is any valuable resource within the hospital. It is further specialized according to the nature of the incidents they may suffer from and those they likely propagate. The following set of subclasses constitute a partition of the concept "Asset" since they have no common instances and that their union completely covers the domain [13].

- **Building** asset (*Building* ⊑ *Asset*): A building asset is a geographical entity that corresponds to the building in which a hospital (or a part of it) is located. An asset building has a variable granularity going from "room" to a "complex of buildings." The concept building asset has two subclasses: a **simple building** asset (a non-divisible location, *SimpleBuilding* ⊑ *Building*) and **complex building** asset that groups multiple simple building assets (*ComplexBuilding* ⊑ *Building*). The domain of the building assets corresponds to the hospital's physical infrastructure.
- **Network** concept (*Network* ⊑ *Asset*): A computer network denotes a communication and data exchange channel linking at least two devices (nodes). Networks connect the components of the hospital's cyber infrastructure.
- **Staff** (*Staff* ⊑ *Asset*): Staff represents any physical person performing regular or occasional tasks within the hospital. In addition to direct employees, the staff includes external stakeholders acting on-site or remotely This concept excludes patients since they are formalized as business processes.
- **Device** concept (*Device* ⊑ *Asset*): Device refers to any tangible equipment, whether associated to computer software with an automatic action (camera, sensor, server) or not (door, lamp). Computer device *ComputerDevice* ⊑ *Device*, building equipment such as doors, chairs, including those with an automated operating process *BuildingDevice* ⊑ *Device* (sensor, camera), medical devices *MedicalDevice* ⊑ *Device* (scanner, pacemaker).
- **Software** concept (*Software* ⊑ *Asset*): Softwares are virtual programs (sequences of computer code) with data processing capabilities. They support a determined business process such as medical acts or human resources. Note that operating systems are also considered software.
- **Data** (*Data* ⊑ *Asset*): Data play a major role in security management since multiple attacks are carried out using or targetting data. we separate two context categories: **patient data**, and **operating data** used to support the hospital processes (access policies, camera flows, metadata..).

Access Point Concept: The access points are the gateways that enable access to asset use and allow the entrance and the occurrence of the incident. An access point can be either physical (door for room) or cyber (a port for network). The access point is an asset with a security access role (*AccessPoint* ⊑ *Asset*).

Controller Concept: Controllers are physical equipments or virtual protocols implementing assets' access restrictions, formalized in predefined policies. Access to the surgery rooms requires a door (access point), supervised by a door access controller. A controller is an asset that identification ensures safe use and anticipates a possible incident occurrence and propagation (*Controller* ⊑ *Asset*).

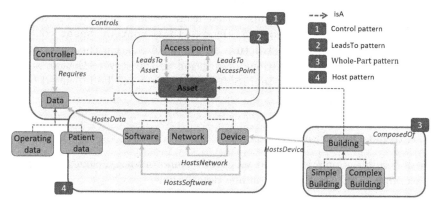

Fig. 2: The asset subontology concepts and relations

4.2   Relationships Identification

The relationships depict how assets interact in the healthcare context and what are their properties. We have identified two families of relations. The first one corresponds to concepts attributes (data properties in OWL): a staff hasRole, a building hasLevel, a software hasVersion, etc. The second family of relations corresponds to concepts interactions (object properties in OWL). They are highlighted in Fig. 2 with solid labeled arrows. We organized these relations in groups matching our propagation channels' analysis to fit the task purposes. This analysis revealed some structural patterns that support reasoning on incident's propagation following their nature. Among patterns we can identify:

Controls pattern allows specifying the conditions and mechanisms for granting or revoking access to cyber or physical assets. The pattern uses the Controller that Controls the Access point. The control patterns requires Data as access policies. For example, a physical Access Control system based on a smart card is composed of three elements: the access rights stored locally or remotely, door readers to check whether data on the card is consistent with the policy and the door (Access point) which would be unlocked when the card is approved.

Leads to pattern captures the access and communication possibilities between assets. This access applies for both physical or cyber flows and is materialized through the Leads To Access and Leads To Asset relations. The access mechanism could be one-way or bidirectional.

The whole-part pattern assumes that if an incident happens on the whole, there could be an impact on its parts. Inversely, if parts are attacked, the whole could also suffer from the impact of the attack. In the healthcare structures, this pattern applies to locations Building assets, that are Composed Of smaller entities . Moreover, the propagation concerns in this case essentially "physical incidents" such as "unauthorized access", "fire" or "flooding". For example, if there is an intrusion on one floor of a hospital, it potentially affects all the rooms.

The hosts-content pattern assumes that if an incident happens on an asset named `host asset` then the content, referred to as `content asset` could be affected by this incident. The structure of the pattern is enriched by rules to enhance the validity of the relations description. If the server (Device) suffers from "fire", the information system (Software) it hosts will be impacted (inaccessible).

4.3   Axioms Definition

A set of formal axioms is defined to specify some ontology elements. Hereby some examples with their corresponding descriptions:

- $simpleBuilding \sqsubseteq building$ (all simple buildings are buildings)
- $AccessPoint \sqsubseteq device \sqcup building \sqcup software \sqcup network$
  (An access point can be either a device, a building, a software or a network)
- $Controller \sqsubseteq \forall controls\ (building \sqcup device \sqcup network \sqcup software)$
- $PhysicalIncident \sqcap CyberIncidents \sqsubseteq \bot$
  CyberIncidents and PhysicalIncidents are totally disjoint concepts
- $data \sqcap requiredby\ Controller \sqsubseteq operatingData$
  Data required by a Controller are exclusively OperatinData
- $leadsToAsset, leadsToAccessPoint \sqsubseteq leadsTo$
  leadsToAsset and LeadsToAccessPoint are subrelations of leadsTo relation
- $ComplexBuilding \sqsubseteq \exists composedof\ Simplebuilding$
  ComplexBuilding is composed of at least one SimpleBuilding

4.4   Implementation

The implementation phase consists of codifying the ontology in a formal language such as OWL. Many software editors allow modeling ontologies and knowledge bases in OWL [3]. We have chosen Protégé[5], an ontology and knowledge base editor that enables the construction of domain ontologies, customized data entry forms to enter data, and comes with visualization and reasoning packages. It offers reasoning tools to support editing inference rules for the impact propagation application. Figure 3 depicts an extract of the SafecareOnto designed in Protégé.

5   Safecare Use Cases

To prove the contribution of our ontology within the SAFECARE project and more particularly for the management of cyber-physical incidents, a vast testing operation have been organized with the project partners. It covered the 12 cyber-physical scenarios. For confidentiality reasons, we will limit ourselves to demonstrate its contribution in two use cases using an illustrative scenario

---

[5] https://protege.stanford.edu/.

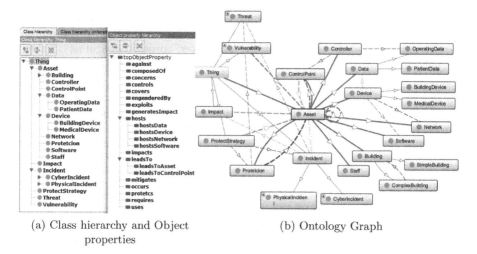

(a) Class hierarchy and Object properties   (b) Ontology Graph

Fig. 3: SafecareOnto implementation in Protégé

related to storing and transporting the Covid vaccines. These latter require particular logistics with a strong constraint regarding the cold chain. This constraint is managed and guaranteed by software connected to temperature sensors placed on the storage freezer for real-time freezer monitoring. This cyber-physical scenario depicts the different actions executed by an attacker to destroy the covid vaccine stocks. The attacker in our use case performs a set of cyber and physical actions: social engineering on a maintenance team member(step 1), sending a spearfishing email (step 2), remotely access the staff member's machine (step 3), identification of the pharmacy room location on the building management software and the freezer temperature sensor (step 4), then going to the hospital, stealing a badge giving access to the pharmacy (step 5), access to the pharmacy room (step 6) and finally placing a fake temperature sensor on the locked fridge (step 7). The direct consequence of this cyber-physical attack is the destruction of the hospital's vaccine stocks, leading to a significant financial loss and the patient vaccination campaign's stopping. This consequence seriously harms the hospital's reputation. It contributes to delaying the health crisis's release, with significant impacts on hospital processes. To show the support offered by our ontology, let us consider two uses cases. The first one concerns the derivation of new facts from the capitalized knowledge using the capabilities of reasoner like Pellet[6]. These new facts may serve asset management task. To show this, using the Semantic Web Rule Language (SWRL), we implemented some inference rules to enrich the knowledge base with additional entries about assets and their protections. The following inference rule is an example :

$$R_1 : hosts(?a, ?d), protects(?p, ?a), against(?p, ?t),$$
$$physicalThreat(?t) \longrightarrow protects(?p, ?d)$$

[6] https://www.w3.org/2001/sw/wiki/Pellet.

Rule R_1 means that a protection p implemented for an office a against physical threat t, will be also a protection for any device d hosted by the office. In our example, the pharmacy hosts a "computer" protected against "unauthorized connection" threat (cyber), using a login system. Besides, protections implemented to secure access to the "pharmacy" ("camera", "door access controller") complete the computer's protections against physical threats ("destruction", "theft"..). SPARQL Queries enable querying the knowledge base to manage assets. While the initial knowledge base state (after raw knowledge acquired), returns only direct facts (login system for computer), the inference rules application extend the results to the derived facts.

The second use case covers the incidents cascading effects. An incident (cyber or physical) that occurs on an asset can lead to several incidents on other connected assets. Through the definition of propagation inference rules, it is possible to identify the extent of reachable assets starting from the initially targeted one. These rules are formulated thanks to the knowledge acquired from security experts (Sect. 3)and generalized considering the ontology. For the Covid scenario example, the following propagation rules draw a "physical intrusion" potential impacts:

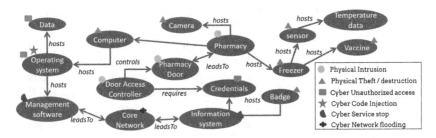

Fig. 4: Incident's impact propagation on scenario assets

(a) Physical access to office may generate devices destruction (physical impact) and computer's data theft (cyber impact).

$R_2 : occurs(?i, ?a), Intrusion(?i), SimpleBuilding(?a), Device(?d)$
$hosts(?a, ?d), Data(?t), hosts(?d, ?t) \longrightarrow impacts(?p, ?t), Impact(?p), Steal(?p)$
$R_3 : occurs(?i, ?a), Intrusion(?i), SimpleBuilding(?a), Device(?d)$
$hosts(?a, ?d) \longrightarrow impacts(?p, ?d), Impact(?p), Destruction(?p)$

(b) Code injection on a cyber system, might generate a network flooding impact.

$R_4 : occurs(?i, ?a), CodeInjection(?i), Software(?a), Network(?n)$
$leadsTo(?a, ?n) \longrightarrow impacts(?p, ?n), Impact(?p), NetworkFlooding(?p)$

The Fig. 4 shows the impacts generated by the rules R_2, R_3 and R_4. Initially, the intrusion incident occurs on the pharmacy door and propagates to pharmacy. Thanks to rule R_2, we can identify that objects hosted by the "pharmacy" (hosts is transitive) are exposed to impact "physical destruction or theft". Furthermore, computer's "data" may suffer from "theft" cyber impact. Applying rule R_3 allows detecting the impact "network flooding" on "core network" asset. This incident would propagate to all cyber assets connected to the network and be predicted thanks to additional rules (not included for space reasons).

6   Related Work

We classified related works into two categories: those dedicated to asset management and analysis, and those contributing to the information system security.

Asset Management and Analysis-Related Works. Several models are presented in the literature for asset analysis ([2,6,23,27,29]). Authors in [29] propose an ontology called OLPIT, and claim that it reflects the layering suggested by the ITIL and CoBit frameworks. It defines hierarchical relationships between the three represented levels: process level, service level, and infrastructure level. [27] represents the assets dependence chain by an oriented graph where the assets are the nodes. They are organized hierarchically into business system layer, information system layer, and system component layer. In [2] dependencies between assets are arranged in a tree-based hierarchy with the "building" asset as the top-level node. The hierarchy links are of two kinds: the "OR" and the "AND" links. In [6], a first version of a metamodel describing the Mission and Asset Information Repository (MAIR) is described. A representative set of assets is depicted with their type of dependencies on other assets through hierarchical layers from [14]: the mission layer for mission and business processes assets, the service layer containing common IT service, and asset layer gathering IT infrastructure assets. For asset analysis, in [23], four dependency layers are defined: the mission, the operational, the application, and infrastructure layers. On another side, we can also mention for the description of assets ArchiMate 2.1, an open and independent Enterprise Architecture modeling language within TOGAF Framework 9.2 and the CIM standard produced by DMTF (formerly known as the Distributed Management Task Force) that is internationally recognized by ANSI (American National Standards Institute) and ISO.

Cyber-Security Related Works. Hospital must hold perfect knowledge of its critical assets to maintain its mission despite incidents occurrence. Assets identification and documentation fall within the scope of the risk analysis, a crucial process in risk management. The latter is described in a wide range of national and international standards: ISO/IEC 27000:2008, ISO/IEC 27002:2013, and NIST SP 800-30. Some resources, like the HITRUST risk management framework [12] are associated with the healthcare sector. To achieve risk analysis, it is possible to rely on existing methodologies based on standards like LIRA [30], CRISRRAM [26] and EBIOS RM [1]. Most of these methodologies give an informal and light descriptions of assets. The literature provides also a plethora of models for security purpose ([7,15–17,31]). Most of them are ontologies. In [31] the proposed ontology is used for attack generation while [16] and [17] ontologies contribute to social engineering analysis and targeted attacks mitigation. These ontologies emphasize the link between the asset concept and the other security concepts. No description neither refinement of the concept "asset" are provided. Only a dependency link is expressed between assets.

7   Conclusion

We proposed, in this paper, an ontology for integrated cyber-physical security in healthcare systems able to support incident propagation and mitigation reasoning. Our modular ontology has been constructed around a central asset module focusing on asset management, and extending to protection and impact modules. The modular ontology structure proved to be very useful as the acquisition of domain knowledge could not be done in one shot given the variety and geographical spread of stakeholders. We showed two use cases, illustrated under a real-attack scenario, enabling asset management and incident propagation analysis features.

Acknowledgements. This research received funding from the European Union's H2020 Research and Innovation Action under SAFECARE Project, grant agreement no. 787002.

References

1. ANSSI: Ebios risk manager - the method (2019). https://www.ssi.gouv.fr/en/guide/ebios-risk-manager-the-method/
2. Breier, J., Schindler, F.: Assets dependencies model in information security risk management. In: Linawati, M.M.S., Neuhold, E.J., Tjoa, A.M., You, I. (eds.) Information and Communication Technology-EurAsia Conference, pp. 405–412. Springer, Heidelberg (2014). https://doi.org/10.1007/978-3-642-55032-4_40
3. Cristani, M., Cuel, R.: A survey on ontology creation methodologies. Int. J. Semant. Web and Inf. Syst. (IJSWIS) 1(2), 49–69 (2005)
4. Depoy, J., Phelan, J., Sholander, P., Smith, B., Varnado, G., Wyss, G.: Risk assessment for physical and cyber attacks on critical infrastructures. In: IEEE Military Communications Conference, pp. 1961–1969 (2005)
5. ENISA: Cyber security and resilience for Smart Hospitals (2016). https://www.enisa.europa.eu/publications/cyber-security-and-resilience-for-smart-hospitals
6. EU PROTECTIVE project: delivrable d4.1 (2017). https://protective-h2020.eu/
7. Fenz, S., Ekelhart, A.: Formalizing information security knowledge. In: Proceedings of the 4th International Symposium on Information, Computer, and Communications Security, pp. 183–194 (2009)
8. Fernández-López, M., Gómez-Pérez, A., Juristo, N.: Methontology: from ontological art towards ontological engineering (1997)
9. Freitas, F., Schulz, S., Moraes, E.: Survey of current terminologies and ontologies in biology and medicine. RECIIS-Electron. J. Commun. Inf. Innov. Health 3(1), 7–18 (2009)

10. Gruber, T.R.: Toward principles for the design of ontologies used for knowledge sharing? Int. J. Hum. Comput. Stud. **43**(5), 907–928 (1995)
11. Grüninger, M., Fox, M.S.: Methodology for the design and evaluation of ontologies (1995)
12. HITRUST: Healthcare sector cybersecurity framework - implementation guide v1.1 (2016). https://hitrustalliance.net/
13. Horridge, M., Knublauch, H., Rector, A., Stevens, R., Wroe, C.: A practical guide to building owl ontologies using the Protégé-OWL plugin and co-ode tools edition 1.0. University of Manchester (2004)
14. Jakobson, G.: Mission cyber security situation assessment using impact dependency graphs. In: 14th International Conference on Information Fusion, pp. 1–8 (2011)
15. Kim, B.J., Lee, S.W.: Understanding and recommending security requirements from problem domain ontology: a cognitive three-layered approach. J. Syst. Softw. **169**, 110695 (2020)
16. Li, T., Wang, X., Ni, Y.: Aligning social concerns with information system security: A fundamental ontology for social engineering. Inf. Syst. 101699 (2020)
17. Luh, R., Schrittwieser, S., Marschalek, S.: TAON: an ontology-based approach to mitigating targeted attacks (2016)
18. Masolo, C., Borgo, S., Gangemi, A., Guarino, N., Oltramari, A.: WonderWeb deliverable d17. Comput. Sci. Preprint Arch. **2002**(11), 74–110 (2002)
19. Noy, N.F., McGuinness, D.L., et al.: Ontology development 101: a guide to creating your first ontology (2001)
20. Pinto, H.S., Martins, J.P.: Ontologies: how can they be built? Knowl. Inf. Syst. **6**(4), 441–464 (2004)
21. Roussey, C., Pinet, F., Kang, M.A., Corcho, O.: An introduction to ontologies and ontology engineering. In: Ontologies in Urban Development Projects, pp. 9–38. Springer, London (2011). https://doi.org/10.1007/978-0-85729-724-2_2
22. Schauer, S., Grafenauer, T., König, S., Warum, M., Rass, S.: Estimating cascading effects in cyber-physical critical infrastructures. In: Nadjm-Tehrani, S. (ed.) CRITIS 2019. LNCS, vol. 11777, pp. 43–56. Springer, Cham (2020). https://doi.org/10.1007/978-3-030-37670-3_4
23. Silva, F.R.L., Jacob, P.: Mission-centric risk assessment to improve cyber situational awareness. Association for Computing Machinery (2018)
24. Staab, S., Studer, R.: Handbook on Ontologies. Springer, Heidelberg (2010). https://doi.org/10.1007/978-3-540-92673-3
25. Suárez-Figueroa, M.C., Gómez-Pérez, A., Fernández-López, M.: The neon methodology framework: a scenario-based methodology for ontology development. Appl. Ontol. **10**(2), 107–145 (2015)
26. Theocharidou, M., Giannopoulos, G.: Risk assessment methodologies for critical infrastructure protection. part II: a new approach (report EUR 27332) (2015)
27. Tong, X., Ban, X.: A hierarchical information system risk evaluation method based on asset dependence chain. Int. J. Secur. Appl. **8**(6), 81–88 (2014)
28. Uschold, M., Gruninger, M., et al.: Ontologies: principles, methods and applications. Technical report University of Edinburgh Artificial Intelligence Applications Institute AIAI TR (1996)
29. vom Brocke, J., Braccini, A.M., Sonnenberg, C., Spagnoletti, P.: Living it infrastructures - an ontology-based approach to aligning it infrastructure capacity and business needs. Int. J. Account. Inf. Syst. **15**(3), 246–274 (2014)

30. White, R., Burkhart, A., George, R., Boult, T., Chow, E.: Towards comparable cross-sector risk analyses: a re-examination of the risk analysis and management for critical asset protection (ramcap) methodology. Int. J. Crit. Infrastruct. Prot. **14**, 28–40 (2016)
31. Wu, S., Zhang, Y., Chen, X.: Security assessment of dynamic networks with an approach of integrating semantic reasoning and attack graphs, pp. 1166–1174 (2018)

Repurpose Image Identification for Fake News Detection

Steven Jia He Lee[1], Tangqing Li[1(✉)], Wynne Hsu[1,2,3], and Mong Li Lee[1,2,3]

[1] Institute of Data Science, National University of Singapore, Singapore, Singapore
{leejiahe,li_tangqing}@u.nus.edu
[2] NUS Centre for Trusted Internet and Community, Singapore, Singapore
[3] School of Computing, National University of Singapore, Singapore, Singapore
{whsu,leeml}@comp.nus.edu.sg

Abstract. The Internet has become the major channel where users gather and disseminate news and information, and has given rise to problems such as fake news and disinformation. This work tackles a class of fake news where visually similar images from past events are purported as visual evidence to exaggerate the severity of a current news event. These images have been repurposed and possess a great affinity to real news, posing a challenge to the fake news detection task. We propose a multi-stage approach that comprises an event type classifier to determine the type of news event, and an image repurpose detector which utilizes a siamese network to detect whether the news is fake and contains a repurposed image. Evaluation on real-world news datasets show that the proposed solution outperforms state-of-the-art methods and is effective in identifying fake news containing repurposed images.

1 Introduction

Online platforms allow users to share pictures and updates on breaking news events and give first-hand witness accounts. However, this also provides a breeding ground for malicious actors to exacerbate the chaos by spreading disinformation. Image content can be manipulated using techniques such as image tampering [6] and repurposing [7,8,15,17]. While image tampering involves content-level operations that alter parts of an image, repurposing pairs an actual image with false contextual information.

Figure 1(a) shows a tweet on the 2019 Amazon rainforest wildfire that uses an outdated image from the 1989 Amazon rainforest wildfire. Figure 1(b) is a tweet that went viral because the image shows people dying on the street from COVID-19. This image is in fact taken from a 2014 art project that remembers the victims of the Nazi's Katzbac concentration camp in Frankfurt. Figure 1(c) shows a protester returning tear gas with a tennis racket. This image, taken from the Yellow Vest Rallies in Europe, is purported to be a scene from the 2019 Hong Kong protest. These images are not digitally manipulated but are shown out of context. The impact can be serious as fake images are found to be shared twice

© Springer Nature Switzerland AG 2021
C. Strauss et al. (Eds.): DEXA 2021, LNCS 12924, pp. 35–47, 2021.
https://doi.org/10.1007/978-3-030-86475-0_4

(a) 2019 Amazon rainforest fire using an outdated photo from 1989 Amazon rainforest fire.

(b) People purported to be dying from coronavirus using an art project image.

(c) 2019 Hong Kong Protest using an image from Yellow Vest Rallies in Europe.

Fig. 1. Example of real world repurposed images.

more than real images [4]. Further, images are deemed to be more credible as they are harder to fabricate compared to textual content.

Detecting repurposed images is a challenge because malicious actors will use visually similar photographs from past events and purport them as visual evidence to exaggerate the severity of a current news event. Repurposed images are often detected manually which requires human expertise, time, and resources. [7] attempts to automate the detection of repurposed images by determining the semantic consistency of an image and its caption, and [17] extended it to include location information. However, the repurposed image in Fig. 1(a) will not be detected because the visual, textual, and location are semantically consistent although the image comes from a different event in the past.

In this work, we introduce a framework called RECAST to detect fake news by identifying repurposed images taken from previous events that have been rebranded with malicious intent. To ascertain the credibility of a piece of news, we classify the type of event that is described by the news and then retrieve the most similar event from a reference base corresponding to the event type. Both the news and the retrieved news are then passed to a repurpose detector to determine credibility. Experiments results indicate the effectiveness of the proposed approach to detect repurpose images in fake news, with improved accuracy over state-of-the-art solutions.

2   Related Work

Existing fake news detection based on news content has focused on common topics, lexical usage, and linguistic style [19], as well as news comments [1]. [18] employs a database of genuine news to identify news whose title disagrees with some previous news. However, this approach only uses textual contents and may be limited in discovering fake news patterns with higher complexity.

[9] proposes a model that takes into account text and image. The authors treat textual content related to social interactions such as tagging other users as social factors and separate them from regular text modality. A recurrent neural network is adopted to jointly learn textual and social features. These features, together with visual features extracted by a pre-trained convolutional neural network model, are fused to determine the veracity of a news. [20] captures written and visual patterns in fake news by extracting text and image features using a pre-trained deep learning model. The joint features are then passed to a fake news detector and an event classifier simultaneously to detect fake news. [11] employs a variational autoencoder to learn the latent features of images and texts, while [22] assesses the semantic coherence between image and text to characterize fake news. These methods assume that the multi-modal features of fake news have a different distribution from that of genuine news. As such, they may not be effective in identifying repurposed images.

Image repurposing is first investigated in [7] which assesses the semantic consistency between an image and its corresponding caption. A reference base of untampered image-caption pairs is used. Semantically inconsistent image-caption pairs are created by swapping the caption of an image randomly with the caption of another image to train a model. This work does not consider the locations and entities of the image-caption pairs. Hence, the model may not detect images that have been repurposed from similar event types which are likely to be semantically consistent with the captions used.

[17] incorporates GPS information and applies named entity recognition on the image captions to identify the location, person, or organisation. However, this approach may not be able to detect if an image has been repurposed from similar past events. [8] develops an adversarial network with a bad actor who produced tampered meta-data and a watchdog who verifies the semantic consistency between images and their corresponding metadata. [15] describes an unsupervised cross-modal consistency verification system that crawls reference images for named entities extracted from the text, and compares the consistency between the query image and the reference image. These works do not distinguish the different events and may not work well for tampered news due to the similarity of the repurposed images from past similar events.

3   Proposed Framework

Figure 2 shows an overview of the proposed RECAST framework comprising of two modules: an event type classifier and a repurpose detector. The input to RECAST is a social media post on an event. An event can be defined as an incident that takes place in some location over a period of time. The location and time of an event will help to differentiate one event from another, e.g., "Wildfire in Amazon rainforest in 2019" vs "Bushfire in Australia in 2020". Events can be classified into different types, e.g., "World Cup 2018" and "European Cup 2016" belongs to the Sports event type.

We parse the content and metadata of the query post to extract the textual content txt, associated image img, event location loc and time $time$. The location

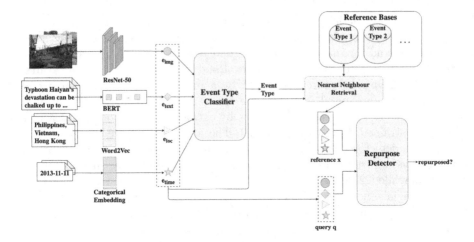

Fig. 2. Overview of RECAST.

is obtained by applying named entity recognition on the post, while the event time is assumed to be the time of the social media post. Then we employ a ResNet-50 [5] architecture pre-trained with ImageNet [2] and fine-tuned over a news corpus to extract an 2048-dimensional feature vector from the image, denoted as e_{img}. A pre-trained BERT model[1] whose last 4 layers has been fine-tuned with the same news corpus is used to obtain a 1536-dimensional word embedding from the text, denoted as e_{txt}. We use the Word2Vec model [14] pre-trained with Google News corpus to obtain a 300-dimensional word embedding of the event location e_{loc}. We perform categorical embedding of the day (1–31), month (1–12) and year (2000–2019) on the event time to obtain a 96-dimensional embedding vector e_{time}. These embeddings are passed to the event type classifier which predicts the event type of the query post. The event type and the embeddings are used to retrieve the most similar tuple from the corresponding reference base.

Recognizing that the characteristics of image, text and spatiotemporal features differ across different event types, we construct a reference base of historical posts for each event type. Each historical post is a tuple $\langle id, e_{img}, e_{txt}, e_{loc}, e_{time} \rangle$. Note that the images in these posts have not been repurposed. The retrieved tuple, together with the embeddings, are sent to the image repurpose detector to determine if the image have been repurposed. For the image repurpose detector, our goal is to retrieve some tuple r in a reference base such that for a given news query q, if the image in r is close to the image in q but the text description of r is different from that of q, then we can conclude that q contains fake information.

3.1   Event Type Classifier

Figure 3 shows the architecture of the event type classifier. It has four modules, one for each modality. Each module comprises of two fully connected layers.

---

[1] https://huggingface.co/transformers/model_doc/bert.html.

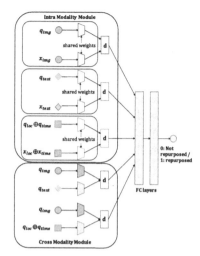

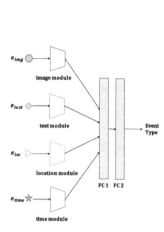

Fig. 3. Event type classifier.   **Fig. 4.** Image repurpose detector.

The embedding vector of each modality is passed to the respective module to obtain the corresponding feature representation. These feature representations are 128-dimensional vectors obtained as follows:

$$f_{img} = \sigma(W_{img} \cdot e_{img})$$
$$f_{txt} = \sigma(W_{txt} \cdot e_{txt})$$
$$f_{loc} = \sigma(W_{loc} \cdot e_{loc})$$
$$f_{time} = \sigma(W_{time} \cdot e_{time})$$

where W_{img}, W_{txt}, W_{loc} and W_{time} are weight matrices of fully connected layers.

These feature representations are then passed through two fully connected layers to obtain the event type. We adopt the multi-class cross entropy loss to be the loss function of our event type classifier.

3.2   Image Repurpose Detector

Our image repurpose detector map the different modalities into a metric space to evaluate their intra-modality and cross-modality distances (see Fig. 4).

We use the reference base $\mathcal{R}$ corresponding to the predicted event type of q to find the nearest tuple x whose image and text are most similar to q's. This is achieved by computing the pairwise L_2 norm distance between the images and texts in q and that of the tuples in $\mathcal{R}$ using FAISS index [10]. The tuple x with the smallest pairwise distance is given by:

$$x = argmin_{r \in \mathcal{R}}(||q.img, r.img||_2 + ||q.text, r.text||_2) \tag{1}$$

We map the different modalities of q and x into a deep metric space and evaluate their intra-modality and cross-modality distances before passing them to the fully connected layers to determine if q has been repurposed.

Intra-modality Module. The intra-modality module uses three siamese neural networks [13]. Each siamese neural network consists of twin networks which accept some modality in the query and reference tuple as inputs. The parameters between the twin networks are tied to ensure that the feature representations of the two inputs are close in the deep metric space. Each network has L layers.

The first twin network takes as inputs the image embedding vectors $q.img$ and $x.img$ and pass them through the L fully connected layers before computing the induced distance vector. The distance vector obtained from the learned feature space is represented as:

$$dist(siamese1(q.img, L), siamese1(x.img, L)) \tag{2}$$

where $dist(.)$ is a distance function, and $siamese1(.)$ denotes the corresponding output vector of the first siamese network.

The second twin network takes as inputs the text embedding vectors $q.txt$ and $x.txt$ and again pass them through L fully connected layers before obtaining the distance vector as follows:

$$dist(siamese2(q.txt, L), siamese2(x.txt, L)) \tag{3}$$

The third twin network looks at the joint space of location and time modality with the distance vector computed as

$$dist(siamese3(q.loc \oplus q.time, L), siamese3(x.loc \oplus x.time, L)) \tag{4}$$

Cross-Modality Module. The cross modality module in Fig. 4 measures the coherence among the text, image, location and time modalities of the query tuple q. For the coherence between image and text, we transform the image and text embeddings of q into a latent feature space through feedforward layers, and obtain the image-text distance vector as follows:

$$dist(W_1 \cdot q.img, W_1 \cdot q.txt) \tag{5}$$

where $dist(.)$ is a distance function, and W_1 denotes the weight matrix of the feedforward layers.

Similarly, we determine the image-spatiotemporal distance vector by

$$dist(W_2 \cdot q.img, W_2 \cdot q.loc \oplus q.time) \tag{6}$$

where W_2 denotes the weight matrix of the feedforward layers.

Finally, we concatenate all the intra- and cross- modality distance vectors and pass it through multiple fully connected layers to predict if q has been repurposed. The output is $\hat{y}$ which is 1 if the model predicts that q has been repurposed, and 0 otherwise.

Learning. We initialize all the network weights in the intra-modality and cross modality fully connected layers to zero-mean and standard deviation as $1/n$ where n is the number of incoming nodes. The training is performed on mini-batches of size 64. For each tuple x in the minibatch, we retrieve the nearest neighbor x'. If the ground truth label of x is 1, indicating that it contains a repurposed image, we set $y(x, x') = 1$, and 0 otherwise. Let $\hat{y}(x, x')$ be the predicted label. We define the loss function as follows:

$$\mathcal{L}_{\text{repurpose}} = y \log(\hat{y}) + (1 - y) \log(1 - \hat{y}) \tag{7}$$

4   Experimental Evaluation

We implemented the proposed framework in PyTorch and carried out experiments to evaluate its effectiveness in identifying fake news. The event type classifier in RECAST is trained for 50 epochs using the Adam optimizer [12]. The learning rate is initially set to be 0.001 and is reduced by half after every 5 epochs. We add a dropout layer before the fully connected layers. The probability parameter of the dropout layer is set to 0.5. The repurpose detector model is trained for 100 epochs with early termination with the Adam optimzier [12]. The learning rate is initially set to be 0.001 and is decayed by 2% each epoch.

4.1   Experimental Datasets

We follow the same protocol described in [7,17] to curate the following datasets:

TamperedNews-Event. This is a subset of TamperedNews dataset [15] where we select news that mentions event entities. The events are extracted from articles in BreakingNews dataset [16] using named entity recognition. We create re-purposed tuples by replacing the images from other events of the same type and further tamper the locations and timestamps of the events.

TamperedWiki-Flickr. We create this dataset from Wiki-Flickr [21] which has both images and text. We utilize the named entity recognition model in [3] to retrieve locations from the text. The time of the post is assumed to be the time of the event. We generate tampered tuples by swapping events with other events of the same type, as well as their location and time.

Table 1 shows the characteristics of the datasets. Note that TamperedNews-Event has more events and fewer news tuples per event.

4.2   Experiments on Event Type Classification

Table 2 shows the performance of RECAST event type classifier. We see that it achieves more than 85% accuracy for both datasets. This is remarkable especially for TamperedNews-Event where the *sports competition* category contains sports events ranging from soccer to tennis which is visually more diverse than the categories such as *flood* and *storm* in TamperedWiki-Flickr.

Effect of Different Modality. Recall that the event type classifier takes as input different modalities. Table 3 shows the effect of these modalities on the accuracy of the classifier. Using image and spatio-temporal information leads to lower accuracy compared to using text and spatio-temporal information.

Table 1. Dataset statistics. *#Events* is the number of unique events in a category.

Dataset	Event type	Category	#Events	#Tuples
TamperedNews-Event	Competition	Competition	16	1199
		Sports competition	91	1326
	Festival	Festival	71	643
		Award	6	395
		Holiday	31	969
		Convention	10	99
	War	War	47	1712
		Shooting	12	105
		Disaster	4	125
TamperedWiki-Flickr	Political	Election	9	812
		Demonstration	15	1767
	Disaster	Earthquake	4	616
		Storm	4	423
		Flood	4	386
	War	War	8	1733
		Terror attack	9	1326

Table 2. Accuracy of event type classifier.

Event type	Accuracy
Competition	0.86
Festival	0.85
War	0.88
Overall	**0.86**

(a) TamperedNews-Event

Event type	Accuracy
Political	0.97
Disaster	0.97
War	0.98
Overall	**0.97**

(b) TamperedWiki-Flickr

Table 3. Effect of modalities on the accuracy of event type classifier.

Dataset	Event type	Image+Text	Image+ST	Text+ST	Image	Text	ST
Tampered News-Event	Competition	0.82	0.77	0.84	0.63	0.83	0.73
	Festival	0.81	0.78	0.83	0.63	0.77	0.71
	War	0.86	0.73	0.86	0.61	0.85	0.73
	Overall	**0.82**	**0.76**	**0.84**	**0.62**	**0.81**	**0.72**
Tampered Wiki-Flickr	Political	0.92	0.85	0.95	0.67	0.91	0.84
	Disaster	0.93	0.83	0.96	0.53	0.92	0.84
	War	0.95	0.88	0.95	0.80	0.93	0.85
	Overall	**0.93**	**0.85**	**0.95**	**0.70**	**0.92**	**0.84**

This suggests that image is less significant than the textual content of news for event type classification. This is because text contains more detailed information of the news event, whereas image only shows a snapshot of the event.

Further, we observe that using text and spatio-temporal information leads to higher accuracy compared to using text and image. One possible reason is that image and text tend to have overlapping information. As a result, including images does not achieve as much improvement as spatial-temporal information.

4.3 Comparative Study

We compare RECAST with the following approaches:

EANN [20]. This method that comprises of three components: a multi-modal feature extractor for images and textual data, a fake news detector that learns discriminative news representations, and an event discriminator. The network is trained adversarially to ensure the learned representation is independent of event-specific features and is transferable to detect fake news on unseen events.

MVAE [11]. This method first applies a pretrained VGG-19 and a recurrent neural network to respectively extract image and text features from the news. A bi-modal variational autoencoder is then trained to encode and reconstruct the extracted visual and textual features. These latent representations are shared with a fake news detection module consisting of fully-connected layers to predict if the news is fake or not.

CCV [15]. This method employs named entity linking to extract event entities mentioned in the text and automatically collects images using from various search engines to construct a reference base for the recognized event entities. The method then calculates the similarity between the news image and images from the reference base for cross-modal consistency verification.

We follow the experiment protocol in [15,20], and report the precision, recall, F1 and AUROC. Table 4 shows the results. Overall, our framework achieves the best performance indicating that it is better able to discriminate fake news from the real ones. Note that the results for RECAST includes the error, if any, made by the event type classifier. We also conducted an experiment where we use the union of the all the event type reference bases to retrieve the nearest neighbor, in other words, we omit the event type classifier. The results show a 5% reduction in precision, 2% reduction in recall, 4% reduction in F1, 4% reduction in accuracy, and 2% reduction in AUROC on average.

Table 4. Results of the comparative study.

Method	TamperedNews-Event				TamperedWiki-Flickr			
	Precision	Recall	F1	AUROC	Precision	Recall	F1	AUROC
EANN	0.60	0.64	0.62	0.63	0.52	0.74	0.61	0.65
MVAE	0.69	0.74	0.72	0.73	0.73	0.61	0.67	0.72
CCV	0.58	0.76	0.66	0.64	0.70	**0.91**	0.78	0.83
RECAST	**0.72**	**0.79**	**0.75**	**0.80**	**0.87**	0.79	**0.83**	**0.91**

MVAE detects fake news by assessing patterns in the linguistic and visual modality of news without relying on external sources of knowledge to verify the credibility of the modalities. When an image in a piece of news is replaced by some image from other similar genuine events, MVAE is unable to detect the fake news based on the modalities of the query news alone. By utilising an external reference base, RECAST is able to utilize the most similar reference news to better predict if the query news is fake.

Although CCV has a high recall of 0.91 in TamperedWiki-Flickr, its precision is low, resulting in a lower F1 score compared to RECAST. Both RECAST and CCV utilize external reference base to verify the truthfulness of news. However, when the retrieved news image is visually similar to the query news, CCV will assign a high credibility score to the query news. As a result, when fake news uses a repurposed image taken from other events that have similar visual scenes to the query event, CCV is not able to detect it as fake news.

4.4 Variants of RECAST

We also examine the effect of different modalities on the performance of the repurpose detector in RECAST. For this experiment, we use the same event type classifier which employ all the modalities to predict the event type, while the repurpose detector uses subsets of modalities as follows:

(I+T) uses image and text information.

(I+S) uses image and spatio-temporal information.

(I+T+S) includes all the modalities.

Table 5 shows the results. We observe that (I+S) produces better AUROC score than (I+T) on both datasets, suggesting that spatio-temporal information contributes more to the detection of repurpose images compared to textual context.

Table 5. Results of the different variants of RECAST.

Method	TamperedNews-Event				TamperedWiki-Flickr			
	Precision	Recall	F1	AUROC	Precision	Recall	F1	AUROC
(I+T)	**0.79**	0.68	0.73	0.76	0.80	0.76	0.78	0.85
(I+S)	0.76	0.70	0.73	0.78	0.81	**0.80**	0.80	0.87
(I+T+S)	0.72	**0.79**	**0.75**	**0.80**	**0.87**	0.79	**0.83**	**0.91**

4.5 Case Study

Finally, we present a case study from TamperedWiki-Flickr to demonstrate the effectiveness of RECAST in detecting fake news containing repurposed images.

Table 6 shows a fake news with a repurposed image taken from the *Tohoku earthquake and tsunami* event to exaggerate the severity of the *Pakistan flood*

Table 6. Fake news detected by RECAST but missed by EANN, MVAE, and CCV.

	Image	Text	Spatial-temporal
Original image source News event: *Tohoku earthquake and tsunami*		A 7.3-magnitude earthquake struck early Saturday morning off Japan's east coast, near the crippled Fukushima nuclear site ...	Tokyo, Fukushima, Japan 2011-03-11
Query news q News event: *Pakistan floods*		Millions of people have been affected by the flood waters in large parts of Pakistan...	Tokyo, Fukushima, Japan 2011-03-11
Retrieved news x News event: *Tohoku earthquake and tsunami*		Hundreds more were reported missing after waves as high as 23 feet swept ashore, according to state broadcaster NHK...	Tokyo, Japan, U.S. 2011-03-11

event. Both the image and textual contents of the query news are taken from historical events and are semantically coherent. As such both EANN and MVAE are unable to detect this news as fake. Since CCV only considers the image modality, it will retrieve multiple images related to the *Pakistan flood* and determine the cosine similarity between the visual features of the query image and reference images. However, the images of the Pakistan flood event happen to be highly similar to that of the *Tohoku earthquake and tsunami*, resulting in CCV's failure to correctly identify the query news as fake news. In contrast, RECAST takes into account visual, textual, and spatio-temporal modalities enabling it to correctly detect this query as fake news.

5   Conclusion

In this work, we have addressed the problem of detecting repurposed images in fake news. We have proposed a framework that takes into account image, text, and spatio-temporal information in the query news. The framework first identifies the event type of the query and retrieves the most similar news from a reference base containing historical news of the same event type. Our repurpose detector considers the cross-modality and intra-modality consistency between the retrieved news and query news. Experimental results on multiple datasets demonstrated that RECAST is effective in detecting repurposed images in news. Future work includes incorporating the proposed framework into a platform for checking fake news at scale.

References

1. Beelen, K., Kanoulas, E., van de Velde, B.: Detecting controversies in online news media. In: ACM SIGIR (2017)
2. Deng, J., Dong, W., Socher, R., Li, L.J., Li, K., Fei-Fei, L.: ImageNet: a large-scale hierarchical image database. In: CVPR (2009)
3. Finkel, J.R., Grenager, T., Manning, C.D.: Incorporating non-local information into information extraction systems by Gibbs sampling. In: ACL (2005)
4. Gupta, A., Lamba, H., Kumaraguru, P., Joshi, A.: Faking sandy: characterizing and identifying fake images on Twitter during hurricane sandy. In: World Wide Web Conference (2013)
5. He, K., Zhang, X., Ren, S., Sun, J.: Deep residual learning for image recognition (2015). arXiv preprint arXiv:1512.03385 (2016)
6. Huh, M., Liu, A., Owens, A., Efros, A.A.: Fighting fake news: image splice detection via learned self-consistency. In: ECCV (2018)
7. Jaiswal, A., Sabir, E., AbdAlmageed, W., Natarajan, P.: Multimedia semantic integrity assessment using joint embedding of images and text. In: ACM MM (2017)
8. Jaiswal, A., Wu, Y., AbdAlmageed, W., Masi, I., Natarajan, P.: AIRD: adversarial learning framework for image repurposing detection. In: CVPR (2019)
9. Jin, Z., Cao, J., Guo, H., Zhang, Y., Luo, J.: Multimodal fusion with recurrent neural networks for rumor detection on microblogs. In: Proceedings of the 25th ACM international conference on Multimedia, pp. 795–816 (2017)
10. Johnson, J., Douze, M., Jégou, H.: Billion-scale similarity search with GPUs. IEEE Trans. Big Data (2019)
11. Khattar, D., Goud, J.S., Gupta, M., Varma, V.: MVAE: multimodal variational autoencoder for fake news detection. In: World Wide Web Conference (2019)
12. Kingma, D.P., Ba, J.: Adam: a method for stochastic optimization. arXiv preprint arXiv:1412.6980 (2014)
13. Koch, G., Zemel, R., Salakhutdinov, R.: Siamese neural networks for one-shot image recognition. In: ICML Deep Learning Workshop, vol. 2 (2015)
14. Mikolov, T., Sutskever, I., Chen, K., Corrado, G.S., Dean, J.: Distributed representations of words and phrases and their compositionality. In: NeurIPS (2013)
15. Müller-Budack, E., Theiner, J., Diering, S., Idahl, M., Ewerth, R.: Multimodal analytics for real-world news using measures of cross-modal entity consistency. In: International Conference on Multimedia Retrieval (2020)
16. Ramisa, A., Yan, F., Moreno-Noguer, F., Mikolajczyk, K.: BreakingNews: article annotation by image and text processing. IEEE Trans. Pattern Anal. Mach. Intell. **40**(5), 1072–1085 (2017)
17. Sabir, E., AbdAlmageed, W., Wu, Y., Natarajan, P.: Deep multimodal image-repurposing detection. In: ACM Multimedia (2018)

18. Su, T., Macdonald, C., Ounis, I.: Ensembles of recurrent networks for classifying the relationship of fake news titles. In: ACM SIGIR (2019)
19. Vo, N., Lee, K.: Learning from fact-checkers: analysis and generation of fact-checking language. In: ACM SIGIR (2019)
20. Wang, Y., Ma, F., Jin, Z., et al.: EANN: event adversarial neural networks for multi-modal fake news detection. In: ACM SIGKDD (2018)
21. Yang, Z., Lin, Z., Kang, P., Lv, J., Li, Q., Liu, W.: Learning shared semantic space with correlation alignment for cross-modal event retrieval. ACM Trans. Multimed. Comput. Commun. Appl. 16(1), 1–22 (2020)
22. Zhou, X., Wu, J., Zafarani, R.: SAFE: similarity-aware multi-modal fake news detection. In: Advances in Knowledge Discovery and Data Mining (2020)

Data and Information Processing

An Urgency-Aware and Revenue-Based Itemset Placement Framework for Retail Stores

Raghav Mittal[1], Anirban Mondal[1], Parul Chaudhary[2(✉)],
and P. Krishna Reddy[3]

[1] Ashoka University, Sonipat, India
raghav.mittal@alumni.ashoka.edu.in, anirban.mondal@ashoka.edu.in
[2] Shiv Nadar University, Greater Noida, India
pc230@snu.edu.in
[3] IIIT, Hyderabad, India
pkreddy@iiit.ac.in

Abstract. Placement of items on the shelf space of retail stores sig-nifcantly impacts the revenue of the retailer. Given the prevalence and popularity of medium-to-large-size retail stores, several research efforts have been made towards facilitating item/itemset placement in retail stores for improving retailer revenue. However, they do not consider the issue of *urgency of sale* of individual items. Hence, they cannot efficiently index, retrieve and place high-revenue itemsets in retail store slots in an *urgency-aware* manner. Our key contributions are two-fold. First, we introduce the notion of urgency for retail itemset placement. Sec-ond, we propose the *urgency-aware* URI index for *efficiently* retrieving high-revenue and urgent itemsets of different sizes. We discuss the URIP itemset placement scheme, which exploits URI for improving retailer rev-enue. We also conduct a performance evaluation with two real datasets to demonstrate that URIP is indeed effective in improving retailer revenue w.r.t. existing schemes.

Keywords: Utility mining · Retail stores · Urgent items · Pattern mining

1 Introduction

Placement of items in the slots of the shelf space of retail stores significantly impacts the revenue of the retailer [6–8,10]. Large retail stores (e.g., Walmart Supercenters, Dubai Mall) with huge retail floor space have become increasingly prevalent. Notably, retail store slots are either *premium* (i.e., slots with high visi-bility/accessibility to customers) or *non-premium* (i.e., low visibility/accessibility slots). Premium slots are those that are near to the eye or shoulder level of cus-tomers and impulse-buy slots at checkout counters; other slots are non-premium. Items placed in premium slots have a significantly higher probability of sale than

© Springer Nature Switzerland AG 2021
C. Strauss et al. (Eds.): DEXA 2021, LNCS 12924, pp. 51–57, 2021.
https://doi.org/10.1007/978-3-030-86475-0_5

items placed in non-premium slots; hence, item placement in premium slots significantly impacts retailer revenue [6,7]. Furthermore, customers often prefer the convenience of *one-stop shopping* i.e., buying sets of items together i.e., *itemsets* [3] (e.g., {*bread, butter, jam*}) instead of buying individual items. *Hence, this work addresses itemset placement in the retail slots to exploit such associations in customer purchase patterns only for premium slots.*

Retailers stock fast-moving consumer goods (FMCG) [4,12], which are characterized by frequent purchases. FMCG include packaged foods, beverages, fashion items (e.g., perfumes, cosmetics, apparel) and consumer electronics. The FMCG market is estimated to reach revenue of \$15,361.8 billion globally by 2025 [1]. Notably, FMCG need to be sold **urgently** due to reasons such as perishability (e.g., for packaged foods), change of fashion styles (e.g., for apparel fashion) and obsoleteness (e.g., for mobile phones); otherwise, the retailer will lose significant revenue. We shall refer to FMCG as **urgent items**.

Existing utility mining approaches identify high-utility itemsets (HUIs). The HUI-Miner algorithm [11] stores itemset utility values in a specialized data structure called *the utility-list*. The MinFHM algorithm [9] uses pruning methods and optimization strategies for extracting MinHUIs (minimal HUIs) i.e., the smallest itemsets with high utility. Utility-based itemset placement approaches for improving retailer revenue consider varied physical sizes of the items [5,7], varied premiumness of the retail store slots [6]. In particular, the TIPDS itemset placement scheme [7] exploits the Slot Type Utility (STU) index [5,7] for quickly retrieving the top-utility itemsets. From different levels of STU, TIPDS allocates itemsets in a round-robin manner until all premium slots are filled. Unlike our proposed URI index, STU does not consider the urgency issue. *Notably, none of the existing works address the issue of* urgency-aware *itemset placement in retail stores for improving retailer revenue.*

We address the problem of *urgency-aware* retail itemset placement for improving retailer revenue. Based on the itemsets extracted from user purchase transactions, the problem is to (a) identify high-revenue itemsets with consideration for their urgency and (b) place such itemsets in a given number of premium slots for improving retailer revenue. We introduce the notion of urgency scores of items to quantify the *urgency of sale* of any given item. Further, we introduce a hybrid urgency revenue score δ to prioritize urgent and high-revenue itemsets for placement. We propose the **U**rgency-aware high-**R**evenue itemset **I**ndexing scheme (**URI**) for efficient indexing and retrieval of urgent and high-revenue itemsets. The core idea of URI is to store *only* the top-λ itemsets based on δ at each index level to reduce candidate itemset generation overhead. We also propose an effective itemset placement scheme, designated as **U**rgency-aware high-**R**evenue **I**temset **P**lacement scheme (**URIP**), for improving retailer revenue. URIP exploits our proposed URI index for strategically placing urgent and high-revenue itemsets. Our key contributions are two-fold:

1. We introduce the notion of *urgency* for retail itemset placement.
2. We propose the *urgency-aware* URI index for *efficiently* retrieving high-revenue and urgent itemsets of different sizes. We discuss the URIP itemset placement scheme, which exploits URI for improving retailer revenue.

We have conducted a performance study with two real datasets to demonstrate that URIP is indeed effective in improving retailer revenue w.r.t. existing schemes. To the best of our knowledge, this is the first work to address *urgency-aware* itemset placement in retail stores.

2 Proposed Framework of the Problem

Consider a finite set Υ of m items $\{i_1, i_2, i_3, ..., i_m\}$. We assume that each item is of the same size and consumes only one slot on the retail store shelves. Each item i_j is associated with a price ρ_{i_j}, a frequency of sales σ_{i_j} and an urgency score θ_{i_j}. Here, $0 < \theta_{i_j} \leq 1$ for urgent items, and 0 otherwise.

Consider a set D of user purchase transactions, where each transaction comprises a set of items from set Υ. We define an itemset of size k as a set of k *distinct* items $\{i_1, i_2, ..., i_k\}$. We assume that each item occurs in a given transaction only once. Notably, this work uses revenue as an example of a utility measure. We shall use the terms revenue, net revenue and utility interchangeably.

The **net revenue** NR_{i_j} of an item i_j is the product of its price and its frequency of sales i.e., $(\rho_{i_j} \times \sigma_{i_j})$. Moreover, the **net revenue** NR_z of an itemset z of size k is the sum of the prices of all k items in z multiplied by the frequency of sales of z i.e., $(\sum_{i=1}^{k} \rho_i) \times \sigma_z$. The **urgency score** θ_z of a given itemset z is the average of the urgency values of all k items in z i.e., $(\sum_{i=1}^{k} \theta_i)/k$. **Hybrid urgency revenue score** δ_z of an itemset z is designed to consider both revenue and urgency. We compute δ_z as $NR_z \times (1 + \theta_z)$. Figure 1 depicts an illustrative example for the aforementioned computations. For example, the net revenue of itemset $\{B,C,E\} = (2 + 6 + 3) * 2$ i.e., 22; the urgency score of $\{A,D\}$ is $(0.8 + 0.1)/2$ i.e., 0.45 and the hybrid urgency-revenue score of $\{A,D\}$ is $24 * (1 + 0.45)$ i.e., 34.8.

Information about Items

Item	A	B	C	D	E
ρ	7	2	6	1	3
σ	3	2	9	5	4
θ	0.8	0.2	0	0.1	0

Item = A to E, σ: Frequency of sales, ρ: Price, NR: Net Revenue, θ : Urgency Score, δ : Hybrid urgency-revenue score				
Itemset	ρ	σ	NR	θ	δ
A,D	(7+1)=8	3	8*3=24	(0.8+0.1)/2=0.45	24*(1+0.45)=34.8
B,C,E	(2+6+3)=11	2	11*2=22	(0.2+0+0)/3=0.06	22*(1+0.06)=23.32

Fig. 1. Example for the computation of NR, θ and δ

Problem Statement: Consider a finite set Υ of m items, where each item i_j is associated with a price ρ_{i_j}, frequency of sales σ_{i_j} and urgency θ_{i_j}. Furthermore, consider a set D of user purchase transactions on the items in Υ. Given N *premium* slots in a retail store, the problem is to place items/itemsets in these slots such that the total revenue of the retailer is improved.

3   URIP: Urgency-Aware Itemset Placement Scheme

We first present our proposed URI indexing scheme, which is used by our proposed URIP itemset placement scheme. Then we discuss URIP in detail.

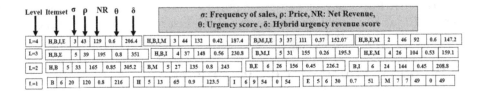

Level	Itemset	σ	ρ	NR	θ	δ		Itemset	σ	ρ	NR	θ	δ		Itemset	σ	ρ	NR	θ	δ		Itemset	σ	ρ	NR	θ	δ
L=4	H,B,I,E	3	43	129	0.6	206.4		H,B,I,M	3	44	132	0.42	187.4		B,M,I,E	3	37	111	0.37	152.07		H,B,E,M	2	46	92	0.6	147.2
L=3	H,B,E	5	39	195	0.8	351		H,B,I	4	37	148	0.56	230.8		B,M,I	5	31	155	0.26	195.3		H,E,M	4	26	104	0.53	159.1
L=2	H,B	5	33	165	0.85	305.2		B,M	5	27	135	0.8	243		B,E	6	26	156	0.45	226.2		B,I	6	24	144	0.45	208.8
L=1	B	6	20	120	0.8	216		H	5	13	65	0.9	123.5		I	6	9	54	0	54		E	5	6	30	0.7	51

σ: Frequency of sales, ρ: Price, NR: Net Revenue, θ: Urgency score, δ: Hybrid urgency revenue score

(M | 7 | 7 | 49 | 0 | 49)

Fig. 2. Illustrative example of the URI indexing scheme

Algorithm 1: Urgency and Revenue based Itemset Placement (URIP)

Input: URI index, N: no. of premium slots, Inventory of each item
Output: Placement of the items/itemsets in N slots

1 Initialize Slots[N] to NULL; CAS=N; ptr p;
2 **while** ($CAS > 1$)
3   **for** (j=1 to λ)
4   **if** (maxL%2==0) **then** p=(maxL/2)+1 **else** p = $\lceil maxL/2 \rceil$
5   **for** (i=0 to $\lceil (maxL/2) \rceil$-1) **if** i%2==0 **then** p=p+i **else** p=p-i
6   Select itemset X from URI[p][j]
7   **if** (inventory of each item in itemset X $\neq$ 0) **then**
8   Place itemset X in slots[N]; CAS = CAS - $|X|$;

URI Indexing Scheme: The core idea of URI is to store the top-λ urgent and high-revenue itemsets of different sizes instead of storing all possible itemsets of any specific size. URI is a multi-level index comprising N levels, where the kth level corresponds to itemsets of size k. Each level of URI corresponds to a hash bucket. For finding the top-λ high-revenue urgent itemsets of a given size k, URI can quickly traverse to the hash bucket corresponding to the kth level. Now, for each level k in URI, the corresponding hash bucket contains a pointer to a linked list of the top-λ high-revenue urgent itemsets of size k. The entries of the linked list are of the form $(its, \rho, \sigma, NR, \theta, \delta)$, where its refers to the given itemset. Here, ρ is the total price of all the items in its, σ is the frequency of sales of its, NR is the net revenue of its, θ is the urgency score of its and δ is the hybrid urgency revenue score of its. The values of NR, θ and δ are computed as discussed in Sect. 2. The entries in the linked list are sorted in descending order of δ to facilitate quick retrieval of the top-λ itemsets of any size k based on both urgency and revenue; any ties are resolved arbitrarily.

URI is built on a level-by-level basis starting with itemsets of size 1 all the way up to the maximum specified level of the index. Itemsets of size 1 are selected based on net revenue and urgency scores. Level 2 of URI is built by combining itemsets of size 1, level 3 is built by combining itemsets of sizes 1 and 2 respectively, and so on. Figure 2 depicts an example of URI with $\lambda = 5$.

Urgency and Revenue Based Itemset Placement (URIP) Scheme: The core idea of URIP is to extract high-revenue and urgent itemsets of different sizes from different levels of URI, starting from the middle level of URI and then branching out progressively to the next higher and the next lower levels. At each level, the extraction starts with the top-1 itemset at that level followed by the top-2 itemset and so on. Observe how URIP prioritizes the placement of itemsets of different sizes. Algorithm 1 depicts the URIP algorithm.

4   Performance Evaluation

Our experiments use two real datasets from the SPMF open-source data mining library [2]. The *Chainstore* retail dataset has 46,086 items and 1,112,949 user purchase transactions, along with item utility (price) values. The *Retail* dataset from a Belgian retail store has 16,470 items and 88,162 transactions. Since *Retail* does not provide utility values, we generated item prices in the [0,1] range by considering ten equal-ranged buckets within this range.

For assigning urgency scores θ to items, we randomly select $U_P\%$ (urgency factor) of all items to be urgent items; for all other items, $\theta = 0$. We assign value of θ to each urgent item for both datasets by considering 8 equal-ranged buckets in the [0,1] range. We assign inventory (number of instances of an item available with the retailer) to items using 8 buckets of equal range in the ranges of [1,750] (for *Retail*) and [1,500] (for *Chainstore*).

We divided each dataset into two parts i.e., training set and test set containing 70% and 30% of transactions respectively. We did placement using training set and evaluated the performance on test set. We measured the execution time (ET) for itemset placement for the training set and the total revenue (TR) of the retailer for the test set. We set the number of top high-utility itemsets per level of the index to 2000 and the total number of slots (T_S) to 6000.

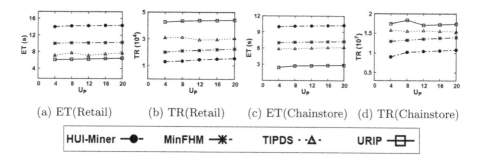

(a) ET(Retail)   (b) TR(Retail)   (c) ET(Chainstore)   (d) TR(Chainstore)

| HUI-Miner ──●── | MinFHM ──✳── | TIPDS ··△· | URIP ──⊟── |

Fig. 3. Effect of variations in Urgency Percentage (U_P)

As reference, we adapt existing schemes, namely MinFHM [9], HUI-Miner [11] and TIPDS [7]. Both MinFHM and HUI-Miner extract high-utility itemsets, but do not perform itemset placement. Using their extracted high-utility itemsets as input, we use our URIP scheme for itemset placement; we shall henceforth refer to these schemes as **MinFHM** and **HUI-Miner** respectively. Since TIPDS itself performs itemset placement, we directly compare w.r.t. TIPDS.

Figure 3 depicts the effect of variations in the percentage of urgent items (U_P) for both datasets. Reference schemes exhibit comparable ET and TR as U_P varies since they are oblivious to the urgency issue. URIP incurs lower ET due to its efficient URI index. URIP significantly outperforms all reference schemes in terms of TR due to its *urgency-awareness*.

5   Conclusion

Given that itemset placement in retail stores significantly impacts retailer revenue, several research efforts have focused on strategically placing itemsets. However, they have not considered the issue of *urgency of sale* of items. In this work, we have introduced the notion of *urgency of sale* of items and proposed the *urgency-aware* URI indexing scheme, which is used by our proposed URIP itemset placement scheme for improving retailer revenue. Our performance study indicates that URIP is indeed effective in improving retailer revenue.

References

1. FMCG Market Size. https://www.alliedmarketresearch.com/fmcg-market
2. SPMF Library. http://www.philippe-fournier-viger.com/spmf/datasets
3. Agrawal, R., Srikant, R.: Fast algorithms for mining association rules. Proc. VLDB **1215**, 487–499 (1994)
4. Aponso, A., Karunaratne, K., Madubashini, N., Gunathilaka, L., Guruge, I.: Analysis and prediction framework: case study in fast moving consumer goods. Int. J. IT Knowl. Manag. **9**, 68–73 (2015)
5. Chaudhary, P., Mondal, A., Reddy, P.K.: A flexible and efficient indexing scheme for placement of top-utility itemsets for different slot sizes. In: Reddy, P.K., Sureka, A., Chakravarthy, S., Bhalla, S. (eds.) BDA 2017. LNCS, vol. 10721, pp. 257–277. Springer, Cham (2017). https://doi.org/10.1007/978-3-319-72413-3_18
6. Chaudhary, P., Mondal, A., Reddy, P.K.: An efficient premiumness and utility-based itemset placement scheme for retail stores. In: Hartmann, S., Küng, J., Chakravarthy, S., Anderst-Kotsis, G., Tjoa, A.M., Khalil, I. (eds.) DEXA 2019. LNCS, vol. 11706, pp. 287–303. Springer, Cham (2019). https://doi.org/10.1007/978-3-030-27615-7_22
7. Chaudhary, P., Mondal, A., Reddy, P.K.: An improved scheme for determining top-revenue itemsets for placement in retail businesses. Int. J. Data Sci. Anal. **10**, 359–375 (2020)
8. Chen, M., Lin, C.: A data mining approach to product assortment and shelf space allocation. Expert Syst. Appl. **32**, 976–986 (2007)

9. Fournier-Viger, P., Lin, J.C.-W., Wu, C.-W., Tseng, V.S., Faghihi, U.: Mining minimal high-utility itemsets. In: Hartmann, S., Ma, H. (eds.) DEXA 2016. LNCS, vol. 9827, pp. 88–101. Springer, Cham (2016). https://doi.org/10.1007/978-3-319-44403-1_6

10. Hansen, P., Heinsbroek, H.: Product selection and space allocation in supermarkets. Eur. J. Oper. Res. **3**, 474–484 (1979)

11. Liu, M., Qu, J.: Mining high utility itemsets without candidate generation. In: Proceedings CIKM, pp. 55–64. ACM (2012)

12. Trihatmoko, R.A., Mulyani, R., Lukviarman, N.: Product placement strategy in the business market competition: studies of fast moving consumer goods. Bus. Manag. Horizon **6**(1), 150–161 (2018)

NV-QALSH: An NVM-Optimized Implementation of Query-Aware Locality-Sensitive Hashing

Zhili Yao, Jiaqiao Zhang, and Jianlin Feng[✉]

School of Computer Science and Engineering, Sun Yat-Sen University,
Guangzhou, China
{yaozhli,zhangjq46}@mail2.sysu.edu.cn, fengjlin@mail.sysu.edu.cn

Abstract. Locality-Sensitive Hashing (LSH) is a popular method for answering c-approximate nearest neighbor queries in high-dimensional spaces. Existing LSH methods are either DRAM-based ones which consume a vast amount of expensive DRAM and are time-consuming to rebuild after programs reboot, or disk-based ones such as the state-of-the-art QALSH (Query-Aware LSH), which suffers from high latency of disk I/O. In this paper, we find that the emerging non-volatile memory (NVM) can be leveraged to solve the above problems. Its economic characteristics and data durability urge us to persist most of the LSH index in NVM to reduce DRAM occupancy; and its byte-addressability and low latency contribute to fast query processing. Since QALSH uses B+-Trees as index data structures and LB-Tree is the state-of-the-art NVM-optimized B+-Tree, we first directly combine QALSH with LB-Tree to get LB-QALSH. However, LB-QALSH shows poor query performance under NVM. To fully utilize the advantages of NVM, we propose an NVM-optimized implementation of QALSH, named NV-QALSH, which is the first NVM-optimized LSH. NV-QALSH adopts three optimization designs to achieve a high query performance. Experiments show that NV-QALSH outperforms LB-QALSH with a 1.5-4.7x speedup. Furthermore, compared with the state-of-the-art DRAM-based LSH, NV-QALSH greatly reduces the DRAM occupancy and the index rebuilt time.

Keywords: Approximate nearest neighbor search · Locality-sensitive hashing · Non-volatile memory

1 Introduction

Finding the nearest neighbor (NN) in high-dimensional spaces is a fundamental problem and has wide applications in various fields such as database, information retrieval, data mining, and artificial intelligence. Due to the curse of dimensionality [17], it is difficult to find exact NN in high-dimensional spaces, hence the alternative problem, called c-approximate NN (c-ANN) search has been widely studied [7]. Locality-Sensitive Hashing (LSH) and its variants [7,18]

© Springer Nature Switzerland AG 2021
C. Strauss et al. (Eds.): DEXA 2021, LNCS 12924, pp. 58–69, 2021.
https://doi.org/10.1007/978-3-030-86475-0_6

are among the most widely adopted methods for answering c-ANN queries in high-dimensional spaces. Intuitively, LSH constructs multiple hash tables forming an index structure to find the c-ANN. Existing DRAM-based LSH methods [2,8,13] place the index structure in DRAM for fast query speed, consuming a vast amount of expensive DRAM and losing the entire index structure upon power failure. Disk-based methods [3–6,10,12,15,16] have a low DRAM occupancy and a short index rebuilt time, but suffer from high latency of disk I/O, resulting in a slow query processing speed.

Non-volatile memory (NVM), an emerging hardware technology, has both disk and DRAM features, i.e., **economy** (cheaper than DRAM), **non-volatility** (data is persisted upon power failure), **byte-addressability** (data accessed in byte rather than disk page I/O) and **low latency** (orders of magnitude faster than disks, but 2-3x slower than DRAM). An intuitive thought is to leverage these properties to combine the advantages of DRAM-based LSH and disk-based ones. Query-Aware LSH (QALSH) [5,6], a state-of-the-art disk-based LSH, uses I/O efficient B+-Trees to index hash tables. Since LB-Tree [9] is the state-of-the-art NVM-optimized B+-Tree, we first combine QALSH with LB-Tree to get a naive implementation of QALSH under NVM, called LB-QALSH. However, LB-Tree is a generic B+-Tree and does not consider any optimizations for QALSH.

To fully utilize the advantages and characteristics of NVM, we propose an NVM-Optimized implementation of QALSH, named NV-QALSH. To the best of our knowledge, NV-QALSH is the first NVM-optimized LSH. NV-QALSH exploits three optimization designs based on the features of NVM and the query phase of QALSH. The first design comes from the DRAM-NVM storage architecture employed by LB-Tree, and we extend this idea to the whole NV-QALSH and utilize a three-level storage architecture. The second optimization is to redesign the leaf nodes to make the NVM B+-Trees more suitable for QALSH. The last optimization is to constrain the collision counting granularity to one leaf node, which takes full advantage of the XPBuffer in NVM. Note that the collision counting is the main operations in the query phase of QALSH.

Experiments show that NV-QALSH achieves a 1.5-4.7x speedup over LB-QALSH, and greatly reduces the DRAM space consumption and the index rebuilt time compared with the state-of-the-art DRAM-based LSH methods.

2 Preliminaries

In this section, we first give the definition of the c-ANN search problem and then briefly introduce how QALSH finds the c-ANN. Then, we discuss the characteristics of NVM which have a great impact on our optimization designs. Finally, we briefly introduce LB-Tree and the implementation of LB-QALSH.

2.1 The c-ANN Search Problem

Given a dataset D which contains n data objects in a d-dimensional space R^d, let $dist(o, q)$ denote a distance metric between two objects o and q, the c-ANN search problem is to construct a data structure that can find an object o such that $dist(o, q) \leq c \times dist(o^*, q)$, where q is any query object in R^d, c is the given approximation ratio that has to be larger than 1, and o^* is the exact NN of q. Intuitively, the c-ANN search problem is to find a near enough object rather than the nearest one for a specific query.

Similarly, the c-k-ANN search problem is to find k objects o_i $(1 \leq i \leq k)$ such that $dist(o_i, q) \leq c \times dist(o_i^*, q)$, where q is any query object in R^d, c is the given approximation ratio, and o_i^* is the exact i-th NN of q.

2.2 The QALSH Method

Based on the novel query-aware bucket partitioning strategy, QALSH [5,6] is proposed to answer c-ANN queries for l_p distance. In this paper, we mainly focus on the l_2 distance, i.e., the Euclidean distance. QALSH adopts the following query-aware LSH function family: $h_a(o) = a \cdot o$. Here o is the vector representation of a data object, and a is a d-dimensional vector where each entry is drawn independently from the standard normal distribution $N(0, 1)$. Intuitively, the LSH function projects an object o onto a random line specified by a. An interesting property of such random projections is that with a high probability, two close objects in the original Euclidean space will remain close on the random line; and similarly, two far objects will stay far on the random line.

To find a c-ANN, QALSH projects the whole dataset onto m random lines. In the query phase, when a query arrives, QALSH locates its projected position on the same m random lines, and constructs m query-centric hash buckets with width w. In this way, objects close to the query will collide into the same bucket with high probability. Then QALSH performs collision counting [3] for the objects in the m buckets, and an object with counting more than l is chosen as a candidate. After that, the bucket width w is enlarged to search for more candidates. QALSH terminates the algorithm when more than β (typically, β is set to be 100 in [6]) candidates are chosen or some candidate is close enough to the query in the original space. Similarly, for c-k-ANN search problem, QALSH terminates the algorithm when more than $(\beta + k)$ candidates are selected or k candidates are close enough.

In practice, QALSH exploits m B+-Trees to index the m random lines, and performs range search in the query phase to find candidates by collision counting. However, to achieve a high query accuracy, QALSH has to search lots of random lines with a gradually increasing bucket width, possibly resulting in a large number of disk I/Os and high query latency.

Besides QALSH, there are other disk-based LSH methods, such as C2LSH [3], SRS [15]. Among them we choose QALSH to optimize in NVM because it is still regarded as a state-of-the-art LSH and has a high query performance especially in high-dimensional spaces. In the tutorials of SIGKDD'19, QALSH is introduced as the representative of LSH. Recently, many variants of QALSH are proposed, including I-LSH [10], R2LSH [11], PDA-LSH [18], VHP [12], indicating that QALSH is still popular and wide studied.

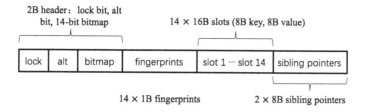

Fig. 1. 256 B leaf nodes of LB-Tree.

2.3 Non-Volatile Memory

Non-Volatile Memory (NVM) is a new storage technique [19]. In this paper, we mainly target the Intel Optane DCPMM[1], a commercially available NVM solution, which has a higher capacity (128–512 GB per DIMM) and a lower price than DRAM. In this paper, we use an NVM-aware filesystem with DAX to manage the Optane memory, and memory-map the data into virtual addresses. Besides, we adopt the App-Direct model of the Optane memory, which allows the CPU to directly access data in NVM and ensures the data persistence.

As indicated in [19], Optane DCPMM is 2–3 slower than DRAM, and the gap between random and sequential reads is 20% for DRAM but 80% for Optane memory. When data are read from the Optane memory to the CPU, 256B data are transferred into an XPLine in the XPBuffer in NVM [19], then 64 B data among them are read to a CPU cacheline. In this paper, we mainly focus on optimizing NVM read because most data operations in the query phase of LSH are read operations.

Recently, some NVM-optimized B+-Trees are proposed, such as FP-Tree [14] and LB-Tree [9]. We follow some useful ideas such as selective persistence and 256B alignment, to reduce the read latency in NV-QALSH.

2.4 LB-Tree and LB-QALSH

LB-Tree is an NVM-optimized B+-Tree which proposes some novel techniques such as entry moving and logless node split, to improve entry insertion performance. As illustrated in Fig. 1, leaf nodes of LB-Tree are 256 B (256 Bytes). The *alt* bit is used to indicate which sibling pointer is valid. The fingerprints are used to speedup the point query.

When implementing LB-QALSH, we mainly make the following two changes in LB-Tree. First, when given the projected value of the ANN query as the target, the lookup operation returns a position p such that the key in p is smaller or equal to the target, while the key in $p + 1$ is larger than the target. In fact, p is the projected position of the query. Second, to support range search from both left and right sides, we use one of the sibling pointers to point to the left sibling node.

---

[1] https://www.intel.com/content/www/us/en/architecture-and-technology/optane-dc-persistent-memory.html.

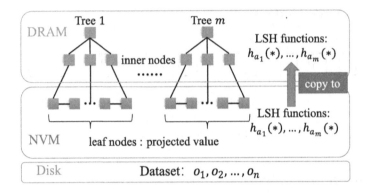

Fig. 2. Three-level storage architecture of NV-QALSH.

Specifically, after a node split, the invalid sibling pointer is set to point to the left sibling node. More details about LB-QALSH will be presented in Sect. 3.3.

3   Optimization Designs

In this section, we thoroughly present the optimization designs of NV-QALSH.

3.1   Three-Level Storage Architecture

NV-QALSH exploits a three-level storage architecture combining the advantages of DRAM, NVM, and disks. As illustrated in Fig. 2, we put in DRAM the "hot" data, i.e., all inner nodes and the m LSH functions. The amount of "hot" data is small, hence they can be recovered rapidly from the data in NVM. Since "hot" data are accessed frequently during the query phase, we place them in faster DRAM to reduce the read latency.

As for the "warm and basic" data, the main part of the whole index structure, are placed in NVM. These data include the leaf nodes which contain all projected values of the dataset. The leaf nodes are used to rebuild the non-leaf part in DRAM, and hence they are the "basic" data of the index. Besides, only a portion of leaf nodes are accessed during the query phase, and hence they are also the "warm" data. Therefore, they are placed in the slightly slower NVM for persistence. The "basic" data also include the m LSH functions to ensure that different queries use the same LSH functions after the program restarts. Besides, all the m LSH functions are used for each query, making them both the "hot" and "basic" data. Therefore, we duplicate them both in NVM and DRAM.

Finally, the "cold" data stored on disks include the whole dataset. The amount of them is enormous, but only a tiny part of them are accessed for original distance calculation for each query. In fact, less than 100 objects are accessed when answering a c-ANN query. Since NVM is still more expensive than disks, these very large but rarely accessed data are placed on cheaper disks.

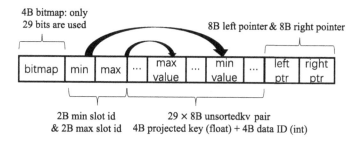

Fig. 3. 256 B leaf nodes of NV-QALSH.

With the three-level storage architecture, NV-QALSH fully utilizes the lowest latency of DRAM, the persistence of NVM, and the economy of disks.

3.2 Leaf Node Optimization

The second optimization is to redesign the leaf node of NV-QALSH based on the query phase. As illustrated in Fig. 3, the leaf node of NV-QALSH is still 256 B.

During the query phase of QALSH, first the projected position of the query is located. Note that the projected value of the query tends to be different from any values in the B+-Trees since the query object is assumed to be different from any objects in the dataset. Since the fingerprints are used for exact point search and have no help of locating projected position, we remove them for saving space.

We follow the original implementation of QALSH to make a key-value pair in a leaf node consist of a 4 B *float* projected value as key and a 4 B *int* data-id as value. Following the idea in [1], entries inside a leaf node are unsorted to reduce NVM writes, and a 4 B bitmap indicates valid data entries and empty slots. During the query phase, the minimum and the maximum hash values are used to determine whether the leaf node is within the searching range. To quickly find the min/max hash values in an unsorted leaf node, two 2 B integers are used to record their slot-ids.

The two 8 B pointers are used to support the range search in the query phase. Since we memory-map the NVM into a virtual address space, each time a new memory-map may change the virtual addresses. Hence, we record in the 8 B pointer the offset of a leaf node to the starting address of the memory-map. Note that the offsets can be easily transformed into virtual memory addresses.

As indicated in Sect. 2, a leaf node of LB-QALSH contains 14 objects, but that of NV-QALSH contains 29 objects. Therefore, with the same dataset size, the number of leaf nodes in NV-QALSH is less than half of that in LB-QALSH. Besides, the number of inner nodes is also reduced, and thus the DRAM occupancy and the index rebuilt time are also reduced. Furthermore, the height of the B+-Tree in NV-QALSH is smaller, and the retrieval speed from the root to leaf nodes will be improved. Finally, all leaf nodes are designed to be aligned with 256 B. Therefore, when some data of a leaf node are read, the whole leaf

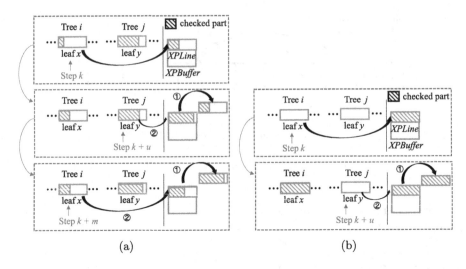

Fig. 4. (a) LB-QALSH counts only one object in a leaf node at a time, causing an XPLine to swap in and out data frequently. (b) NV-QALSH counts all objects in a leaf at a time, avoiding data swapping in and out in an XPLine frequently.

node is brought into exact one XPLine in the XPBuffer of NVM, contributing to the next optimization design.

3.3 Collision Counting Granularity Optimization

The third optimization design is to constrain the collision counting granularity to a leaf node, i.e., counting all the objects in a leaf node at a time.

As described in Sect. 2, in the query phase, QALSH needs to perform collision counting. In practice, QALSH starts from the projected position of the query in a B+-Tree and performs a range search towards the left and the right. In the original implementation, to check the objects on all lines fairly, QALSH will check one object for each line circularly. LB-QALSH takes this approach but causes XPLine in XPBuffer to swap in and out data frequently. As illustrated in Fig. 4a, when searching leaf x in Tree i at step k, the whole leaf x is brought into an XPLine in XPBuffer, and only one object is checked. Next at step $k + u$, the leaf y in Tree j is checked, and happens to be switched into the XPLine where leaf x is accommodated before. Hence, the leaf x is switched out and the leaf y is swapped in. But the next time the Tree i is checked again at step $k + m$, since leaf x still contains unchecked objects, the XPLine has to replace leaf y with leaf x. In this example, to check all objects in leaf x, an XPLine requires **at least two** data swapping in and out operations, whcih leads to frequent internal data transmission in NVM, resulting in a high average read latency.

To remedy this issue, NV-QALSH counts all the objects in a leaf node at a time. In this way, when checking all objects in a leaf node, NV-QALSH causes **at most one** data swapping in and out operation in an XPLine. As illustrated in Fig. 4b, when Tree i is checked, all the objects in leaf x are checked. Although leaf x is swapped out for leaf y in the XPLine later, it will not be switched in the XPLine again since all the objects are checked. In this manner, NV-QALSH greatly reduces the number of data swapping in and out of an XPLine, and improves the query processing performance.

4 Experiments

4.1 Experiment Setup

All methods are implemented in C++ and are compiled with gcc 8.3 using -O3 optimization. All experiments are running on a machine equipped with two Intel Xeon(R) Gold 5222 CPUs, a 128 GB DRAM, and a 512 GB Optane DCPMM consisting of 4 × 128 GB Optane NVDIMM. The NUMA effect is removed by first placing the NVM data on the node where CPU0 is located, and then binding CPU0 to the experimental program using *taskset* instruction.

4.2 Datasets and Queries

The four real-world datasets used in experiments are listed below.

- **Mnist**[2]. The Mnist dataset contains 60,000 50-dimensional feature vectors. The 100 queries are chosen randomly from its test set. Note that we follow [3,6] to only consider the top-50 dimensions with the largest variance.
- **Gist**[3]. We use 100,000 960-dimensional vectors of Gist as datasets, and choose 100 queries randomly from its corresponding test set.
- **CIFAR-10**[4]. We use 50,000 3072-dimensional vectors of CIFAR-10 as datasets, and choose 100 queries randomly from its corresponding test set.
- **P53**[5]. This dataset contains 31,059 5408-dimensional vectors and a query set of 100 vectors. We follow QALSH [6] to remove the objects with missing values and normalize the coordinates to be integers in a range of [0,10000].

4.3 Evaluation Metrics

We adopt following metrics to evaluate the performance given a specific accuracy.

- **Overall Ratio** [16] is used to measure the accuracy of an LSH method. For a c-k-ANN query q, it is defined as $\frac{1}{k} \sum_{i=1}^{k} \frac{dist(o_i,q)}{dist(o_i^*,q)}$. Here o_i is the i-th object returned by the LSH method while o_i^* is the exact i-th NN. Intuitively, a smaller overall ratio means higher accuracy.
- **Query Time** is used to evaluate the efficiency of an LSH method.
- **DRAM Occupancy** is the DRAM space consumed by the LSH index.
- **Index Rebuilt Time** is the time spent on rebuilding the whole LSH index.

---

[2] http://yann.lecun.com/exdb/mnist/.
[3] http://corpus-texmex.irisa.fr/.
[4] http://www.cs.toronto.edu/~kriz/cifar.html.
[5] http://archive.ics.uci.edu/ml/datasets/p53+Mutants.

4.4 Benchmark Methods

The methods we evaluate are listed below:

- **NV-QALSH**. An NVM-optimizied implementation of QALSH we proposed.
- **LB-QALSH**. A naive implementation of NVM-QALSH.
- **disk-based QALSH**[6] (simply **D-QALSH**). The original QALSH [6] is designed for external memory, which is I/O latency-sensitive.
- **DRAM-based QALSH**[7] (simply **M-QALSH**). This is the memory version of QALSH, which is also a state-of-the-art DRAM-based LSH method.
- **DRAM-based SRS**[8] (simply **SRS**). The memory version of SRS [15] is also a state-of-the-art DRAM-based method.
- **LCCS-LSH**[9]. Recently, LCCS-LSH [8] proposes a novel search framework to improve LSH performance. It is also a state-of-the-art DRAM-based method.

Parameters Setting. For the c-k-ANN search problem, we set k to 100 by default. All the LSH methods are fine-tuned for the best performance. For QALSH-based methods, we vary the approximation ratio c and let the other parameters be computed automatically as described in [6]. For the disk-QALSH, the page size is set to 64 KB by default. For the SRS, we fine-tune the number of hash functions m, approximation ratio c, maximum number of objects checked t. For LCCS-LSH, we tune the number of hash tables L.

4.5 Results and Analysis

NVM to Disk Speedup Ratio. In this experiment, we set the approximate ratio $c = 4$ and run D-QALSH, LB-QALSH and NV-QALSH so that they all lead to a same query accuracy. Then we compare the improvement of performance of LB-QALSH and NV-QALSH over D-QALSH. As illustrated in Fig. 5, NV-QALSH always has a higher speedup ratio than LB-QALSH. Generally, NV-QALSH is 1.5-4.7x faster than LB-QALSH, which shows that not only the NVM hardware advantages but also our optimization designs that accelerate the query phase. Besides, when the dimension increases, both LB-QALSH and NV-QALSH achieve a higher speedup because the number of hash tables (random lines) increases, and D-QALSH needs more disk I/Os to find ANNs. In other words, the improvement brought by NVM is more obvious in high-dimensional spaces.

Query Time. The second experiment is to compare the query time of different LSH methods given a specific accuracy. As shown in Fig. 6, the DRAM-based methods achieve faster query speed than the two NVM-based methods because DRAM is still 2–3 faster than NVM. However, NV-QALSH achieves a near-DRAM query time when the given overall ratio is high. Besides, when the query

---

[6] https://github.com/HuangQiang/QALSH.

[7] https://github.com/HuangQiang/QALSH_Mem.

[8] https://github.com/DBWangGroupUNSW/SRS.

[9] https://github.com/1flei/lccs-lsh.

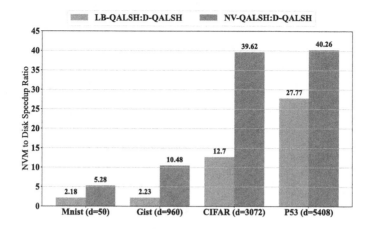

Fig. 5. NVM to disk speedup ratio.

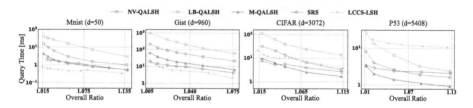

Fig. 6. Query time.

accuracy is higher, the query time gap between LB-QALSH and NV-QALSH becomes larger because more random lines are needed and more objects on the random lines are checked, which causes higher XPLine contention in LB-QALSH. However, since NV-QALSH optimizes the collision counting granularity, the XPLine contention still remains low and the query time does not grow as fast as that of LB-QALSH. Note that M-QALSH has the shortest query time for the last two high-dimensional datasets, indicating that QALSH is still a state-of-the-art LSH method, especially for high-dimensional spaces.

DRAM Occupancy. Although DRAM-based LSH methods have better query performance, they consume lots of expensive DRAM space. As shown in Fig. 7, we evaluate the DRAM occupancy of all the LSH methods except SRS because it does not record the memory usage information. For all four datasets, NV-QALSH occupies negligible DRAM space and is much more cost-saving than the DRAM-based LSH, making it a more practical LSH method. The DRAM occupancy of NV-QALSH is slightly lower than that of LB-QALSH. Although it is not clear in the figure, NV-QALSH saves up to 26%–41% DRAM space compared with LB-QALSH, resulting from the leaf node redesign optimization. Besides, compared with DRAM-based methods, when achieving a high query accuracy, NV-QALSH saves 90%–94% of DRAM space.

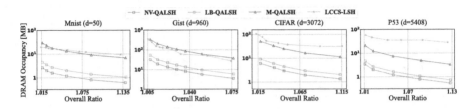

Fig. 7. DRAM occupancy.

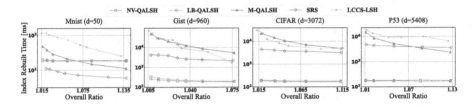

Fig. 8. Index rebuilt time. LB-QALSH and NV-QALSH have a similar short index rebuilt time and their curves stick together.

Index Rebuilt Time. As illustrated in Fig. 8, for high-dimensional datasets and high query accuracy, NV-QALSH and LB-QALSH achieve a shortest index rebuilt time due to the lower DRAM occupancy. For the Mnist dataset, the index rebuilt time of the NVM-based methods are slightly longer than that of SRS. Compared with LB-QALSH, NV-QALSH has a slightly faster index rebuilt time although it is not clear in the figure. In general, NV-QALSH achieves a stable and fast index rebuilt speed.

5   Conclusion

In this paper, we proposed NV-QALSH, an NVM-optimized implementation of QALSH, which leverages NVM to not only reduce the high query latency suffered by disk-based LSH methods, but also reduce the high consumption of expensive DRAM space suffered by DRAM-based LSH methods.

Acknowledgements. The corresponding author of this work is Jianlin Feng. This work is partially supported by China NSFC under Grant No. 61772563.

References

1. Chen, S., Gibbons, P.B., Nath, S., et al.: Rethinking database algorithms for phase change memory. In: CIDR, vol. 11, p. 5 (2011)
2. Datar, M., Immorlica, N., Indyk, P., Mirrokni, V.S.: Locality-sensitive hashing scheme based on p-stable distributions. In: Proceedings of the Twentieth Annual Symposium on Computational Geometry, pp. 253–262 (2004)

3. Gan, J., Feng, J., Fang, Q., Ng, W.: Locality-sensitive hashing scheme based on dynamic collision counting. In: Proceedings of the 2012 ACM SIGMOD International Conference on Management of Data, pp. 541–552 (2012)
4. Gong, L., Wang, H., Ogihara, M., Xu, J.: iDEC: indexable distance estimating codes for approximate nearest neighbor search. Proc. VLDB Endow. **13**(9), 1483–1497 (2020)
5. Huang, Q., Feng, J., Fang, Q., Ng, W., Wang, W.: Query-aware locality-sensitive hashing scheme for l_p norm. VLDB J. **26**(5), 683–708 (2017). https://doi.org/10.1007/s00778-017-0472-7
6. Huang, Q., Feng, J., Zhang, Y., Fang, Q., Ng, W.: Query-aware locality-sensitive hashing for approximate nearest neighbor search. Proc. VLDB Endow. **9**(1), 1–12 (2015)
7. Indyk, P., Motwani, R.: Approximate nearest neighbors: towards removing the curse of dimensionality. In: Proceedings of the Thirtieth Annual ACM Symposium on Theory of Computing, pp. 604–613 (1998)
8. Lei, Y., Huang, Q., Kankanhalli, M., Tung, A.K.: Locality-sensitive hashing scheme based on longest circular co-substring. In: Proceedings of the 2020 ACM SIGMOD International Conference on Management of Data, pp. 2589–2599 (2020)
9. Liu, J., Chen, S., Wang, L.: Lb+ trees: optimizing persistent index performance on 3dxpoint memory. Proc. VLDB Endow. **13**(7), 1078–1090 (2020)
10. Liu, W., Wang, H., Zhang, Y., Wang, W., Qin, L.: I-LSH: I/O efficient c-approximate nearest neighbor search in high-dimensional space. In: 2019 IEEE 35th International Conference on Data Engineering (ICDE), pp. 1670–1673. IEEE (2019)
11. Lu, K., Kudo, M.: R2LSH: a nearest neighbor search scheme based on two-dimensional projected spaces. In: 2020 IEEE 36th International Conference on Data Engineering (ICDE), pp. 1045–1056. IEEE (2020)
12. Lu, K., Wang, H., Wang, W., Kudo, M.: VHP: approximate nearest neighbor search via virtual hypersphere partitioning. Proc. VLDB Endow. **13**(9), 1443–1455 (2020)
13. Lv, Q., Josephson, W., Wang, Z., Charikar, M., Li, K.: Multi-probe LSH: efficient indexing for high-dimensional similarity search. In: Proceedings of the 33rd International Conference on Very Large Data Bases, pp. 950–961 (2007)
14. Oukid, I., Lasperas, J., Nica, A., Willhalm, T., Lehner, W.: FPTree: A hybrid SCM-DRAM persistent and concurrent B-tree for storage class memory. In: Proceedings of the 2016 International Conference on Management of Data, pp. 371–386 (2016)
15. Sun, Y., Wang, W., Qin, J., Zhang, Y., Lin, X.: SRS: solving c-approximate nearest neighbor queries in high dimensional Euclidean space with a tiny index. In: Proceedings of the VLDB Endowment (2014)
16. Tao, Y., Yi, K., Sheng, C., Kalnis, P.: Efficient and accurate nearest neighbor and closest pair search in high-dimensional space. ACM Trans. Database Syst. (TODS) **35**(3), 1–46 (2010)
17. Weber, R., Schek, H.J., Blott, S.: A quantitative analysis and performance study for similarity-search methods in high-dimensional spaces. VLDB **98**, 194–205 (1998)
18. Yang, C., Deng, D., Shang, S., Shao, L.: Efficient locality-sensitive hashing over high-dimensional data streams. In: 2020 IEEE 36th International Conference on Data Engineering (ICDE), pp. 1986–1989. IEEE (2020)
19. Yang, J., Kim, J., Hoseinzadeh, M., Izraelevitz, J., Swanson, S.: An empirical guide to the behavior and use of scalable persistent memory. In: 18th {USENIX} Conference on File and Storage Technologies ({FAST} 2020), pp. 169–182 (2020)

NCRedis: An NVM-Optimized Redis with Memory Caching

Jiaqiao Zhang, Zhili Yao, and Jianlin Feng[✉]

School of Computer Science and Engineering, Sun Yat-Sen University,
Guangzhou, China
{zhangjq46,yaozhli}@mail2.sysu.edu.cn, fengjlin@mail.sysu.edu.cn

Abstract. Non-volatile memory (NVM) has byte-addressability and data-durability. Redis, a popular in-memory kv-store system, can persist data when replacing DRAM with NVM. However, to implement NVM Redis, we need to use general NVM allocators to obtain NVM and guarantee the data consistency of Redis. There are two problems in NVM Redis. First, it is expensive to directly use NVM allocators with numerous metadata modification and logging. Second, logging which is also used to guarantee the data consistency of NVM Redis leads to write amplification and degrades the run-time performance of Redis. In this paper, we find that these two problems can be solved by exploiting memory caching inside Redis under NVM. Firstly, memory caching like the well-known Linux Slab will cache freed memory, reducing expensive NVM allocation/deallocation. Secondly, by recording all allocated NVM, caching can handle the persistent memory leak of Redis, which is the only inconsistent state induced by failures. Thus, the use of logging in NVM Redis can be avoided by using memory caching. In this paper, we propose an NVM caching called LFSlab (Log-Free Slab), while the conventional Slab needs logging to guarantee its consistency under NVM. Using LFSlab, we propose NCRedis (NVM Caching Redis) under NVM. In the experiment with Optane persistent memory, Redis with LFSlab outperforms the naive implementation of NVM Redis with no caching by 1.52-2.65x and DRAM Redis with data backup in disks by 1.27x, and gets at least 94% performance of DRAM-only Redis, while Optane persistent memory is 2-3x slower than DRAM at a 39% cost savings.

Keywords: Non-volatile memory · Redis · Slab · Logging

1 Introduction

Intel Optane persistent memory[1] is now commercially available. This novel non-volatile memory (NVM) device is byte-addressable, data-durable and cheaper than DRAM. Redis [4] is a popular in-memory kv-store system and stores all

---

[1] https://www.intel.com/content/www/us/en/arch-itecture-and-technology/optane-dc-persistent-memory.html.

C. Strauss et al. (Eds.): DEXA 2021, LNCS 12924, pp. 70–76, 2021.
https://doi.org/10.1007/978-3-030-86475-0_7

transient key-value pairs in DRAM. Replacing DRAM with Optane, Redis can persist data without DRAM and disks, saving the hardware cost.

To implement NVM programs, we can use Intel's PMDK (Persistent Memory Development Kit) [7] or other allocators [1,8]. However, Using NVM allocators is expensive. These allocators use multiple persistent metadata to manage the NVM space [1,3,7,8]. Modification of these metadata takes much latency under NVM and logging is used to guarantee the failure-atomicity of data modification in NVM allocators [1,7]. NVM Redis is also responsible to guarantee its own data consistency, handling all possible failure-induced inconsistent states. Conventional logging can be used to guarantee the consistency [5,6] of Redis under NVM, but it will degrade the run-time performance of Redis. Based on 8-byte write atomicity of NVM [6], there is no other inconsistent state except the failure-induced persistent memory leak. **Persistent memory leak** means user programs lose the address of NVM and cannot access the NVM anymore. To improve the run-time performance of NVM Redis, we need to tackle slow NVM allocation and inefficient logging.

In this paper, we find that embedding memory caching in NVM Redis is a good solution to the two issues. Firstly, using caching, the memory freed by programs will be cached in the caching manager for future reuse, reducing the expensive NVM allocation/deallocation of NVM allocators. Secondly, memory caching can handle the only failure-induced inconsistent state of persistent memory leak in NVM Redis. In this way, the use of logging for guaranteeing the data consistency in NVM Redis is not needed. Caching can handle persistent memory leak because of its work mechanism. Programs using caching, allocated NVM is first recorded by the caching manager and then used by user programs. Thus, the memory used by user programs is the subset of the memory recorded in caching manager, and the difference set is the memory that is not accessible and used by user programs, i.e. the free/leaked memory. We can find out the leaked memory and reuse it by comparing the two memory sets.

Memory caching is necessary for NVM Redis, and we can implement the well-known Linux Slab [2] under NVM. The conventional Slab organizes its metadata in a linked list. However, failure might induce inconsistent states when Slab operates a node in the middle of a linked list. Adding/deleting a middle node is non-atomic, including two steps of breaking the list and reconnecting it. Thus, NVM Slab needs logging in logic. In this paper, we propose a LFSlab (Log-Free Slab) without logging embedded under NVM. Using atomic interfaces of NVM allocators [7], LFSlab can operate a tail node atomically. The LFSlab is restricted to the operation of the tail in a linked list in caching management. We name our NVM Redis the NCRedis (NVM Caching Redis) because there is LFSlab memory caching embedded. To the best of our knowledge, NCRedis is the first NVM-optimized Redis.

2   Implementation of NCRedis

2.1   Architecture of NCRedis

In this paper, we focus on improving the run-time performance of Redis when using Intel's PMDK under Optane persistent memory of Intel, and we propose NCRedis. As illustrated in Fig. 1, NCRedis has three main components: the

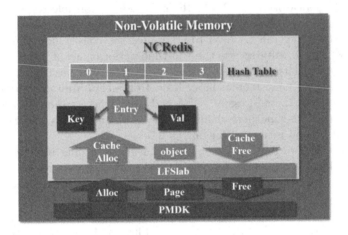

Fig. 1. Architecture of NCRedis.

hash table index, our LFSlab caching and the PMDK allocator. **Hash table** is the same as the one in the original Redis. A hash bucket in the hash table is a linked list storing entries with the same hash value. A hash entry has two string objects storing key/value and a *next* pointer pointing to the next entry. **LFSlab** is an NVM caching implemented in this paper. As illustrated in Fig. 1, LFSlab is embedded in NCRedis and works as a connector between NCRedis and PMDK. NVM allocated from PMDK is first recorded in LFSlab, and then used by NCRedis. **PMDK** allocator is an NVM allocator used in our implementation. The LFSlab allocates/deallocates NVM from PMDK at the granularity of a page. Although PMDK uses the 16-byte address to locate data, we only use the last 8-byte to locate the data, i.e. 8-byte address, and keep the front 8-byte the same.

2.2   Log-Free Designs of LFSlab

The architecture of LFSlab is similar to that of original Slab. The main difference is that LFSlab guarantees the data consistency of caching management under NVM without logging, which conventional Slab needs. In the caching management of the Slab mechanism, the caching manager will allocate memory pages from allocators and install them in a linked list. Then a page is divided into several memory objects used by user programs. Addresses of free memory objects are stored in freelists. When a memory object is freed by user programs,

---

Algorithm 1: Log-Free Set Operation in NCRedis

---

1 **Function** Set(*hash_table, Key, Value*):
2   index = **GetKeyIndex**(*hash_table, Key*)
3   head_entry = hash_table[index]
4   new_entry = **CacheAlloc**(*ENTRY_SIZE*)
5   **InitEntry** (*new_entry, Key, Value, head_entry*)
6   **SetFirstEntry**(*new_entry*)

---

it will be put back to freelists and cached again. When a page is completely free and should be deallocated, the caching manager needs to uninstall the page and remove the specific node recording the page address from the linked list.

However, the linked list will be broken into two pieces temporarily when inserting/removing a node in the middle of the list, and then the two pieces will be reconnected later. If a failure occurs before the reconnection of the breaked linked list, we will lose half the piece of the linked list behind the operated node and there is an inconsistent state. That is why conventional Slab needs logging under NVM. Instead, we redesign the caching management and guarantee the data consistency without logging in our LFSlab. Note that PMDK guarantees that a node can be atomically installed or uninstalled in the tail of a linked list by its atomic allocation/deallocation interfaces [7]. And no inconsistent state occurs. So LFSlab is confined to operating the tail node of the linked list only. Installing/uninstalling a page in a node ressembles operating a tail node.

2.3   Handling Persistent Memory Leak by LFSlab

Redis needs to handle the crash-enabled persistent memory leak under NVM, which is an inconsistent state. The crash-enabled persistent memory leak happens when a piece of NVM is allocated successfully but fails to be installed by programs. Conventionally, we can use logging to guarantee the data consistency but degrade the run-time performance of Redis under NVM. Instead, we find that memory caching of LFSlab can also handle the persistent memory leak. Using LFSlab, NVM is first allocated from allocators and recorded by LFSlab at the granularity of page. Afterward, a memory object is separated from a page and used by user programs. Thus, NVM recorded in LFSlab is the universal set and the NVM used by user programs is the subset. By comparing the two sets, we can find and reuse the leaked memory, solving the persistent memory leak.

2.4   Log-Free Designs of NCRedis

Based on LFSlab and 8-byte atomic write in NVM, we design the key-value operations in NCRedis as log-free. As illustrated in Algorithm 1, we design the log-free *Set* operation to insert a new key-value pair. In the beginning, we will initialize the new entry and pre-set the address of its next entry. Then the new entry is inserted into the linked list of a hash bucket. Based on 8-byte write

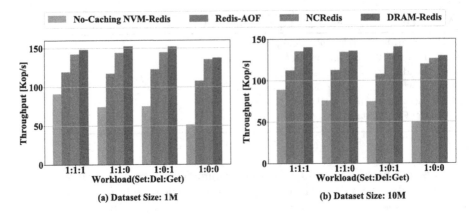

Fig. 2. Memtier Benchmark Test of NCRedis (Higher is better).

atomicity of NVM, the address modification in Line 6 runs atomically without inconsistent states such as incomplete modification. If a failure happens before Line 6 in Algorithm 1, the new entry has been allocated but fails to be installed in the hash table. And the NVM of the new entry is leaked. This inconsistent state can be solved with LFSlab so we don't need logging. Similarly, other operations in NCRedis are also log-free. Owing to the length of the article, we choose not to elaborate the recovery, in which NCRedis only compares its used memory with the one recorded in LFSlab to handle the persistent memory leak.

3   Evaluation

3.1   Experimental Setup

Experiments are conducted on a machine equipped with two Intel Xeon(R) Gold 5222 CPUs. And there is a total of 128 GB DRAM and 512 GB Optane persistent memory with the app-direct mode in the machine. There exists 256 GB Optane persistent memory in each socket of the machine. The machine runs Linux with 4.19.0-13 kernel.

To compare NCRedis with the naive NVM Redis implemented without caching, we implement the **No-Caching NVM-Redis** under NVM, which directly uses PMDK and micro logging [6] to handle the inconsistent state described in Sect. 2.4. To compare NCRedis with conventional DRAM Redis with disks, we run **Redis-AOF** that writes logging to the appended-only file in disks. In addition, we use the performance of **DRAM-Redis** as the upper bound of NVM Redis's performance, because Optane persistent memory is 2-3x slower than DRAM.

3.2   Memtier Benchmark Test

To test the run-time performance of NCRedis, we use the Memtier benchmark of RedisLabs. The Memtier program simulates the real workload and operates

Redis with Set/Del/Get operations. The keys and values are 32-byte random strings. In the test, the Redis will firstly be warmed up with 1M/10M data. Then we vary the ratio of Set:Del:Get and operate the Redis with the 1M/10M operations, testing the throughput of Redis.

As illustrated in Fig. 2, our NCRedis outperforms the naive No-Caching NVM-Redis by 1.52-2.65x and the conventional Redis-AOF with DRAM and disks by up to 1.27x. Because Optane is cheaper than DRAM, NCRedis has a better run-time performance than Redis-AOF at a cost savings of 39%. Meanwhile, NCRedis gets 94%–99% performance with the one of DRAM-Redis. NCRedis can outperform the No-Caching NVM-Redis mainly on account of our LFSlab, which reduces the overhead of using NVM allocators and enables the log-free design of NCRedis. NCRedis outperforms Redis-AOF, because NCRedis can persist data with only NVM and don't need to backup data in slow disks. We improve Redis under NVM a lot by exploiting memory caching and make NVM Redis outperform Redis under DRAM and disks in this paper.

4   Conclusions

In this paper, we optimize Redis under NVM through the exploitation of memory caching. A log-free Slab manager called LFSlab is implemented under NVM. Then LFSlab is embedded into Redis, reducing the run-time overhead of NVM allocation/deallocation and handling the inconsistent state of persistent memory leak of Redis under NVM. LFSlab enables the design of NVM Redis to be log-free and with this program embedded, the NCRedis is implemented. In the experiment with Optane persistent memory, NCRedis outperforms naive NVM Redis with no caching by 1.52-2.65x and DRAM Redis-AOF with data backup in disks by 1.27x. NCRedis yields at least 94% performance of DRAM-only Redis. With Optane persistent memory, NCRedis saves the hardware cost of 39%, compared with DRAM Redis.

Acknowledgements. We thank NetEase for providing the machine with Optane persistent memory. The corresponding author of this work is Jianlin Feng. This work is partially supported by China NSFC under Grant No. 61772563.

References

1. Bhandari, K., Chakrabarti, D.R., Boehm, H.J.: Makalu: fast recoverable allocation of non-volatile memory. ACM SIGPLAN Not. **51**(10), 677–694 (2016)
2. Bonwick, J., et al.: The slab allocator: An object-caching kernel memory allocator. In: USENIX Summer, vol. 16. Boston (1994)
3. Cai, W., Wen, H., Beadle, H.A., Kjellqvist, C., Hedayati, M., Scott, M.L.: Understanding and optimizing persistent memory allocation. In: Proceedings of the 2020 ACM SIGPLAN International Symposium on Memory Management, pp. 60–73 (2020)
4. Carlson, J.: Redis in Action. Simon and Schuster, New York (2013)

5. Liu, J., Chen, S., Wang, L.: Lb+ Trees: optimizing persistent index performance on 3dxpoint memory. Proc. VLDB Endow. **13**(7), 1078–1090 (2020)
6. Oukid, I., Lasperas, J., Nica, A., Willhalm, T., Lehner, W.: FPTree: a hybrid SCM-DRAM persistent and concurrent b-tree for storage class memory. In: Proceedings of the 2016 International Conference on Management of Data, pp. 371–386 (2016)
7. Scargall, S.: Programming Persistent Memory: A Comprehensive Guide for Developers (2020)
8. Schwalb, D., Berning, T., Faust, M., Dreseler, M., Plattner, H.: nvm malloc: memory allocation for NVRAM. ADMS@VLDB **15**, 61–72 (2015)

A Highly Modular Architecture for Canned Pattern Selection Problem

Marinos Tzanikos[(✉)], Maria Krommyda[(✉)], and Verena Kantere[(✉)]

ECE, NTUA, Athens, Greece
el13147@mail.ntua.gr, {mariakr,verena}@dblab.ece.ntua.gr

Abstract. Due to the continuously increasing rate of data production from multiple sources, especially from social media, data analysis techniques are focusing on identifying patterns in the formed graphs and extracting knowledge from them. Most techniques till now, begin with given patterns and calculate the coverage in the graph. Here, we propose a graph mining architecture that focus on finding small sub-graph patterns, referred to as canned pattern, from a database of graphs without any domain knowledge of the graph. These patterns can be used to expedite the query formulation time, increase the domain knowledge and support the data analysis. The canned pattern should maximize coverage and diversity over the graph database while minimizing the cognitive-load of the patterns. The approach presented here is based on an innovative modular architecture that combines state-of-art techniques to extract these patterns and validate the extracted result.

Keywords: Canned patterns · Graph mining · Pattern mining

1 Introduction

Graph pattern matching and mining is an NP-hard problem, which is defined using subgraph isomorphism. Graph pattern mining and graph pattern matching have a lot practical use cases and have been a subject of research in the last decades. The problem is based on invertible functions, one-to-one correspondence, that have been proven to be too restrictive for patterns in real world applications.

Practically, the term pattern matching is used to describe the methodology needed to find the number of occurrences of a small sub-graph inside a graph. Most of the pattern mining algorithms focus on receiving potential small sub-graph patterns, most of the time representing complex queries, and extracting the information about the number of times this pattern is available in the graph.

The reverse approach, extracting popular patterns from graphs without any previous knowledge, is called canned pattern selection problem and was introduced in Catapult [7]. The canned pattern selection problem is defined over a given a graph database D and a pattern budget b as the selection of a set of patterns P that are satisfying b and that maximize coverage and diversity while

© Springer Nature Switzerland AG 2021
C. Strauss et al. (Eds.): DEXA 2021, LNCS 12924, pp. 77–83, 2021.
https://doi.org/10.1007/978-3-030-86475-0_8

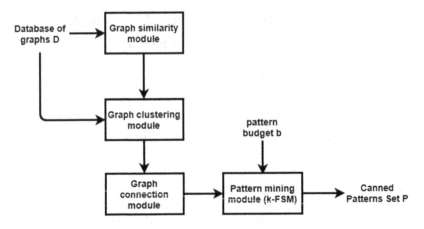

Fig. 1. System architecture.

at the same time minimize the cognitive load of P. The main advantage of this methodology is that users no longer have to design the pattern resulting in the improvement of the query formulation time. In addition, it eliminates the need for specific domain knowledge over the given database, a work-intensive task that requires extensive training.

Contributions. We are presenting here a modular architecture that allow us to provide a solution for the canned pattern selection problem. The architectural design offers some significant challenges, which are:

- Each module can utilize custom solutions and current state-of-art techniques in order to offer the optimal results for the task. In addition, abstract interfaces between the modules ensure that the underlying implementation can be easily interchangeable.
- Division of the canned pattern selection problem into independent tasks that can be optimised and adapted as needed.
- The machine learning neural solution for graph similarity offers quick addition of new graph sets and fast graph similarity score between two graphs.

2 System Architecture

In Fig. 1 we present the proposed architectural design, that is composed of four main components. For a given database of graphs, first the similarity score between the graphs is computed. Then this score is used to divide the graphs into clusters, which are then merged in one continuous graph. Finally, the continuous graphs are used to extract the patterns. In the following subsections we are describing in detail the functionalities provided by each module.

2.1 Graph Similarity Module

In the proposed architectural design, the graph similarity module is responsible for providing a similarity score between any two graphs from our graph dataset.

Graph similarity computation is a well known problem with many applications in graph community. Early algorithms that evaluated graph similarity through distance, such as Graph Edit Distance (GED) [3] and Maximum Common Subgraph (MCS) [4] are high demanding both in search space and time-complexity and thus fail to scale for graphs with more than a few modes. Aiming to reduce complexity by pruning the search space, heuristic algorithms like Hungarian [9] and VJ [5] were proposed to calculate the GED approximately. Many machine learning solutions have been tested with the graph neural models gaining in accuracy and popularity.

Currently, the SimGNN [1] solutions is proposed for the graph similarity module, that uses graph convolutional networks and graph embeddings to calculate a similarity score on two given graphs G_1 and G_2. The SimGNN has been selected due to the key features that ensure a high degree of accuracy for the prediction of the similarity score. These are:

- The features and structural properties of the graph nodes are taken into consideration.
- An embedding function generates a vector for each graph, using an attention mechanism.
- The function is trained by using GED distance values as ground-truth similarity scores. These GED values are the minimum approximations between three heuristic GED algorithms.
- Both the node-level and graph-level embeddings of two graphs are examined for the computation of the score.

For the graph similarity module, the SimGNN model is trained over the available dataset, following the proposed optimization to tune the hyperparameters. Some hyperparameters that can be tuned are rate of the dropout layers, epochs, batch size, histograms and learning rate. The trained model is tasked with providing an accurate similarity score between two given graphs that is then used as a clustering property in the graph clustering module. We define the similarity score provided between graph G_i and G_j as $s(G_i, G_j)$. When the score is equal to zero it indicates a perfect match.

2.2 Graph Clustering Module

The graph clustering module is responsible for partitioning the graph database D into a set of clusters $C = \{C_1, C_2, ..., C_k\}$, in a way that ensures both graph homogeneity within each cluster and graph heterogeneity between clusters. For this module, a custom solution that is based on a graph-based version of the classic k-means clustering algorithm [6] is proposed.

In detail, the proposed solution begins with selecting k random graphs from the database, the representatives of each cluster. For the remaining graphs, the

---

Algorithm 1: Pattern Clustering Module

Input: A database of graphs D
Output: A set of clusters $C = \{G_1, G_2, ..., G_k\}$
1. k graphs are chosen randomly from the graph database D and are the initialized as representatives of each cluster.
2. The remaining graphs are assigned to the closest cluster based on the graph similarity score $s(G_i, G_j)$ from the each graph and the cluster representative.
3. The new representative (set median graph) of each cluster is calculated.
4. Steps 2 and 3 are repeated until the cluster representatives no longer change.

---

distances from each cluster representative is calculated using the graph similarity module. Next, a new representative for each cluster is computed using the set median graph of each cluster. The process is repeated till the stabilization of the representatives. The detailed steps are presented in Algorithm 1.

The set median graph as defined in [11] provides a method for calculating representative from a set of graphs. Given a random cluster of graphs $C_r = \{G_1, G_2, ..., G_k\}$, the set median graph $\tilde{g}$ is given by the following formula:

$$\tilde{g} = \underset{G \in C_r}{\operatorname{argmin}} \sum_{G_i \in C_r} s(G, G_i)$$

where s is the similarity score, in our case obtained by the graph similarity module. The set median graph $\tilde{g}$ is a graph that belongs to the cluster C_r and represents the overall cluster suitably.

2.3   Graph Connection Module

This module is responsible for connecting all the graphs that belong in a cluster as frequent sub-graph mining techniques require for the graph to be connected. It take as input the generated clusters of the previous module and outputs one connected graph per cluster.

In order to achieve that, one node with a specific, pre-defined label is added to the cluster and then this node is connected with only one edge to all graphs in the cluster. A custom solution is proposed here, that aims to identify the node of each graph where this edge will be added in a way that will not affect the importance of the nodes in the graph and will minimize the domination of the addition of the edges. For this, the PageRank algorithm [10,12] is used to select and connect to the unique node only the node of each graph that has the lowest centrality. The detailed steps are presented in Algorithm 2.

2.4   Pattern Mining Module

The pattern mining module is responsible for generating a large amount of patterns and filtering them based on a pattern score. The problem of finding all

---

Algorithm 2: Pattern Connection Module

Input: A set of clusters $C = \{C_1, C_2, ..., C_k\}$
Output: A set of graphs $CG = \{CG_1, CG_2, ..., CG_k\}$
for *each cluster C_i* **do**
  $CG_i \longleftarrow \emptyset$;
  for *each graph G_j in cluster C_i* **do**
    $Vscores_j = G_j.\text{pagerank}()$;
    $V_{min} = Vscores_j.\text{min}()$;
    if $CG_i = \emptyset$ **then**
      $CG_i \longleftarrow G_j$;
      $CG_i.\text{addnode}(label\_unique, id_s)$;
      $CG_i.\text{addedge}(id_s, V_{min})$;
    else
      $CG_i \longleftarrow \text{disjoint\_union}(CG_i, G_j)$;
      $CG_i.\text{addedge}(id_s, V_{min})$;

---

labeled patterns with k edges that are frequent in graph G is called the k-Frequent Subgraph Mining problem (k-FSM). The frequency of such a pattern is usually calculated with the minimum node image (MNI) metric [2]. While there are many pattern mining frameworks available, for the pattern mining module the Peregrine framework [8] is proposed as it is scaleable, with good performance even for large graphs. To the best of our knowledge, Peregrine outperforms other solutions because of its pattern aware approach and its ability to minimize total matches, canonicality computations and sub-graph isomorphism calculations. Additionally, Perigrine provides an API for the motif counting, k-FSM and pattern matching graph problems, making it the best solution for the evaluation of the exported patterns.

The pattern mining module takes as input the connected graphs generated from each cluster $CG_1, CG_2, ..., CG_k$ and a pattern budget b = (a_{min}, a_{max}, c) where a_{min} is the minimum number of nodes of a canned pattern, a_{max} is the minimum number of nodes of a canned pattern and b is the total number of canned patterns that are requested. Output is the set of canned patterns P that satisfy the pattern budget requirements. The detailed steps are presented in Algorithm 3. In order for pattern comparison to be possible a pattern score metric is defined as score(p) = $\frac{MNI(p,G) \cdot div(p,P\backslash p)}{cog(p)}$ where:

- $MNI(p, G)$ is the minimum node image of a pattern p in a graph G
- $div(p, P \backslash p) = \min GED(p, p_i)$ where $p_i \in P$ and GED( ) is the graph edit distance operator
- Given a pattern p with V_p nodes and E_p edges $cog(p) = \frac{2 \cdot |E_p| \cdot |E_p|}{|V_p| \cdot (|V_p| - 1)}$

Algorithm 3: Pattern Mining Module

Input: A set of graphs $CG_1, CG_2, ..., CG_k$ and a pattern budget
$\quad\quad b = (a_{min}, a_{max}, c)$
Output: Canned pattern set P
$P' \longleftarrow \emptyset$;
for *each graph* CG_j **do**
$\quad$ **for** *i from a_{min} to a_{max}* **do**
$\quad\quad P' \longleftarrow P' \cup$ peregrineFSM(i,CG_j);
$\quad\quad$ //Above line generates all patterns of size i from CG_j that are above a
$\quad\quad$ given MNI.

CalculatePatternScores(P');
$P \longleftarrow$ SelectBestPatterns(P',c);

---

3   Conclusions

A modular architecture has been presented here as a solution to the canned pattern selection problem. Due to the independence of the modules, and the usage of abstract interfaces between them, it is possible to perform extensive experimental analysis, to identify the optimal technique for each module as well as potential bottlenecks and mitigation strategies. For example, different approaches to the formulation of the continuous graph can be evaluated.

References

1. Bai, Y., Ding, H., Bian, S., Chen, T., Sun, Y., Wang, W.: Simgnn: a neural network approach to fast graph similarity computation. In: ACM ICWSDM (2019)
2. Bringmann, B., Nijssen, S.: What is frequent in a single graph? In: PAKDD (2008)
3. Bunke, H.: What is the distance between graphs. Bull. EATCS **20**, 35–39 (1983)
4. Bunke, H., Shearer, K.: A graph distance metric based on the maximal common subgraph. Pattern Recogn. Lett. **19**(3), 255–259 (1998)
5. Fankhauser, S., Riesen, K., Bunke, H.: Speeding up graph edit distance computation through fast bipartite matching. In: Jiang, X., Ferrer, M., Torsello, A. (eds.) GbRPR 2011. LNCS, vol. 6658, pp. 102–111. Springer, Heidelberg (2011). https://doi.org/10.1007/978-3-642-20844-7_11
6. Galluccio, L., Michel, O., Comon, P., Hero, A.O.: Graph based k-means clustering. Signal Process. **92**(9), 1970–1984 (2012)
7. Huang, K., Chua, H.E., Bhowmick, S.S., Choi, B., Zhou, S.: Catapult: data-driven selection of canned patterns for efficient visual graph query formulation. In: Proceedings of the 2019 International Conference on Management of Data. SIGMOD 2019, pp. 900–917. Association for Computing Machinery (2019)
8. Jamshidi, K., Mahadasa, R., Vora, K.: Peregrine. In: Proceedings of the Fifteenth European Conference on Computer Systems, April 2020. https://doi.org/10.1145/3342195.3387548
9. Kuhn, H.W.: The Hungarian method for the assignment problem. In: Jünger, M., et al. (eds.) 50 Years of Integer Programming 1958-2008, pp. 29–47. Springer, Heidelberg (2010). https://doi.org/10.1007/978-3-540-68279-0_2

10. Langville, A., Meyer, C.: A survey of eigenvector methods of web information retrieval. SIAM Rev. **47**(1), 135–161 (2004). https://doi.org/10.1137/S0036144503424786
11. Munger, A., Bunke, H.: On median graphs: properties, algorithms, and applications. IEEE Trans. Pattern Anal. Mach. Intell. **23**, 1144–1151 (2001). https://doi.org/10.1109/34.954604
12. Page, L., Brin, S., Motwani, R., Winograd, T.: The pagerank citation ranking: bringing order to the web. Technical Report 1999–66, Stanford InfoLab (1999)

AutoEncoder for Neuroimage

Mingli Zhang[1(✉)], Fan Zhang[2(✉)], Jianxin Zhang[3(✉)], Ahmad Chaddad[4],
Fenghua Guo[5], Wenbin Zhang[6], Ji Zhang[7], and Alan Evans[1]

[1] Montreal Neurological Institute, McGill University, Montreal, Canada
mingli.zhang@mcgill.ca
[2] Shandong Future Intelligent Financial Engineering Laboratory, Yantai, China
[3] Dalian Minzu University, Dalian, China
[4] School of Artificial Intelligence, Guilin University of Electronic Technology,
Guilin, China
[5] Shandong University, Jinan, China
[6] Carnegie Mellon University, Pittsburgh, USA
[7] University of Southern Queensland, Darling Heights, Australia

Abstract. Variational AutoEncoder (VAE) as a class of neural networks
performing nonlinear dimensionality reduction has become an effective
tool in neuroimaging analysis. Currently, most studies on VAE consider unsupervised learning to capture the latent representations and
to some extent, this strategy may be under-explored in the case of heavy
noise and imbalanced neural image dataset. In the reinforcement learning point of view, it is necessary to consider the class-wise capability
of decoder. The latent space for autoencoders depends on the distribution of the raw data, the architecture of the model and the dimension of
the latent space, combining a supervised linear autoencoder model with
variational autoencoder (VAE) may improve the performance of classification. In this paper, we proposed a supervised linear and nonlinear
cascade dual autoencoder approach, which increases the latent space discriminative capability by feeding the latent low dimensional space from
semi-supervised VAE into a further step of the linear encoder-decoder
model. The effectiveness of the proposed approach is demonstrated on
brain development. The proposed method also is evaluated on imbalanced neural spiking classification.

1 Introduction

Modeling brain age is critical for the diagnosis of neuropsychiatric disorder.
Investigations on brain age have benefited from the development of advanced
magnetic resonance imaging (MRI) [3] and from large-scale initiatives such as
the Pediatric Imaging, Neurocognition, and Genetics (PING) [6] studies. One of
the simplest ways to model brain age is predicting participant age from magnetic
resonance imaging data through machine learning and statistical analysis. One
of the important topics in neuroscience is about designing an effective encoding
model and applying it to neural spiking prediction [4]. A shared variance component analysis method was applied for the estimation of the neural population's
variance reliably encoding a latent signal [4].

© Springer Nature Switzerland AG 2021
C. Strauss et al. (Eds.): DEXA 2021, LNCS 12924, pp. 84–90, 2021.
https://doi.org/10.1007/978-3-030-86475-0_9

Supervised learning in the context of neural networks is commonly used into a variety of neuroimaging tasks. Unsupervised VAE are commonly used in learning complex distribution of the dataset [1]. *autoencoders* (AEs) based supervised regression is proposed in [8] and a supervised linear AEs is proposed and applied to brain age prediction [6]. Autoencoders can be treated as a truncated Principal Component Analysis (PCA) [2] or analysis and synthesis dictionary learning [6,7].

We propose a framework to integrate supervised linear AE [6] and nonlinear VAE based regression [8] into cascade dual autoencoders. Intuitively, a supervised linear autoencoder will be fit to variational autoencoder based regression. The dual autoencoder can tight the latent representation with their high-dimensional input dataset. Establishing a more robust latent representation and a discriminative linear combination, with setup the relationship between the preliminary latent representation and their high-dimensional input dataset. In addition, in the supervised linear autoencoder, we adopt class-wise information estimation to make a more discriminative output for the final classification and realize supervised learning.

The major contributions of this work are as follows:

- We propose a novel cascade dual autoencoder for generating discriminative and robust latent representations, which is trained with the variational autoencoder based regression and class-wise linear autoencoder.
- We present a joint learning framework to embed the inputs into a discriminative latent space with variational autoencoder of the cascade dual autoencoder, and assign them with initial input to the ideal distribution by class-wise linear autoencoder.
- The proposed non-linear and linear cascade dual autoencoder framework is more efficient and robust for different kinds of datasets than current AEs.
- Empirical experiments on two different datasets demonstrate the effectiveness of our proposed approach that outperforms state-of-the-art methods.

2   The Proposed Approach

The proposed approach is composed of two important components: 1) The variational autoencoder based regression [8] with encoder network E, the decoder network D and the regression estimation R of label L; 2) The linear class-wise autoencoder with $\mathbf{P} = [\mathbf{P}_1, \cdots, \mathbf{P}_k, \cdots, \mathbf{P}_K]$ as the linear encoder and $\mathbf{D} = [\mathbf{D}_1, \cdots, \mathbf{D}_k, \cdots, \mathbf{D}_K]$ as the linear decoder [6], where $\mathbf{P}_k \in d \times m$ and $\mathbf{D}_k \in m \times d$. We treat the initial input samples as $\mathbf{X} = [\mathbf{X}_1, \cdots, \mathbf{X}_k, \cdots, \mathbf{X}_K]$ from all the K classes, and the data in k-th class is denoted as $\mathbf{X}_k \in \mathbb{R}^{d \times N_k}$, where d is the dimension of training samples and N_k is the number of samples of class k.

2.1   Variational AutoEncoder Based Regression

In VAE based regression, the encoder E provides the reduced latent representation of the input $\mathbf{X}$, latent representation $\mathbf{Z}$, the decoder network D can reconstruct $\mathbf{X}$ from $\mathbf{Z}$ as $\hat{\mathbf{X}}$, and $\mathbf{Z}$ is associated with labels L. Our goal is applied the label into the variational autoencoder based regression, then

$\mathbf{Z} = [\mathbf{Z}_1, \cdots, \mathbf{Z}_k, \cdots, \mathbf{Z}_K] \in \mathbb{R}^{d_1 \times N_k}$ and $\mathbf{X}$ with L as the input of the class-wise linear autoencoder for more accurate and robust classification. We first train the variational autoencoder based regression of the cascade dual autoencoder framework.

Generative Model. For the input features $\mathbf{X}$, there is a latent representation $\mathbf{Z} \in \mathbb{R}^{d_1 \times N}$ associated with the labels $L \in \mathbb{R}^{1 \times N}$, where d_1 is the dimension of latent representation and N is the total number of training samples, it is not one dimension but related with 1D label. To well guarantee the quality of latent representation. we construct a deep generative model and the generative process of x as $p(x, z, l) = p(x|z)p(z|l)p(l)$, where $p(z) = \mathcal{N}(z|0, \mathbf{I})$, $p_\theta(x|z) = f(x; z, \theta)$, where $f(x; z, \theta)$ is a suitable likelihood function (e.g., Gaussian or Bernoulli distribution when x is binary), $p(z|l)$ is a label related prior on latent representation, $p(z|l)$ is a linear generator model, $p(z|l) \sim w^T l + \|w\|_1$, $s.t.\, w^T w = \mathbf{I}$, where $\| \ \|_1$ is l_1 norm, and $p(l)$ is prior on label. $p_\theta(x|z)$. The estimated samples from the posterior distribution over $p(z|x)$ are used to predicting the label l.

Inference Model. To have a scalable and tractable variational inference and parameter learning, we adopt the standard variational inference and an auxiliary function $q((z, l)|x)$, we omit the parameter φ for $q_\varphi((z, l)|x)$, to approximate the posterior, $p((z, l)|x)$. $\log p(x)$ is the sum of the divergence between $q((z, l)|x)$ and $p((z, l)|x)$. we assume there is $q((z, l)|x) = q(z|x)q(l|x)$. We follow the variational principle by applying a lower bound objective function to guarantee the accuracy of the posterior approximation. The lower bound objective model can be written as

$$\mathcal{J}_1 := -D(q(l|x), p(l)) + \lambda \mathbb{E}_{q(z|x)}[\log p(x|z)] - \mathbb{E}_{q(l|x)}[D(q(z|x), p(z|l))] \tag{1}$$

where λ is a hyper-parameter that controls the relative weight of the discriminative and generative learning. The higher λ is, the more weight for decoded reconstruction from latent representation to estimated input. $D(q(.), p(.))$ is a divergence function, such as KL-divergence. $q(l|x)$ is formulated as a univariate Gaussian distribution. $q(z|x)$ as the probabilistic encoder enforces the input to latent space with multivariate normal distribution [8]. The last term of Eq. (1) forces $q(z|x)$ as close as possible to the label-prior $p(z|l)$.

2.2 Supervised Linear Autoencoder

To have much accurate classification results than VAE based regression, we introduce class-wise linear autoencoder with the $\mathbf{X}_k$ and latent representation $\mathbf{Z}_k$ from variational autoencoder as input features, setting as $\mathbf{S}_k = \{\mathbf{X}_k, \mathbf{Z}_k\}$, where label is the class number k, the projective equation and reconstructive equation are as follows

$$\mathbf{A}_k = \mathbf{P}_k \mathbf{S}_k + \mathbf{n}_0 \quad \hat{\mathbf{X}}_k = \mathbf{D}_k \mathbf{A}_k + \mathbf{n}_1 \tag{2}$$

where, $\mathbf{S}_k \in \mathbb{R}^n$ denotes the input features, $\mathbf{A}_k \in \mathbb{R}^d$ is the d dimensional hidden latent variables, $\hat{\mathbf{X}}_k \in \mathbb{R}^n$ is the estimated input $\mathbf{S}_k$ from the $\mathbf{A}_k$. $\mathbf{P}_k$ is linear encoder transforming the input features into the $\mathbf{A}_k$. $\mathbf{D}$ is linear decoder, back-project the $\mathbf{A}_k$ to estimated outputs. $\mathbf{n}_0$ and $\mathbf{n}_1$ are bias. The class-wise linear

Table 1. The total amount of spike in each spike group.

Group	1	2	3	4	5	6	7	8
#	25390	3168	1523	339	69	15	1	2

autoencoder modeled by minimizing the expected squared reconstruction error as

$$\mathcal{J}_2 := \left\{ \underset{\mathbf{P},\mathbf{D}}{\arg\min} s \quad \sum_{k=1}^{K} \|\mathbf{S}_k - \mathbf{D}_k\mathbf{P}_k\mathbf{S}_k\|_F^2 + \lambda\|\mathbf{P}_k\bar{\mathbf{S}}_k\|_1 \right\}_{\mathbf{S}_k=\{\mathbf{X}_k,\mathbf{Z}_k\}}, \quad (3)$$

where the Frobenius norm is fidelity term, the term with strict sparsity l_1 norm is to force the samples of other classes $\bar{\mathbf{S}}$ not to fit into the modeling of the current class and hence ensure the model to be class-wise discriminative. This supervised linear autoencoder can be solved same as the former works [6,7].

Therefore, the overall loss of the proposed supervised cascaded autoencoder network is as follows,

$$\min_{\theta,\varphi,\mathbf{P},\mathbf{D}} \quad \mathcal{J}_1 + \mathcal{J}_2, \quad (4)$$

2.3   Implementation Details

The supervised variational autoencoder takes data $(\mathbf{X}, L)$ as input then outputs $(\hat{\mathbf{X}}, \hat{L})$, where $\hat{L}$ is the predicted label corresponding to the label posterior. Keras is applied to train $q(l|x)$ with three convolutional layers, two max pool layers and one softmax layer, then dropout and ReLU activation. When training $q(z|x)$, three convolutional layers, and two fully connected layers with batch normalization, dropout and ReLU activation are applied. For $p(x|z)$ and $p(x|l, z)$, a fully connected layer with three ReLU activation and then the sigmoid for the output. The linear supervised autoencoder is optimized with an alternating direction method of multipliers (ADMM).

3   Experiments

The proposed cascade dual autoencoder framework is evaluated on modeling brain age from 3 to 21 years old with the cortical thickness from PING dataset, the details of the subjects are listed in [6]. The proposed framework is also evaluated on modeling single-neuron spiking activity with calcium imaging of 30507 slides. For each sequence, the firing rate and power spectrogram are computed. The distribution of spikes is presented in Table 1. The cortical calcium image is in time-sequence format, each spike is not only related to the corresponding cortical calcium image but also their neighbors, we proposed using a non-local image mean based approach and applied it to the spike calcium imaging to generate a number of non-local mean calcium imaging sequences. For detailed about the dataset, please refer to the materials in [5] and Table 1. 5-fold cross-validation

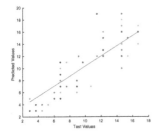

Fig. 1. The predicted age with ground-truth testing $R^2 = 0.7730$

Table 2. Classification results on the brain age.

	RMSE	MAE	ACC
RF	3.5503	2.6186	0.7200
NDPL	3.806	2.799	0.7001
SVM	4.1641	3.3645	0.5525
VAE	3.6832	2.6850	0.6771
Ours	3.5077	2.529	0.7336

is applied on these experiments. To measure performance in terms of prediction accuracy (ACC), root means square error (RMSE) and mean absolute error (MAE) are applied in this paper.

Prediction of Brain Age : We first demonstrate the proposed approach's performance by predicting the brain age from 3 to 21 years old with the cortical thickness of T1 structure MRI on the PING database. We measure the contribution of the proposed method by comparing results against the similar approaches NDPL [6] and VAE [8]. Figure 1 and Table 2 give the prediction accuracy in terms of correlation (R2), RMSE, MAE and accuracy (ACC). As one can see, the proposed method achieved the best accuracy compared with other baselines. Figure 1 shows the general relationship between cortical thickness and age which reflects a high correlation ($R^2 = 0.773$) between age and predicted age (brain age). As expected, a lower prediction accuracy is observed therein on both sides of the age range due to the challenging regression problem of 'regression to mean'.

Prediction of Spiking : The performance of the proposed approach is demonstrated on predicting the single-neuron spiking, based on calcium imaging. Here, we evaluate our method in a classification setting by applying a sliding window on the 30507 spikes for the data preprocessing scheme. The 5-fold cross-validation is applied on these experiments. To measure performance in terms of prediction overall accuracy (oACC), balanced accuracy(bACC). $oACC = N_c/N_t$, where, N_c is total number of all correctly classified subjects and N_t is number of all test subjects. $bACC = \frac{1}{K}\sum_{k=1}^{K} \frac{N_c^k}{N_t^k}$, where, N_c^k is total number of all test subjects in class k. N_t^k is the total number of all test subjects in class k.

Table 3 compares the RMSE, MAE, oACC and bACC obtained by our approach to the recently proposed method NDPL [6], variational autoencoders (VAE) and random forest (RF) for evaluating the proposed model. We see that the proposed approach outperforms the state-of-the-art method. *NLM* is the proposed model with non-local calcium image mean as input features. From Table 3, we can find the non-local frame mean as input features have lower residual classification errors (RMSE and MAE), compared with the calcium image features as input directly. With non-local calcium image features as input, the proposed achieved the best performance, yielding improvements of about 0.1805 in RMSE, 0.0855 in MAE and 0.0367 in bACC.

Table 3. Classification results on the single neuron spikes with 5-fold CV.

	RMSE	MAE	oACC	bACC
RF(NLM)	0.9932	0.7834	0.8400	0.3100
NDPL(NLM)	1.2549	1.0749	0.5608	0.2377
SVM(NLM)	1.4891	1.8575	0.4525	0.1727
VAE	1.1182	0.8700	0.6371	0.2718
Ours	1.0755	0.8840	0.8418	0.3213
Ours(NLM)	0.8950	0.7985	0.8070	0.3580

4 Conclusion

We proposed an efficient and robust cascade dual autoencoder framework model brain development and spikes prediction with calcium image and behavior video frame. Compared with the conventional methods, this approach learns discriminative features by imposing both variational autoencoder and class-wise linear autoencoder, and a l_1 sparsity constraint on coefficients of non-current-class. Experiments on the tasks of predicting the brain age and modeling spikes showed the benefit of our approach compared to state-of-the-art methods for these tasks. Furthermore, our approach can be used in understanding the influence of gender on brain development.

Acknowledgements. This work was partially supported by the NSFC (61902220, 61972062), the Young and Middle-aged Talents Program of the National Civil Affairs Commission, the Fonds de recherche du Québec-Santé (FRQS 271636,298507), the Science and Technology Innovation Program for Distributed Young Talents of Shandong Province Higher Education Institutions under Grant No. 2019KJN042.

References

1. Benou, A., Veksler, R., Friedman, A., Raviv, T.R.: De-noising of contrast-enhanced MRI sequences by an ensemble of expert deep neural networks. In: Deep Learning and Data Labeling for Medical Applications, pp. 95–110. Springer, New York (2016)
2. Bzdok, D., Eickenberg, M., Grisel, O., Thirion, B., Varoquaux, G.: Semi-supervised factored logistic regression for high-dimensional neuroimaging data. In: Advances in Neural Information Processing Systems. pp. 3348–3356 (2015)
3. Cole, J.H., Franke, K.: Predicting age using neuroimaging: Innovative brain ageing biomarkers. Trends Neurosci. 40(12), 681–690 (2017)
4. Stringer, C., Pachitariu, M., Steinmetz, N., Reddy, C.B., Carandini, M., Harris, K.D.: Spontaneous behaviors drive multidimensional, brainwide activity. Science **364**(6437), 255–255 (2019)

5. Xiao, D., et al.: Mapping cortical mesoscopic networks of single spiking cortical or sub-cortical neurons. Elife **6**, e19976 (2017)
6. Zhang, M., et al. : Brain status modeling with non-negative projective dictionary learning. NeuroImage **206**, 116226 (2020)
7. Zhang, M., Guo, Y., Zhang, C., Poline, J.-B., Evans, A.: Modeling and analysis brain development via discriminative dictionary learning. In: Knoll, F., Maier, A., Rueckert, D., Ye, J.C. (eds.) MLMIR 2019. LNCS, vol. 11905, pp. 80–88. Springer, Cham (2019). https://doi.org/10.1007/978-3-030-33843-5_8
8. Zhao, Q., Adeli, E., Honnorat, N., Leng, T., Pohl, K.M.: Variational autoencoder for regression: application to brain aging analysis. In: Shen, D., Liu, T., Peters, T.M., Staib, L.H., Essert, C., Zhou, S., Yap, P.-T., Khan, A. (eds.) MICCAI 2019. LNCS, vol. 11765, pp. 823–831. Springer, Cham (2019). https://doi.org/10.1007/978-3-030-32245-8_91

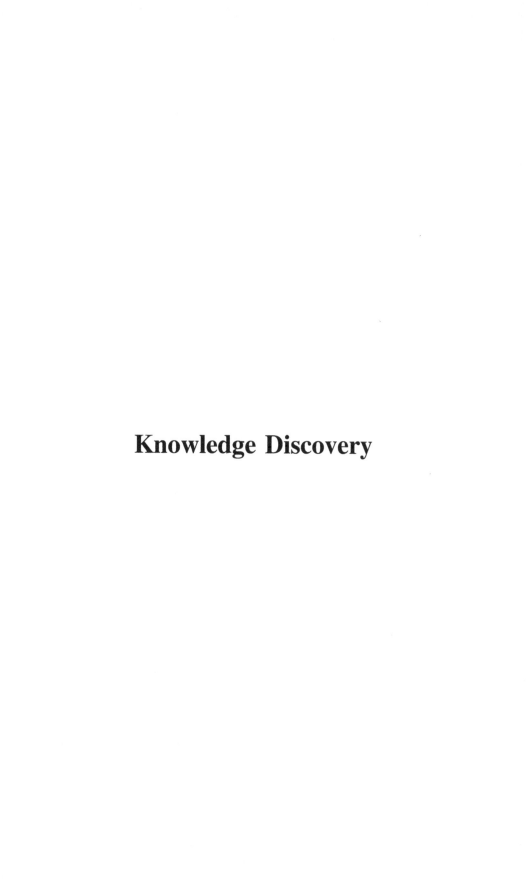

Knowledge Discovery

Towards New Model for Handling Inconsistency Issues in DL-Lite Knowledge Bases

Ghassen Hamdi[(✉)] and Mohamed Nazih Omri

MARS Research Laboratory, University of Sousse, Sousse, Tunisia
mohamednazih.omri@eniso.u-sousse.tn

Abstract. The lightweight description logic (DL-lite) represents one of the most important logic specially dedicated to applications that handle large volumes of data. Managing inconsistency issues, in order to effectively query inconsistent DL-Lite knowledge bases, is a topical issue. Since assertions (ABoxes) come from a variety of sources with varying degrees of reliability, there is confusion in hierarchical knowledge bases. As a consequence, the inclusion of new axioms is a main factor that causes inconsistency in this type of knowledge base. Often, it is too expensive to manually verify and validate all assertions. In this article, we study the problem of inconsistencies in the DL-Lite family and we propose a new algorithm to resolve the inconsistencies in prioritized knowledge bases. We carried out an experimental study to analyze and compare the results obtained by our proposed algorithm, in the framework of this work, and the main algorithms studied in the literature. The results obtained show that our algorithm is more productive than the others, compared to standard performance measures, namely precision, recall and F-measure.

Keywords: DL-Lite · Ontology · Inconsistency · Prioritized knowledge bases

1 Introduction

DL-Lite [1] is a family of tractable DLs designed for applications that deal with large quantities of data and where the most relevant reasoning function is to respond to queries. DL-Lite ensures a low level of computational complexity and it is considered as especially well suited to Ontology-Based Data Access (OBDA) [2]. To reflect generic knowledge, Description logic (DL) use only concepts and role inclusions in logical formulas (called axioms). A DL knowledge base (KB) is divided into two parts: a terminological base (called TBox), considered "generic" or "global", true in all models and for all individuals, and an assertional base (called ABox), this information is "specific" or "local", true for certain particular individuals.

How to deal with inconsistency is a crucially important question that emerges in OBDA [3,4]. With respect to some assertions that contradict the terminology, inconsistency is defined in such a setting. Usually, a TBox is typically checked and confirmed, while the assertions may be supplied by different and inaccurate sources in vast amounts and can contradict the TBox. Furthermore, manually checking and validating all of the assertions is often prohibitively expensive. This is why, in OBDA, reasoning in

© Springer Nature Switzerland AG 2021
C. Strauss et al. (Eds.): DEXA 2021, LNCS 12924, pp. 93–99, 2021.
https://doi.org/10.1007/978-3-030-86475-0_10

the face of inconsistency is important. Several works (*e.g.* [5]), inspired fundamentally by database approaches. or propositional logic approaches. treat with inconsistency in DLs, by adapting many inconsistency tolerant inference techniques, named semantics. These are focused on the assertional (or ABox) repair principle, which is closely related to the maximally consistent subset concept used in propositional logic. A repair of an ABox is merely a maximum assertional subbase that is consistent with a given TBox.

In this paper, the key current inconsistency-tolerant reasoning methods for prioritized KBs are first discussed. Interestingly enough, our approach is appropriate for the DL-Lite environment in the sense that, by producing a single preferred assertional repair, they allow effective handling of inconsistency.

The remainder of this article is arranged as follow. In Sect. 2 we present the main approaches that tried to solve the problem of repairing Inconsistencies in Dl-Lites Knowledge Bases. Section 3 will be dedicated to the presentation of an overview L-Lite ontology and management of inconsistencies. Section 4 details our proposed approach and presents the used algorithm. It also provides the experimental study conducted in this work and analyse the obtained results. In the final Sect. 5, we present a summary of the work as well as some ideas for future projects.

2   Related Works

Several works (*e.g.* [5]) attempted to deal with DL-Lite inconsistency by adapting several inference methods that accept inconsistency, based on database approaches (*e.g.* [6]). Priorities play an important role in the management of inconsistency, and they have been extensively discussed in the literature in the sense of propositional logic (*e.g.* [7]). In querying inconsistent databases or DL KBs, numerous works examined the notion of priority (*e.g.* [8]). Regretfully, only a few works are available in the OBDA format, such as the one given in [10, 11] for dealing with reasoning under the prioritized DL-Lite ABox. In a recent line of research, the inconsistency in lightweight ontology is being investigated. In particular, the writers [5], look at the issue of KB inconsistency by computing a collection of consistent subsets of assertions known as repairs, which recover ontology consistency, and then using them to address queries. In addition, in [13, 14], polynomials algorithms are proposed by the authors to pick a single preferred repair from a prioritized inconsistent DL-Lite KB, allowing for efficient query answering once the repair has been decided. Specifically, the writers in [13] propose a new approach focused on a single preferred repair solution. However, sequential inference strategies based on the selection of a single coherent assertional basis are proposed by the authors in [14]. In [15–17], the authors propose a new algorithm for answering queries without requiring access to Web databases.

3   DL-Lite Ontology and Management of Inconsistencies: An Overview

Inconsistency Treatment for Prioritized DL-Lite Assertional Bases: A DL-Lite knowledge base with priority is the set of assertions denoted by: $\mathcal{A} = \{S_1, ..., S_n\}$. The sets S_i

are called layers and each layer S_i contains the set of assertions with the same priority level i and are considered to be the most reliable than those present in a layer S_j when $j > i$. As a consequence, S_1 includes the most relevant assertions, while S_n contains the less important assertions. In the following, we use the DL-Lite notation $\mathcal{K} = \langle \mathcal{T}, \mathcal{A} \rangle$ to link to a prioritized DL-Lite knowledge base of the kind: $\mathcal{A} = \{S_1, ..., S_n\}$.

Linear-Based Repair: Let $\mathcal{K} = \langle \mathcal{T}, \mathcal{A} \rangle$ be an inconsistent and priority DL-Lite knowledge base. The linear based repair of $\mathcal{A}$ [14] is indicated by: $\ell(\mathcal{A}) = \{S'_1, ..., S'_n\}$ such that

- i) if i = 1 then

$$S'_1 = \begin{cases} Sl & \text{if } \langle \mathcal{T}, S_l \rangle \text{ is consistent} \\ \emptyset & Otherwise \end{cases} \tag{1}$$

- ii) for i = 2, ..., n

$$S'_i = \begin{cases} S_i & \text{if } \left\langle \mathcal{T}, S'_1 \cup ... \cup S'_{(i-1)} \cup S_i \right\rangle \text{ is consistent} \\ \emptyset & Otherwise \end{cases} \tag{2}$$

Clearly, $\ell(\mathcal{A})$ is acquired by dismissing an S_i layer when its facts contradict with the previous layers. Indeed, the subbase $\ell(\mathcal{A})$ is unique and it is consistent with $\mathcal{T}$.

Non-Defeated Repair: Another way to achieve a preferred repair is to interractively recover, layer by layer, all of the free elements free $(\mathcal{A})$.

We consider $\mathcal{K} = \langle \mathcal{T}, \mathcal{A} \rangle$ a DL-Lite knowledge base with priority. We denote by $free(\mathcal{A})$ the set of assertions relate to A that aren't to blame for any inconsistencies in $\langle \mathcal{T}, \mathcal{A} \rangle$.

The Non-Defeated Repair [14], denoted by $nd(\mathcal{A})$ is a sequence $nd(\mathcal{A}) = S'_1 \cup ... \cup S'_n$, such that:

$$\forall i = 1 ... n, S'_i = free(S_1 \cup ... \cup S_i)$$

Namely, $nd(\mathcal{A}) = free(S_1) \cup free(S_1 \cup S_2) \cup ... \cup free(S_1 \cup ... \cup S_{S_i})$.

4 Most-Possible Repair Proposed Approach

4.1 Most-Possible Repair Algorithm

Our algorithm is presented in this way: First, we initialize M-PR$(\mathcal{A})$ on an empty set. Then we sort the layers according to their cardinalities in a decreasing way (according to the layers that contain the largest number of assertions). If two sources have the same cardinality, we apply FIFO (First In First Out): that means, the first in is the first out. Then, if an assertion of the form of a concept and an assertion of the form of a role are inconsistent, we remove the assertion of the form of a role and keep the assertion of the form of a concept. Otherwise, if two assertions of the form of a concept are inconsistent, we keep the first and delete the second.

Algorithm 1. Most-Possible Repair proposed Algorithm

Data: $\mathcal{K} = \langle \mathcal{T}, \mathcal{A} \rangle$ with $\mathcal{A} = \{S_1, ..., S_n\}$
Result: A set of consistent assertions $M\text{-}PR(\mathcal{A})$
$M\text{-}PR(\mathcal{A}) \leftarrow \emptyset$
for $i = 1$ *to* n **do**
    $posmax \leftarrow i$
    for $j = i + 1$ *to* n **do**
        if $S_j > Sposmax$ **then**
        | $posmax \leftarrow j$
        end
    end
    $temp \leftarrow Sposmax$
    $Sposmax \leftarrow S_i$
    $S_i \leftarrow temp$
end
for $i = 1$ *to* n **do**
    if *A role and a concept are inconsistent* **then**
    | Delete (role)
    else if *Two concepts $C1$ and $C2$ are inconsistent* **then**
    | Delete ($C2$)
    $M\text{-}PR(\mathcal{A}) \leftarrow M\text{-}PR(\mathcal{A}) \cup C_i \cup R_i$ ($C_i \cup R_i$ is a set of assertions in concept or role form)
end
return $M\text{-}PR(\mathcal{A})$

4.2   Experimental Study and Results Analysis

The ontology used in this article describes the workflow in a development IT company. Our ontology is encoded by a TBox which contains 43 axioms and an ABox provided by 100 separate sources $\mathcal{A} = (S1,..., S100)$ represented as characteristics of the used ABox data collection : 97 Concept assertions and 56 Role assertions.

The tests are to be applied on ABox which contain *22, 44* and *66* sets of conflicts respectively. Our algorithm shows a calculation of correct assertions and makes our ontology consistent.

We are interested in the performance indices used to evaluate our algorithm. In our case, we classify ABox assertions into two classes: consistent and inconsistent.

To evaluate our approach, we consider the precision, the recall, and F-measure defined as follows:

The *Precision* P is calculated by dividing the total number of assertions by the number of consistent assertions. If it's big, it means the method has the most consistent assertions and can be called "precise". The *Recall* R of the number of consistent assertions found to the total number of consistent assertions is known as the Recall. The *F-measure* F is the harmonic average of the precision P and the recall R to give the performance of the system.

$$P = \frac{CE}{CE + IE} \tag{3}$$

$$R = \frac{CE}{CE + CNE} \tag{4}$$

$$F = \frac{2 * (P * R)}{(P + R)} \tag{5}$$

Where CE denotes the number of consistent assertions returned after the algorithms have been applied. CNE is the number of consistent assertions that were not returned after the algorithms were applied. And IE stands for the number of inconsistent assertions that exist before applying the algorithms.

Table 1. Linear-Based Repair algorithm evaluation results.

Linear-Based Repair $\ell(\mathcal{A})$			
Conflict size	Precision (P)	Recall (R)	F-measure (F)
22	84,61	77.07	80.66
44	66.41	69.04	67.70
66	48.03	53.04	50.41
Average value	66.35	66.38	66.25

Table 2. Non-Defeated Repair algorithm evaluation results.

Non-Defeated Repair nd($\mathcal{A}$)			
Conflict size	Precision (P)	Recall (R)	F-measure (F)
22	84.93	78.98	81.84
44	70.66	84.12	76.81
66	58.75	81.73	68.36
Average value	71.44	81.61	75.67

Table 3. Most-Possible Repair algorithm evaluation results.

Most-Possible Repair M-PR($\mathcal{A}$)			
Conflict size	Precision (P)	Recall (R)	F-measure (F)
22	86.16	87.26	86.70
44	73.49	96.82	83.56
66	62.92	97.39	76.45
Average value	74.19	93.82	82.23

By analyzing the results obtained in the series of experiments, there is a difference in efficiency between the various approaches. Indeed, the measurements gathered in Tables 1, 2 and 3 indicate that our **Most-Possible Repair** algorithm is the best among the other algorithms. In addition, we find that the Linear-Based Repair and Non-Defeated Repair algorithms do not return a maximum number of consistent assertions. In fact, the repair returned by our algorithm is the largest and most productive than that returned by the Linear-Based Repair and Non-Defeated Repair algorithms. We also noticed that our algorithm is the best compared to the other algorithms.

5   Conclusion and Prospects

In this paper, we have proposed a new approach for dealing with inconsistencies in prioritized DL-Lite knowledge bases. To do this, we started by studying the main approaches to dealing with inconsistencies in DL-Lite ontologies existing in the literature. At the end of a synthetic assessment, which we carried out, to compare these approaches according to a certain number of criteria, we were able to propose and detail our algorithm to deal with the inconsistency. We have carried out experiments studies on our approach, using standard performance measures, namely precision, recall and F-measure. The experimental study and analysis have shown that our **Most-Possible Repair** solution is more productive.

As future work, several perspectives remain possible. We cite two of them which are short-term: a first contribution consists in applying our approach to a larger ontology in order to confirm its reliability and robustness. A second possible prospect consists in adding a module to automatically manage the inconsistency when it appears when adding new assertions, through a graphical interface allowing this addition.

References

1. Artale, A., Calvanese, D., Kontchakov, R., Zakharyaschev, M.: The DL-lite family and relations. Comput. Res. Reposit. (CoRR) Volume abs/1401.3487 (2014)
2. Lenzerini, M.: Ontology-based data management, In: Proceedings of the 6th Alberto Mendelzon International Workshop on Foundations of Data Management, vol. 866, pp. 12–15, ACM, Glasgow (2011)
3. Hamdi, G., Omri, M.N., Benferhat, S., Bouraoui, Z., Papini, O.: Query answering DL-lite knowledge bases from hidden datasets. Ann. Math. Artif. Intell. **89**, 271–299 (2021)
4. Hamdi, G., Omri, M.N., Papini, O., Benferhat, S., Bouraoui, Z.: Querying DL-lite knowledge bases from hidden datasets. In: International Symposium on Artificial Intelligence and Mathematics, Fort Lauderdale, Florida (2018)
5. Bienvenu, M., Rosati, R.: Tractable approximations of consistent query answering for robust ontology-based data access. In: Proceedings of the 23rd International Joint Conference on Artificial Intelligence. pp. 775–781, IJCAI/AAAI, Beijing (2013)
6. Bertossi, L.E.: Database Repairing and Consistent Query Answering, Morgan & Claypool Publishers, San Rafael (2011)
7. Benferhat, S., Dubois, D., Prade, H.: How to infer from inconsistent beliefs without revising? In: Proceedings of the Fourteenth International Joint Conference on Artificial Intelligence, pp. 1449–1457, Morgan Kaufmann, Montréal Québec (1995)
8. Staworko, S., Chomicki, J., Marcinkowski, J.: Prioritized repairing and consistent query answering in relational databases. Ann. Math. Artif. Intell. **64**, 209–246 (2012)
9. Du, J., Qi, G., Shen, Y.: Weight-based consistent query answering over inconsistent over inconsistent SHIQ knowledge base. Knowl. Inf. Syst. **34**, 335–371 (2013)
10. Hamdi, G., Telli, A., Omri, M.N.: Querying of several DL-Lite knowledge bases from various information sources-based polynomial response unification approach. J. King Saud Univ. Comput. Inf. Sci. (2020)
11. Telli, A., Hamdi, G., Omri, M.N.: Lexicographic repair under querying prioritised dl-lite knowledge bases. Sci. J. King Faisal Univ. Basic Appl. Sci. **22** (2021)
12. Baral, C., Kraus, S., Minker, J.: Combining multiple knowledge bases. IEEE Trans. Knowl. Data Eng. **3**, 208–220 (1991)
13. Benferhat, S., Bouraoui, Z., Tabia, K.: How to select one preferred assertional-based repair from inconsistent and prioritized DL-Lite knowledge bases? In: Proceedings of the Twenty-Fourth International Joint Conference on Artificial Intelligence, pp. 1450–1456 (2015)
14. Telli, A., Benferhat, S., Bourahla, M., Bouraoui, Z., Tabia, K.: Polynomial algorithms for computing a single preferred assertional-based repair. Kunstliche Intelligenz **31**, 15–30 (2017)
15. Boughammoura, R., Omri, M.N.: Querying deep web data bases without accessing to data. In: 13th International Conference on Natural Computation, Fuzzy Systems and Knowledge Discovery, pp. 597–603, IEEE, Guilin (2017)

16. Boughammoura, R., Omri, M., Hlaoua, L.: Information retrieval from deep web based on visual query interpretation. Int. J. Inf. Res. Rev. **2**, 45–59 (2012)

17. Boughammoura, R., Hlaoua, L., Omri, M.N.: G-Form: a collaborative design approach to regard deep web form as galaxy of concepts. In: 12th International Conference of Cooperative Design, Visualization, and Engineering, pp. 170–174, Springer, Mallorca, Spain (2015). https://doi.org/10.1007/978-3-319-24132-6_20

ContextWalk: Embedding Networks with Context Information Extracted from News Articles

Chaoran Chen[1,2(✉)] ⬭, Mirco Schönfeld[3] ⬭, and Jürgen Pfeffer[4] ⬭

[1] Department of Biosystems Science and Engineering, ETH Zürich,
Basel, Switzerland
chaoran.chen@bsse.ethz.ch
[2] Swiss Institute of Bioinformatics, Lausanne, Switzerland
[3] University of Bayreuth, Bayreuth, Germany
mirco.schoenfeld@uni-bayreuth.de
[4] Bavarian School of Public Policy, Technical University of Munich,
Munich, Germany
juergen.pfeffer@tum.de

Abstract. Extracting meaningful social networks from a large corpus of news articles is challenging. This paper presents a network embedding technique which uses the content of texts as context information for relationships among persons referenced in texts. The goal of our approach is to automatically extract social networks from news articles and to identify structures in these networks. *ContextWalk* is an algorithmic approach extending random walk-based techniques by adapting a context-aware path sampling. We demonstrate the functionality and performance of the algorithm on an 8,000 nodes network extracted from a large German news dataset. Comparison with state-of-the-art algorithms shows that using the context to identify similar types of relationships significantly improves the algorithm's ability to detect fine-grained semantic groups.

Keywords: Social networks · Representation learning · Context information · News

1 Introduction

For the field of network science, the core assumption when analyzing social or technological systems is that these systems should not only be understood as a set of elements but that their connections and interactions are even more important [20,29]. But how can we retrieve information about interpersonal relationships? On the one hand, there are typical social science data collection methods such as interviews and surveys [21] as well as technology-supported solutions such as voluntary tracking of mobile devices (e.g., [5,22,34]). A downside of these approaches is that it requires the willingness to cooperate with scientists.

© Springer Nature Switzerland AG 2021
C. Strauss et al. (Eds.): DEXA 2021, LNCS 12924, pp. 100–114, 2021.
https://doi.org/10.1007/978-3-030-86475-0_11

Consequently, such approaches are not always practical to investigate politics or other elites since, e.g., top politicians are unlikely to tell scientists about their relationships and share their phone or movement records.

A secondary data source is newspaper articles. Using news as a dataset has many advantages due to the high availability, currentness, and wide range of coverage. News articles are particularly interesting for analyzing elites, since politicians, top managers, sport stars, etc. are in the focus of journalistic reporting. Several approaches to extract social networks from texts exist [4,9,13,24]. These traditional approaches for extracting networks from texts rely on co-occurrence of entities within texts, paragraphs, or sentences. The co-occurrence networks—technically they are one-mode networks constructed from person/source two-mode networks—differ in respect of their structural properties from social networks constructed from interviews and surveys. For instance, prominent actors can have a very high degree, or summary articles mentioning many actors will form a complete sub-graph connecting all of these actors.

Network embedding techniques try to reduce the dimensionality of representations of complex systems and have been applied for many different data sources [3,10,18]. They aim to find vector representations of nodes or edges that preserve the structure and relevant properties of a network—which properties are considered as relevant depends on the investigated network. The main argument of this paper is that current network embedding algorithms need to be extended to better work with social networks extracted from news articles by utilizing context information extracted from the texts.

Problem Statement. The study aims to develop a network embedding algorithm that is suited for dense and noisy networks such as co-reference networks extracted from news datasets. The algorithm takes as input a social network in the form of a graph $G = (V, E)$ and a set of documents (news articles) D which are mapped to the nodes V (the persons mentioned in the articles). The edges $E \to \mathbb{N}$ are undirected and weighted by the number of documents containing references to both nodes: $E \to \mathcal{P}(D)$, where $\mathcal{P}$ is the power set function. Furthermore, let $\mathcal{N}(v)$ be the neighbors of v and $deg(v)$ be the degree (number of neighbors) of v. The network embedding algorithm $\Phi : V \to \mathbb{R}^d$ uses a context embedding method $\Psi : \mathcal{P}(D) \to \mathbb{R}^c$, where d is the dimension of the node embedding (which should be much smaller than $|V|$) and c is the dimension of the context embedding.

Contributions. This study makes two contributions, both aiming to support social scientists in finding structures in large networks. Firstly, it discusses the challenges and the potential of news articles to serve as a data source for social network research. A network was generated from over three million German articles and made public [8]. By only using a few common NLP techniques and openly available data sources, the results of this work are easy to reconstruct, transfer, and extend. Secondly, this paper presents *ContextWalk*, a network embedding technique that is especially suited for clustering. It treats the news articles as the context of the referenced people. Only by incorporating additional information besides the network typography, it was possible to turn an otherwise noisy network into a valuable data source. The presented algorithm is efficient with a runtime complexity of $|E| + |V| \ log(|V|)$.

The paper is organized as follows: Sect. 2 gives an overview of related works. Section 3 presents the dataset used in the study. The *ContextWalk* algorithm is explained in Sect. 4. Finally, Sect. 5 shows experimental results and comparison with other algorithms. In Sect. 6, we discuss limitations and generalizability of our approach.

2   Related Work

The main challenge of this work is to extract meaningful networks of people from text sources. Network text analysis has developed several approaches to identify agents and their social networks in large collections of texts [4,9,13,24]. The network extraction process typically consists of two steps. First, persons are identified (entity extraction). Secondly, relationships among persons are derived from the co-occurrence of two or more persons within one text, paragraph, sentence, or within a sliding window of n words.

Network embedding is the technique of finding a representation of elements of a network—nodes or edges—in a vector space [3,10,18]. This work is related to the many random walk-based network embedding methods that have their roots in the Skip-Gram algorithm [26]. Developed in the field of computer linguistics to embed words, Skip-Gram expects a text corpus as input and uses a neural network with one hidden layer which will be trained to predict the surrounding words of a given word. The idea is that those words with a similar meaning can be used synonymously and will often have a similar neighborhood, thus, being placed close to each other in the vector space. The algorithm has proven to be very effective in capturing the relationships between words; a famous example is the following equation: $\Phi(king) - \Phi(man) = \Phi(queen) - \Phi(woman)$. DeepWalk [31] adapts the idea for network embeddings by using fixed-length random walks in the graph instead of the text strings. The method has proven to be quite effective and the concept of combining Skip-Gram with sampling from the graph was reused by other algorithms such as LINE [35] and Node2Vec [19].

An aspect that has remained unclear is the actual semantic of a random walk compared to consecutive words in a sentence. The original Skip-Gram takes sequences of tokens that, firstly, are guaranteed to be meaningful and, secondly, consist of different types of words that fulfill different roles (e.g., subject, predicate, object). DeepWalk, on the other hand, feeds the Skip-Gram algorithm with one type of token: In the case of a social network, a random walk would be a sequence of person names. Furthermore, DeepWalk does not regard the overall semantics of a random walk. It is only defined that two consecutive nodes are connected but not whether the nodes on a walk make up a meaningful composition altogether. *ContextWalk* addresses this question by using a more selective sampling technique.

While DeepWalk itself is designed for networks with unweighted edges, Node2Vec has proposed an efficient extension for weighted edges using alias sampling [37]. *ContextWalk* uses the extended DeepWalk as a baseline.

In the real world, it is often possible to obtain more information about a network and its entities than those found in an adjacency matrix. TADW [40], Paper2vec [17], and TriDNR [30] use text data that can be associated to the nodes. They are essentially mixing network analysis with classical profile-based techniques but, unlike *ContextWalk*, they do not use the texts to gain insights into the relationships between the nodes. CANE [36] has a similar motivation as this paper and recognizes that a node could act differently when interacting with different neighbors. The solution it offers is, however, very different from this study: instead of generating a single embedding with a context-aware method, it proposes a different embedding for each context. For tasks that require one definitive embedding such as clustering, CANE averages the context-specific embeddings mitigating the main feature of the approach.

Further, there are several community-aware network embedding methods (e.g., [6, 39, 41]). Even though, ContextWalk creates embeddings that are well-suited for the task of community detection, the method itself is not community-aware and does not optimize the modularity or any other clustering-defining scores.

3   Dataset

For this study, a network from texts was built. With persons being the nodes, two nodes are connected if the persons have been mentioned in one text and the edges are weighted by the number of co-mentioned. Additionally, the mentions' contexts are also considered as contexts of the edges. The corpus that was used for this study consists of news articles from 8 German newspapers with a total amount of 3.8 million documents. While most of the papers are operating nationwide, few regional papers were included as well to increase the diversity and the number of covered people. The articles range from 1946 to June 2019, although most were written in the past few years. While we have focused on currently active actors in our experiments, it made sense to include older data since a political career can last many decades.

To find the names of persons mentioned in the articles, the StanfordNLP named entity recognition software [15] is used. The names are then matched with person names in Wikidata [38], this reduces the number of false positives and ensures that only the persons with at least a certain degree of prominence are included. Unambiguous names are not resolved and only mentions of full names are accepted. The latter is assumed to not cause a significant loss of person recognitions because inspections of samples from the selected sources showed that in almost all articles, the full name of every person is mentioned at least once.

A co-mention network is extracted from this dataset. The nodes of the network are the persons in the articles and two persons are connected with an edge if there are articles in which both persons are mentioned. The edges are

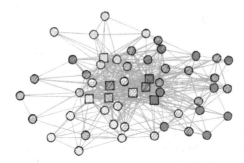

Fig. 1. Subgraph with 54 German politicians (394 edges), members of the government in the eight states of Germany with the highest populations in February 2020. Colors represent states, the rectangles are heads of governments. The network is dense and has a very uneven degree distribution.

weighted by the number of articles the two persons share. To reduce the network size and to only keep those persons for whom sufficient data exists, 8000 were chosen: they are the 5000 persons with the most articles supplemented with the 3000 most-mentioned politicians who are not in the first 5000. Then, edges with a weight of less than five were removed with the consequence that 714 nodes became disconnected and got dropped. The articles in which two persons are mentioned are considered as the context of the edge that connects them.

3.1 Challenges

One major challenge of applying typical social network algorithms on the given network is that it is not a social network with respect to many key characteristics. For instance, it does not show patterns of a small world [28]. Instead of the usual six degrees of separation, the median shortest path is only 2—the network has a very high density. The degree distribution is very uneven with 10% of the edges belonging to 25 nodes. These 25 nodes also constitute a clique and the subgraph of the 50 nodes with the highest degrees possesses 90% of all possible edges. Angela Merkel – the person with the highest degree – is directly connected to 3768 other persons which is much larger than Dunbar's number of around 150 [14]. The diameter of the graph is 7, the average eccentricity is 4.8 (median: 5), the smallest eccentricity is 4. This topology is untypical for a social network and for networks in general, thus, making it challenging to find structures with existing techniques. Looking for cliques, for example, might not be the best approach, since the high density also implies a large number of cliques. Figure 1 visualizes a small subset of the network.

The derived network does not describe actual social interactions. Being mentioned in a news article requires an action of a journalist but not of the affected persons. To study the actors, it is, therefore, necessary to gain an understanding of the meaning of the edges first.

The generated network consists of about 7,300 people who were mentioned in German news. As will be shown in Sect. 5, it is easy to divide them into larger clusters, mostly by their professions but existing methods fail to determine more fine-grained groups, for example, within the politicians. Latter is indeed quite difficult due to the problem that in many cases, there is no single, conclusive way to cluster them, and a plain network graph is not able to reveal the semantic of the relationships between the actors. Only by including context information, it becomes possible to identify finer structures.

4   Algorithm

4.1   Context Embedding

A context-aware application is understood as a system that knows and uses information regarding the situation of the user that is relevant for the task which the user wants to perform [12]. Such information often includes the location, time, and identity of the user – the same information that is considered as significant for the orientation of a person [2]. Also persons and objects in close (physical) proximity are of interest [33].

This concept can be fluently transferred to the social network analysis. News articles are thereby a very good source as a report usually contains exactly this information. For simplicity, this paper will only use the keywords of the articles as context. The underlying idea hereby is that a relationship is meaningful if there is a small set of keywords that connects the two persons and that appears repeatedly in their conversations or, as in our case, in articles reporting about them. The context vector is the normalized distribution of keywords in the set of articles associated with the edge:

$$\Psi_{keyword} : \mathcal{P}(D) \rightarrow [0, 1]^c \tag{1}$$

This way, each entry of the embedding vector has a clear meaning. The entries with the highest values show the dominant keywords that connect two persons, and the distribution of the values indicates the character of their relationship. As an example, in the used dataset, the context of the tie between Angela Merkel and Barack Obama was inspected. Prominent keywords in their relationship include USA, Europe, Berlin, Washington, Russia, China, and NATO. The sum of the 30 largest values is 0.22 – a clear sign that the news reports about them are far from random. A randomly picked edge has only a score of approximately 0.01. Comparing Merkel's relationship to Obama with hers to Nicolas Sarkozy shows that the latter contains more mentions of France, Europe, Brussels, Greece, and banks but less about Wladimir Putin, Russia, Syria, and Iran. These observations were used to develop *ContextWalk*.

4.2   ContextWalk

Part 1: Context-Aware Sampling. This method is inspired by DeepWalk [31] which consists of two components: a sampling mechanism and the usage of Skip-Gram [27]. As DeepWalk, the algorithm creates a "corpus" by sampling

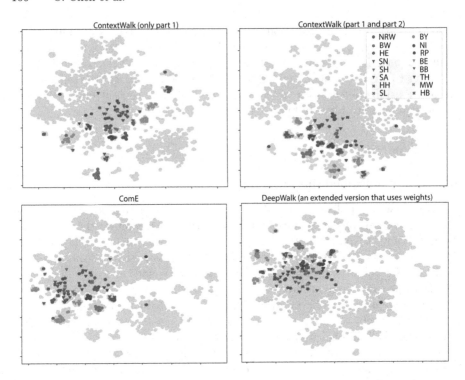

Fig. 2. Embeddings of the network with different methods after dimensionality reduction with t-SNE. The bright-red circles represent the nodes that are not in the subset A. The other colors and symbols represent the communities in subset A.

from the network and feeds it to Skip-Gram but it uses a more complex and context-aware sampling technique. In either case, sampling is based on a random walk. A walk w of length k is a list of vertices $(v_0, ..., v_{k-1})$ where v_i is a neighbor of v_{i-1} (for all $0 < i < k$). The aim is to find an embedding function Φ which maximizes the probability of vertices of a walk appearing given a vertex:

$$\max_{\Phi} Pr(w \backslash v | \Phi(v)) \tag{2}$$

Each vertex is used as v_0 for a fixed defined number of times. The basic version of DeepWalk as described in [31] only deals with unweighted edges and v_i (where $i \neq 0$) is uniformly sampled from the neighbors of v_{i-1}. Assuming $v_i \in \mathcal{N}(v_{i-1})$:

$$Pr[v_i \in w] = \frac{1}{deg(v_{i-1})} \tag{3}$$

An easy extension to that approach which takes the edge weights into accounts is to sample v_i with a probability proportional to the weight:

$$Pr[v_i \in w] = \frac{f(v_{i-1}, v_i)}{\sum_{v_j \in \mathcal{N}(v_{i-1})} f(v_{i-1}, v_j)} \tag{4}$$

The sampled data is passed to the Skip-Gram method: The set of sampled walks is equivalent to a corpus in natural language processing, a single walk is essentially considered as a sentence, and a node is a token. Transferring the concept of Skip-Gram to network science has been proven to be quite successful as many works have adopted the idea. However, word embeddings rely on actual, real-world texts while network embeddings use randomly constructed walks. There is no guarantee that it makes any sense to group the nodes of a walk together. This algorithm addresses this issue and strives to generate walks that have some semantic meaning and, therefore, come closer to a real (text) corpus.

This algorithm incorporates the context into the sampling process. The idea is that the probability of a walk to "make sense" is higher if the nodes' relationship to each other can be characterized by few and repeatedly occurring keywords. Using the keyword mapping context $\Psi_{keyword}$ ($=: \Psi$), each entry of a context vector corresponds to a keyword. Let t be the number of top keywords that should be considered and $h_t(x)$ the indices of the entries in x with the t largest values:

$$h_t : \mathbb{R}^c \rightarrow \mathbb{N}^t \tag{5}$$

After collecting random walks as in 4, the walks are added to the corpus with a probability depending on the sum of the t top keywords' values in the averaged context. For a given walk w of length k, let e be the average of the context vectors of the edges between the nodes in w:

$$e = avg(\{\Psi(g(v_i, v_j))|v_i, v_j \in w, (v_i, v_j) \in E\}) \tag{6}$$

The probability of being added to the corpus W can then be expressed as:

$$y = \sum_{i \in h_t(e)} e_i \tag{7}$$

$$Pr[w \in W] \propto y, \qquad \text{if } y > l \tag{8}$$

$$Pr[w \in W] = 0, \qquad \text{otherwise} \tag{9}$$

where l represents the minimal expected amount of similarity between the relationships of the nodes in a walk. In other words, the algorithm favors walks that contain persons who are mentioned with each other in the context of certain specific keywords.

Part 2: Two-Mode Embedding. Consider the sentence "Söder Seehofer Herrmann Hofreiter". Even though the words of this "sentence" have a semantic connection – these are the names of four top politicians from Bavaria, Germany – it is far away from being as meaningful as a real sentence. It only contains subjects but misses a predicate. Verbs describe the relationship between the subjects and, therefore, would transform the network into a normal social network.

However, further aiming to generate a more natural language-like corpus, the idea of this approach is to add keywords as objects to the sentence. Given a sampled walk w from Algorithm 1, w' will be constructed as follows:

$$w' = (v_0, u_0, v_1, ..., u_{k-2}, v_{v-1}), \quad \text{where } u_i \in h_t(v_i, v_{i+1}) \tag{10}$$

In other words, persons and keywords are embedded simultaneously. u_i is sampled uniformly from $h_t(v_i, v_{i+1})$ for performance reasons but it might also be interesting to depend the sampling probability on its value in the context vector.

4.3 Complexity

The runtime complexity of the sampling algorithm is in $O(|E|)$ and does not increase compared to DeepWalk for weighted graphs. The computation of the embedding after sampling can be achieved in $O(|V| \, log(|V|))$ by using the techniques of hierarchical softmax and negative sampling [26, 31].

5 Experiments

We chose community detection [7, 16, 23, 32] as the downstream task for an evaluation since it is of great relevance in the social network analysis. Considering the density of the network and the large variety of covered topics – thereby also communities – in news articles, detecting real-world communities is highly challenging. The used dataset is described in Sect. 3.

5.1 Compare Clusterings

The aim is to assess whether persons who are in the same real-world group are placed close to each other. Since it is unfeasible to define a meaningful clustering for over seven thousand persons, only subgraphs will be clustered at one time. Embeddings of the whole network will be computed but clustering will be performed on subsets of the network which contains persons that can be divided into non-overlapping or barely-overlapping groups. We have defined three subsets that contain well-known people in real groups. The members of these groups typically know and interact with each other, even though they are not always connected in the given network.

- Subset A contains members of German state governments (Landesregierungen) as of July 2020. The clusters have different sizes not only because of the different number of ministers in the states but also because not everyone is in the dataset. The clusters are non-overlapping. (n = 155, k = 16)
- In subset B, there are the football players of the Bundesliga who can be clustered by their club memberships. Even though the clusters do not overlap at any given point in time, the players change their clubs from time to time. Since the network does not rely on data for a short period of time but for many decades, a perfect clustering in the sense of the defined ground truth is probably not possible. (n = 261, k = 18)

– Subset C consists of members of Germany's federal parliament (Bundestag) who are members of the five biggest committees (Ausschüsse). Similar to subset B, the data is noisy for this task considering that it is not uncommon for politicians to change their fields of interest and join different committees after re-election. Furthermore, it is expected from a member of the parliament to be familiar with a wide range of topics and to comment on them. There are also six members who are in two or more of the selected five committees. $(n = 172, k = 5)$

Table 1. The NMI of the three clusterings

	Subset A	Subset B	Subset C
ContextWalk (only part 1)	0.7887	0.6730	0.0866
ContextWalk (part 1 + 2)	**0.8435**	**0.7084**	**0.0947**
DeepWalk (extended)	0.6592	0.5444	0.0891
ComE	0.7647	0.3900	0.0742
LINE (1st)	0.2216	0.2650	0.0894
LINE (2nd)	0.2553	0.2328	0.0840

ContextWalk was compared with an extended version of DeepWalk [31] which considers the edge weights, with ComE [6] that explicitly trains for community-aware embeddings, and with LINE [35]. After computing the embeddings, the clustering was performed with k-Nearest-Neighbors. To compare with the true clustering, the normalized mutual information (NMI) values [11] were calculated. The number of dimensions of the vectors was set to 100.

The results are listed in Table 1 and Fig. 2 visualizes the embeddings obtained with ContextWalk, DeepWalk and ComE for subset A. For subset A and B, *ContextWalk* outperformed every other method and the second part of *ContextWalk* proved to make a significant contribution to the performance. LINE showed by far the worst results which, however, was not unexpected: LINE looks at the first respectively second proximity of a network, and with the average shortest path being only two, this approach is not viable for this network. Also not surprising are the bad performances of all methods for subset C: The members of Bundestag are highly interconnected with many different and semantically meaningful clusterings – not only by their committees but, for example, by their party memberships or state.

Using the context data of the edges between nodes of a cluster, it is possible to gain insights into the nature of a group. This is demonstrated in Table 2 which shows three clusters for each of the three subsets. Reusing the idea of 7, the second column presents the sums of the scores of the 30 most common keywords within each cluster. A large value suggests that the relationships between the members share a number of keywords that regularly come up. The third column lists the three most important keywords. All clusters have relatively high values for the top 30 keywords – in comparison: the respective value for a set of a hundred randomly selected edges is approximately 0.002. For subset A and B, the top three keywords are distinctive for the clusters while, for clusters in subset C, the keywords are very general.

Table 2. Most weighted context information of ties within clusters; sum = sum of top 30 keywords.

Cluster name	Sum	Top 3 keywords
Subset A		
NRW	0.3927	CDU/0.0295, Nordrhein-Westfalen/0.0246, FDP/0.0217
BY	0.3940	CSU/0.0308, Neu!/0.0294, Politik/0.0192
BW	0.4573	CDU/0.0424, Baden-Württemberg/0.041, Stuttgart/0.0393
Subset B		
1. FC Köln	0.4273	Köln/0.0387, 1. FC Köln/0.0375, Spiele/0.0345
1. FC Union Berlin	0.5603	Berlin/0.0500, Union Berlin/0.0456, Spiele/0.0426
Fortuna Düsseldorf	0.5262	Düsseldorf/0.058, Fortuna Düsseldorf/0.0564, Spiele/0.0496
Subset C		
Recht und Verbraucherschutz	0.3031	Deutschland/0.0210, Politik/0.0204, Politiker/0.0204
Inneres und Heimat	0.2970	FDP/0.015, CSU/0.0148, SPD/0.0148
Auswärtiges	0.2940	Politik/0.017, Politiker/0.017, SPD/0.0169

With an exception for LINE, the figures in Fig. 2 show a number of larger clusters. These clusters mostly represent fields of professions. The main clusters that are clearly separated are German politicians, foreign politicians, different groups of athletes and, entertainers (musicians or actors).

5.2 Network and Embedding Distances

Last but not least, the relationships between the distances in the actual network and the distances in the computed embeddings were analyzed. Random nodes were selected and their distances calculated. The edge weights were transformed into distances so that an edge with a large weight has a small distance value. While the correlations are rather small for all algorithms, it is to note that, in comparison, it is larger for the new algorithm than for DeepWalk and ComE. It is a little surprising since no direct effort was undertaken to increase the emphasis on the network distances. An imaginable explanation for this phenomenon could be that certain correlations between the edge weights and edge contexts exist.

6   Discussion

This paper presented *ContextWalk* - a simple approach to incorporate context information into network embeddings. By adopting a context-aware sampling and simultaneously embedding the nodes and the context information, *ContextWalk* group nodes with similar relationships to construct a semantically meaningful corpus. The method was demonstrated on a dataset consisting of 3.8 million German news articles and a network with over 7,000 persons. Using community detection as the downstream task, the evaluation showed that the contexts indeed contribute to finding semantic groups.

Limitations. Using only keywords defined by the journalists, this embedding highly relies on their way of classification. While good interpretability is an advantage, some facets of the contents of the text documents are entirely ignored. It might be interesting to make use of more sophisticated NLP techniques in the future and include further attributes such as the publication date and platform, author, and type (e.g., report, opinion, or interview). An alternative approach could be to directly embed the underlying text documents individually for which a state-of-the-art technique is Paragraph Vector [25]. Paragraph Vector is supposed to be able to capture the semantics of texts of any length through a neural network and is also used in other network embedding techniques such as SubVec [1] where good results were reported. A limitation of this idea, however, is that it produces a black box: The meanings of the dimensions of the computed vectors are less interpretable and could not directly be used to gain insights into the relationships in the network.

Generalizability. The presented techniques were developed with a particular dataset in mind but they can also be applied to a wide range of scenarios. In general terms, this paper provides a context-incorporating network embedding algorithm. It is especially useful for networks that are connected to a collection of text documents due to the unstructured but information-rich nature of texts, thus, possessing a high potential for analysis.

Given, for example, a dataset containing all the letters, emails, and chat messages of a group of persons – not only to each other but also to members outside of the group, even though it is a social network describing actual social interactions, it offers some challenges. The network would be quite dense and consisting of many overlapping clusters since a person is usually in a large number of groups. Considering that it is not unusual to have hundreds or even thousands of "friends" on Facebook, the degree of a node in such a communication network would be probably much higher than around 150 as Dunbar's number suggests. With the methods described in this paper, a context could be defined as metadata such as date and platform of communication together with the actual text content of the messages. Since the topics and the way of discussions with, for example, the family, friends, colleagues, and customers usually differ, the presented techniques could help to distinguish between different types of communities.

References

1. Adhikari, B., Zhang, Y., Ramakrishnan, N., Prakash, B.A.: Distributed Representations of Subgraphs. In: International Conference on Data Mining Workshops (ICDMW). pp. 111–117. IEEE, New Orleans, LA, USA (2017)
2. Berrios, G.E.: Disorientation states and psychiatry. Compreh. Psychiatr. **23**(5), 479–491 (1982)

3. Cai, H., Zheng, V.W., Chang, K.C.: A comprehensive survey of graph embedding: problems, techniques, and applications. IEEE Trans. Knowl. Data Eng. **30**(9), 1616–1637 (2018)
4. Carley, K.M., Diesner, J., Reminga, J., Tsvetovat, M.: Toward an interoperable dynamic network analysis toolkit. Decis. Supp. Syst. **43**(4), 1324–1347 (2007)
5. Cattuto, C., den Broeck, W.V., Barrat, A., Colizza, V., Pinton, J.F., Vespignani, A.: Dynamics of person-to-person interactions from distributed RFID sensor networks. PLoS ONE **5**(7), e11596 (2010)
6. Cavallari, S., Zheng, V.W., Cai, H., Chang, K.C.C., Cambria, E.: Learning community embedding with community detection and node embedding on graphs. In: Proceedings of the Conference on Information and Knowledge Management, pp. 377–386. New York (2017)
7. Chakraborty, T., Dalmia, A., Mukherjee, A., Ganguly, N.: Metrics for community analysis: a survey. ACM Comput. Surv. **50**(4), 54:1–54:37 (2017)
8. Chen, C.: Social network from German news (2021). https://doi.org/10.7910/DVN/5MGEUY, https://doi.org/10.7910/DVN/5MGEUY
9. Corman, S.R., Kuhn, T., Mcphee, R.D., Dooley, K.J.: Studying complex discursive systems. Hum. Commun. Res. **28**(2), 157–206 (2002)
10. Cui, P., Wang, X., Pei, J., Zhu, W.: A survey on network embedding. IEEE Trans. Knowl. Data Eng. **31**(5), 833–852 (2019)
11. Danon, L., Díaz-Guilera, A., Duch, J., Arenas, A.: Comparing community structure identification. J Stat. Mech. Theor. Exp. **2005**(09), P09008–P09008 (2005)
12. Dey, A.K.: Understanding and using context. Personal Ubiquitous Comput. **5**(1), 4–7 (2001)
13. Diesner, J.: From texts to networks: detecting and managing the impact of methodological choices for extracting network data from text data. KI - Künstliche Intell. **27**(1), 75–78 (2013)
14. Dunbar, R.I.M.: Neocortex size as a constraint on group size in primates. J. Hum. Evol. **22**(6), 469–493 (1992)
15. Finkel, J.R., Grenager, T., Manning, C.: Incorporating non-local information into information extraction systems by Gibbs sampling. In: Proceedings of the 43rd Annual Meeting on Association for Computational Linguistics, pp. 363–370 (2005)
16. Fortunato, S., Hric, D.: Community detection in networks: a user guide. Phys. Rep. **659**, 1–44 (2016)
17. Ganguly, S., Pudi, V.: Paper2vec: combining graph and text information for scientific paper representation. In: Jose, J.M., et al. (eds.) ECIR 2017. LNCS, vol. 10193, pp. 383–395. Springer, Cham (2017). https://doi.org/10.1007/978-3-319-56608-5_30
18. Goyal, P., Ferrara, E.: Graph embedding techniques, applications, and performance: a survey. Knowl.-Based Syst. **151**, 78–94 (2018)
19. Grover, A., Leskovec, J.: Node2Vec: scalable feature learning for networks. In: Proceedings of the 22nd International Conference on Knowledge Discovery and Data Mining (KDD 2016), pp. 855–864. ACM, New York (2016)
20. Hennig, M., Brandes, U., Pfeffer, J., Mergel, I.: Studying Social Networks. A Guide to Empirical Research. Campus Verlag, Frankfurt (2012)
21. Hollstein, B., Töpfer, T., Pfeffer, J.: Collecting egocentric network data with visual tools: a comparative study. Netw. Sci. **8**(2), 223–250 (2020)
22. Hong, H., Luo, C., Chan, M.C.: SocialProbe: Understanding Social Interaction Through Passive WiFi Monitoring. In: Proceedings of the 13th International Conference on Mobile and Ubiquitous Systems: Computing, Networking and Services (MOBIQUITOUS 2016), pp. 94–103. ACM, New York, November 2016

23. Javed, M.A., Younis, M.S., Latif, S., Qadir, J., Baig, A.: Community detection in networks: a multidisciplinary review. J. Netw. Comput. Appl. **108**, 87–111 (2018)
24. Johnson, J.C., Krempel, L.: Network visualization: "the bush team" in Reuters news ticker 9/11-11/15/01. J. Soc. Struct. **5**(1) (2004)
25. Le, Q., Mikolov, T.: Distributed representations of sentences and documents. In: Proceedings of the 31st International Conference on International Conference on Machine Learning, vol. 32, pp. II-1188–II-1196 (2014)
26. Mikolov, T., Chen, K., Corrado, G., Dean, J.: Efficient estimation of word representations in vector space. In: Bengio, Y., LeCun, Y. (eds.) 1st International Conference on Learning Representations (ICLR 2013), Scottsdale, Arizona, USA, May 2-4, 2013, Workshop Track Proceedings (2013)
27. Mikolov, T., Sutskever, I., Chen, K., Corrado, G., Dean, J.: Distributed representations of words and phrases and their compositionality. In: Proceedings of the 26th International Conference on Neural Information Processing Systems (NIPS 2013), vol. 2, pp. 3111–3119. Curran Associates Inc., Red Hook (2013)
28. Milgram, S.: The small world problem. Psychol. Today **2**(1), 60–67 (1967)
29. Newman, M.: Networks - An Introduction. Oxford University Press, Oxford (2010)
30. Pan, S., Wu, J., Zhu, X., Zhang, C., Wang, Y.: Tri-party deep network representation. In: Proceedings of the Twenty-Fifth International Joint Conference on Artificial Intelligence (IJCAI 2016), pp. 1895–1901. AAAI Press (2016)
31. Perozzi, B., Al-Rfou, R., Skiena, S.: DeepWalk: Online learning of social representations. In: Proceedings of the 20th ACM SIGKDD International Conference on Knowledge Discovery and Data Mining (KDD 2014), pp. 701–710. ACM, New York (2014)
32. Rossetti, G., Cazabet, R.: Community discovery in dynamic networks: a survey. ACM Comput. Surv. **51**(2), 35:1–35:37 (2018)
33. Schilit, B., Theimer, M.: Disseminating active map information to mobile hosts. IEEE Netw. **8**(5), 22–32 (1994)
34. Shoval, N., Ahas, R.: The use of tracking technologies in tourism research: the first decade. Tour. Geogr. **18**(5), 587–606 (2016)
35. Tang, J., Qu, M., Wang, M., Zhang, M., Yan, J., Mei, Q.: LINE: large-scale information network embedding. In: Proceedings of the 24th International Conference on World Wide Web (WWW 2015) , pp. 1067–1077. Republic and Canton of Geneva (2015)
36. Tu, C., Liu, H., Liu, Z., Sun, M.: CANE: context-aware network embedding for relation modeling. In: Proceedings of the 55th Annual Meeting of the Association for Computational Linguistics (vol. 1: Long Papers), pp. 1722–1731. Association for Computational Linguistics, Vancouver, Canada, July 2017. https://doi.org/10.18653/v1/P17-1158
37. Vose, M.: A linear algorithm for generating random numbers with a given distribution. IEEE Trans. Softw. Eng. **17**(9), 972–975 (1991)
38. Vrandečić, D., Krötzsch, M.: Wikidata: a free collaborative knowledgebase. Commun. ACM **57**(10), 78–85 (2014)

39. Wang, X., Cui, P., Wang, J., Pei, J., Zhu, W., Yang, S.: Community preserving network embedding. In: Thirty-First AAAI Conference on Artificial Intelligence, February 2017
40. Yang, C., Liu, Z., Zhao, D., Sun, M., Chang, E.: Network representation learning with rich text information. In: Twenty-Fourth International Joint Conference on Artificial Intelligence, June 2015
41. Yang, L., Cao, X., He, D., Wang, C., Wang, X., Zhang, W.: Modularity based community detection with deep learning. In: IJCAI. vol. 16, pp. 2252–2258 (2016)

FIP-SHA - Finding Individual Profiles Through SHared Accounts

Carolina Nery, Renata Galante, and Weverton Cordeiro[✉]

Institute of Informatics, Federal University of Rio Grande do Sul, Porto Alegre, Brazil
{carolina.nery,galante,weverton.cordeiro}@inf.ufrgs.br

Abstract. Many recommendation systems rely on users' account history (such as visited/purchased/classified items) to predict which other items they may be interested in. In practice, family members or friends can share a single account. For this reason, deriving a single user profile from an account's history can lead to imprecise item suggestions. In this work, we propose to identify individual profiles behind shared accounts to better customize the suggestions of items for the person who is currently logged in. In short, the problem is solved by identifying online sessions on a platform and afterward, clustering these sessions to identify the profiles of the users behind the (potentially) shared account.

Keywords: Shared account · Clustering · Item similarity · User profile

1 Introduction

Various entertainment, e-commerce, and media outlet companies, seeking to offer the best user experience possible, work hard on recommending items (like movies, products, or news) their customers most likely will be interested in [2]. To compute item recommendations, these companies' platforms often take advantage of user accounts, to which actions (movie viewing, product browsing, purchases, and ratings) are registered; these actions are then fed to recommendation systems, which attempt to identify other items those users may also like.

For convenience, various persons (e.g., family members or friends) may share a single account on those platforms. For example, various family members might share a single account in an e-commerce platform, as it may be simpler to manage purchases in the same (saved) bank card and deliver those purchases to the family's address. Consequently, recommendations based on the entire account history might fail to capture the person's real interests. In a talk, Rastogi [13] mentioned that handling multiple personas behind single customer accounts is one of the research challenges pursued by Amazon.com.

Despite the potentialities of existing investigations on the topic [1,6,9,17], item recommendation in platforms subject to account sharing remains an open problem in the literature. One main limitation is that existing work either approach content sharing in the context of platforms that handle *homogeneous* items

© Springer Nature Switzerland AG 2021
C. Strauss et al. (Eds.): DEXA 2021, LNCS 12924, pp. 115–126, 2021.
https://doi.org/10.1007/978-3-030-86475-0_12

only (e.g., shoes [13], videos [1,17], to which item metadata is available. Zhang et al. [17], for example, explored user movie rating preferences to identify multiple users behind an account. Jiang et al. [6], in turn, approached the issue in the context of online streaming services, and relied on node embedding on heterogeneous graphs whose edges represent relationship between items (music), item and metadata (e.g., artist singing a music), and between metadata (e.g., album from an artist). In contrast, metadata information in scenarios beyond streaming services might be difficult to capture or might not be available at all (like items available in a multi-department online store). Even if available, item metadata might be inaccurate and/or change over time.

We argue that a fully-fledged solution for unveiling multiple persona's profiles behind shared accounts must handle *heterogeneous* items, i.e., items that share very few characteristics (e.g., books, electronics, auto, clothing, music, home and kitchen, and personal care), and to which no metadata information is available. To bridge this gap, we propose a approach to identify users behind shared accounts in platforms having heterogeneous items (e.g., an e-commerce platform). The challenge we approach in this paper is to identify one or more profiles that are present in a shared account, considering a platform handling heterogeneous items, and assuming as input a stream of users' actions like user identifier and anonymized numeric identifiers of visited items (unlike related work, which approached the issue in the context of homogeneous items of which metadata is available). By working with numeric identifiers for users and visited items, we ensure that our solution remains applicable to heterogeneous items, without relying on assumptions related to product type or user preferences. We also provide evidence, from an experimental evaluation, that our approach performs effectively even with minimal user input available and without training.

The remainder of the paper is structured as follows. Section 2 describes the background. Section 3 provides a review of the literature. In Sect. 4 we discuss the method to identify users in shared accounts and showing the functional steps and components. Section 5 presents the experiments to implement the proposed method along with the acquired results, while discusses these results and their application to actual setups. Section 6 presents the results obtained in each experiment. Section 7 concludes the paper, providing directions for future work.

2   Background

There are three core concepts upon which our work is based: *item similarity*, *word embedding*, and *clustering*. The first one involves analyzing the similarity of items that are part of a user's online session (for example, movies browsed in a streaming platform), and are mainly used in the context of collaborative filtering. Item similarity is often assessed based on user interactions – if any two people share the same interest on a given item, they are more likely to express the same interest on other items too. In this context, items are classified and indexed according to their similarity concerning an item matrix. Cosine measure [11] is one option to compute similarity.

The second concept, word embedding, captures semantic similarity between tokens or pixels and projects them into user-defined vector space [8]. Therefore, one can argue that embedding is low-dimensional spaces that can project a high-dimensional vector. We faced the dimensionality problem in our work, as a generic model for unveiling shared accounts in platforms dealing with heterogeneous items is unable to extract relevant patterns from users' item browsing in the platform. Therefore, we use t-Distributed Stochastic Neighbor Embedding (t-SNE) to reduce the dimensionality of word vector space [7,10].

To find non-Euclidean plans in a space of n dimensions, t-SNE uses Machine Learning to converge to the 2D plane that best represents the set of characteristics that produce greater relevance to the data set [12]. In summary, it uses local relationships between points to create a low-dimension mapping, which enables capturing a non-linear structure. The probability distribution is created using the Gaussian distribution, which defines the relationships between the points in the high-dimensional space. t-SNE uses the Student t-distribution to recreate the probability distribution in the small space avoiding the problem of agglomeration. Finally, t-SNE optimizes the fitting using a gradient descent method. In the first step, creates a probability distribution that suggests the relationships between neighboring points [5]. In the second step, t-SNE tries to recreate a small space that follows this probability distribution in the best possible way.

The last core concept of our work is the clustering technique. In our context, we use clustering to group sessions of a shared account by their similarities, so each cluster that contains sessions is associated with a user profile. We use Affinity Propagation (AP) [4] to find the number of clusters automatically, which are represented by the element that best generalizes all the elements within the cluster. AP is based on the exchange of messages between items until exemplars are found for each cluster. The similarities between each item are received as input and at each iteration, two types of messages are faced: the responsibilities and availability, which are calculated according to the similarities. The number of exemplars, i.e. of clusters, is influenced by the input preferences value and, the process ends as soon as it reaches a specified number of iterations or when the cluster structure stabilizes with a certain number of iterations [3]. The AP algorithm's clustering performance and convergence problem mainly depend on two parameters: preference and damping (λ). The damping factor λ influences the convergence performance of the AP algorithm. In the grouping process, the adequate value of the damping factor avoids the optimum local condition, based on the different convergence speeds of search from the center of the ideal cluster at the different stages of the algorithm. Considering these points, we did some experiments to prove this challenge and choose the best value for our data sets.

3   Related Work

In [17] the authors presented the first study in the identification of shared accounts in the context of film platforms. The goal is to identify whether a particular account is shared and to group the actions of users who share it.

The authors developed a model for shared accounts based on linear subspace unions and used the clustering technique to perform the identification. Following the same line, Bajaj et al. [1] proposed a method of grouping channels based on similarity to group similar channels for accounts and use the *Apriori* algorithm to decompose the actions of online TV accounts into different people who share the same login by analyzing the viewing characteristics and individualizing the experience of each persona. After that, the authors used personal profiles to recommend additional channels to the account. Yang et al. [16] also analyzed the similarity of each type of item over a period of time to detect whether a sequence is generated by the same user, to make recommendations to the identified users.

In [13], the authors introduced an array of latent variables to capture multiple personas. Addressing the issue of recommending items to users through shared accounts on Amazon.com in the context of recommending product size for items such as clothing and shoes. The problem of product size recommendations is addressed based on the customer's purchase and return data, which results in high rates of return by the customer buying incorrect sizes.

The authors of [15] presented a study of top-N recommendations for shared accounts in the absence of contextual item-based information. The data is represented as a preference matrix in which the rows represent the users and the columns the items. It is referred to as $N - N$ recommendation, based only on positive binary data, for a user, this recommendation system finds KNN(j), the k items most similar to j, for each preferred item j using a similarity measure. They are unable to identify the profile of each user behind shared accounts.

In [6], the authors proposed SHE-UI, a framework to differentiate users' preferences and group sessions per user in the domain of multimedia streaming. Using a structure based on feature learning, unsupervised learning, and normalized random walks, authors can identify a set of users behind a shared account, from the streaming of music. In addition, given a new streaming session for an account, its structure is capable of identifying a persona. The authors' solution depends on resource extraction and the technique (random walk) of homogeneous items, which limits its applicability in an e-commerce scenario. SHE-UI aims to identify multiple users on a shared account from the logs passed from requesting songs on each account. The idea is to find different patterns of items requested in each session and associate each pattern with a different user.

4   FIP-SHA

In our work, we approach the problem of identifying user profiles behind shared accounts by clustering users' online sessions belonging to a single account. In this case, each cluster would materialize the online preferences and interests of a single user behind the shared account. We only assume as input a stream of actions that users performed in the platform: action timestamp, user identifier, and numeric identifier of visited item. Our goal is to classify these actions into potential profiles behind an account having a given user identifier.

Our work is based on the hypothesis that *online sessions from the same user are similar (for example, the set of items visited or categories of items) and, therefore, they can be grouped into user profiles.* To explore this hypothesis, we focus on the following research questions: **RQ1**. What are the most relevant similarities between the items and how to identify the beginning and end of each session? **RQ2**. Is it possible to use the clustering technique to identify shared accounts in scenarios without item metadata available? **RQ3**. What is the effectiveness of using clustering to classify user sessions into profiles? In the following sections, we provide an answer to these questions.

4.1 Session Representation

Figure 1 presents an overview of the solution approached in this paper. In each online session at some platform, a person navigates (browses through and/or purchases) a set of items; these items can be reached through item categories, custom searches, or previously recommended/related items. The order in which items are visited (which can be seen as a directed graph) might reveal one person's browsing pattern/behavior/preference. Our main idea is to build profiles based on the similarity of those online sessions (or item browsing patterns).

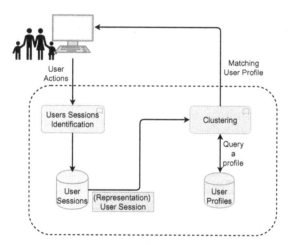

Fig. 1. FIP-SHA overview

In our work, we adopted the following pipeline: (*i*) identify users' online sessions; (*ii*) represent those sessions; and (*iii*) cluster them into users' profiles. The first one required a mechanism able to capture—as a single user session in the platform—browsing actions made by the same person in a given period. The challenge is deciding when some session ends and another one starts. One trivial case is a long period of inactivity. A more complex case, however, is when a person (who was browsing for new age music) gives a seat to another one (who now listens to heavy metal). Previous investigations [6,14] have considered duration and/or idle period as a delimiter for sessions. In our work, we considered item-item similarity, in addition to idleness, to delimit browsing sessions. We do so by computing an average *distance* between visited items; once it exceeds a threshold, it indicates that someone else is now browsing (a new session).

Algorithm 1 provides an overview of cut sessions strategy. We generate a Features Matrix from the log of users' activities. To detect what is relevant in a matrix with such dimensions, we used t-SNE dimensionality reduction technique to calculate through existing dimensions. Then, we compared the standard deviation in pairs, obtaining the plan that best represents these contents. In this way, each item in the data set receives a coordinate corresponding to the generated space. Then, to find the centroid of the groupings, we used Affinity Propagation (AP) for the purpose of finding its coordinates. The criterion used to find which grouping each item belongs to was the centroid that obtained the shortest Euclidean distance. In summary, we separated the groups of content into groupings and infer whether the analyzed content belongs to the same topic or not, if they are grouped close together it means that they can be represented by the same topic. In this way, we can analyze the frequency of content for each user, and when a different content is observed from the profile, a single session is split into a new session (likely from a different user).

Algorithm 1: FIP-SHA - Split Sessions

Input: L = < u,s,i >
Output: split of sessions into shared accounts
1 constant CUTOFF_THRESHOLD
2 read features
3 generate tsne file
4 **Function** *get_cluster_centers*:
5 | clustering = AffinityPropagation.fit_transform
6 | cluster_centers_indices = clustering.cluster_centers_indices_
7 | labels = clustering.labels_
8 | **return** cluster_centers, labels
9 get random labels
10 get corresponding centroid labels
11 map euclidean from centroid
12 **Function** *cut_sessions_based_on_cutoff_threshold*:
13 | **for** *session* ∈ *sessions* **do**
14 | | **if** *distance* > *CUTOFF_THRESHOLD* **then**
15 | | | cut current session into a new one;
16 | | **end**
17 | **end**
18 | **return** sessions

For the second aspect, we represented sessions through vectors of terms and their frequencies that describe the items visited. Finally, the third aspect required techniques for session clustering into users' profiles. Algorithm 2 introduces our strategy for clustering sessions into user profiles. It mainly comprises two steps: (i) similarity calculation, where the sessions of each account are compared in pairs using the similarity metric by Cosine, and the result is the Similarity Matrix between the sessions in each account. Afterwards, the similarity between the sessions is used in the AP algorithm to perform the (ii) grouping of sessions and generate the clusters that best represent the users' profiles.

Algorithm 2: FIP-SHA - Cluster Sessions

Input: split of sessions into shared accounts
Output: profiles
1 **for** *account* ∈ *dataset* **do**
2 | fetch list of sessions as items lists;
3 | obtain item matrix from each session;
4 | similarities = similarity(session_item_lists)
5 | clustering = AffinityPropagation.fit(similarities)
6 **end**
7 **return** clusters;

---

5 Experimental Evaluation Setup and Metrics

We carried out a series of tests that combine similarity between items and clustering algorithm. These experiments assessed the overall performance of our solution to unveiling individual profiles behind a shared account. The source code of our project, along with evaluation scripts, data used to carry out the experiments, and raw results obtained, are publicly available on GitLab[1]. For the experiments, we simulated "fake" shared accounts using datasets from Globo.com[2] and music listening sessions from Last.fm[3]. Table 1 presents additional information on the datasets used. The Globo.com dataset contains the logs of user interactions from a news portal, including clicks, distributed in more than 1 million sessions of 320,000 users. The Last.fm dataset contains tuples <user, timestamp, artist, song> collected using the Last.fm API.

Table 1. Characterization of the datasets used.

	Period	Session/Visited items	Number of users
Globo.com	October 1 to 16, 2017	1.048.594	322.897
Last.fm	May, 2010	907.887	992

In our work, we artificially created "fake" shared accounts by grouping together user accounts from the original dataset. In particular, we created shared accounts that merge two, three, and four accounts from the original dataset. The accounts chosen for merging into a shared account were chosen randomly.

To evaluate the effectiveness of our solution in revealing the user profiles behind the shared account, we use ARI, FMI, NMI, and AMI scores to assess the quality of clusters grouping user sessions[4]. These metrics are the same as used in related work. We also define in the context of this work a *user separation* metric, which quantifies how much each cluster *actually is* an individual behind our

---

[1] https://gitlab.com/carolinaNery94/accountprofiles-mscproject.

[2] https://www.kaggle.com/gspmoreira/news-portal-user-interactions-by-globocom.

[3] http://ocelma.net/MusicRecommendationdataset/lastfm-1K.html.

[4] https://scikit-learn.org/stable/modules/clustering.html.

"fake" shared account. We compute this metric per shared account, as described next. For each cluster grouping the sessions from the shared account, we count the highest number of sessions in the cluster that belong to a same user, and divide it by the size of the cluster. This operation gives us a highest proportion index. A value of 1 (optimum) indicates that all sessions in the cluster belong to a same user. We then compute the average and standard deviation of the highest proportion index computed for each cluster obtained from the "fake" shared account. We also computed a *weighted user separation* metric. To this end, we compute the weight of each cluster as its size (number of sessions in it) divided by the total number of sessions in the shared account. We then sum the highest proportion index of each cluster weighted by its size.

We split our tests into two categories: Analysis of similarity resulting in splitting the sessions and Clustering those sessions into user profiles.

Cut-Off Sessions. To evaluate the session cut-off step we set up the t-SNE library from *sklearn.manifold* with parameters perplexity = 50, n_iter = 1000, and init = random. The perplexity is related to the number of nearest neighbors. Larger data sets usually require a larger perplexity. n_iter refers to the maximum number of iterations for the optimization and the init refers to the initialization of word embedding. We tested the combinations of perplexity for the data sets and choose the value that obtained the best result and after getting the cluster centers as shown in Algorithm 1, resulting in cut off the sessions if needed.

Clustering. The experiments regarding the clustering follow a similar configuration used in cut-off sessions setup. AP needs to obtain the cluster center through continuous iteration, which leads to the high time complexity of the algorithm, the clusters and exemplars returned are obtained with the influence of the damping parameter. This value is responsible for maintaining a balance between convergence and oscillation. So, before using AP in clustering step presents in Algorithm 2, we need to perform a test to determine the damping value.

6   Results

Next we present the results obtained in the tests defined in the previous section.

6.1   Cut Off Sessions

From a set of sessions representing a shared account with 2, 3, or 4 users, and given a defined cut-off point, we can split a given session into another. In addition to analyzing the sessions and breaking them. We also need to adjust the t-SNE to reduce the dimensionality, t-SNE has a *perplexity* parameter and this value needs attention before we start work. The idea is to ensure that the algorithm runs long enough to stabilize. We find the cluster's center, using AP, and so building the fictitious labels for each point, representing the values of coordinates.

After calculating the coordinates, we start the third stage, calculating the Euclidean distance of each session's line, as soon as this distance reaches a certain cutoff, the current session is split into a new one. Tables 2 show the results.

Table 2. Example of cut off session for Globo.com dataset

	User Id	Session Id	Click Article Id	x centroid	y centroid	Distance
0	15275	1506875407922788	101192	−23.93	−19.91	0.00
1	15275	1506875407922788	102738	4.27	−4.04	32.36
2	15275	1506875407922788	102701	4.27	−4.04	0.00
3	15275	1506875407922788	102692	16.82	−11.57	14.63
4	15275	1506875407922788_000	101193	−23.93	−19.91	**41.59**
5	15275	1506875407922788_000	102696	4.27	−4.04	32.36
6	15275	1506875407922788_000	102661	0.05	1.41	6.90
7	15275	1506875407922788_000	102686	4.27	−4.04	6.90
8	15275	1506875407922788_000	101194	−23.93	−19.91	32.36
9	15275	1506875407922788_000	102668	4.27	−4.04	32.36
10	15275	1506875407922788_000	102718	−18.14	3.25	23.58
11	15275	1506875407922788_000	102669	4.27	−4.04	23.58
12	15275	1506875407922788_000	101495	−23.93	−19.91	32.36
13	15275	1506875407922788_000	104173	4.27	−4.04	32.36
14	15275	1506889585128598	101195	−23.93	−19.91	32.36

6.2 Clustering

The *damping* value in the AP algorithm influences the number of clusters, and also plays a decisive role in the convergence speed. The inadequate choice of this value can lead to the oscillation, making it easy to not converge and, finally, influencing the grouping. As a result of the experiments, the *damping* value 0.9 performed better compared to the other values tested in both data sets. We used this value in the next experiments carried out.

In summary, for each set, shared accounts with 2, 3, and 4 users per account were created and then, for each generated user account, the cosine similarity metric between sessions was applied and AP clustering algorithm was performed having as input value the similarities between sessions, and the profiles containing the sessions were output.

The first 4 accounts of 1,703 were analyzed. Table 3 presents the results of the analyzed cases for each Cluster (noted as C in the Table). The estimated value of the ARI is above 0.5, the NMI average was 0.8, the lowest AMI was 0.59, and the FMI 0.81. For the Last.fm dataset, the analysis is similar, however, due to the volume of data per shared account, the number of groups/profiles was much higher. Table 4 presents the results of the analyzed cases, for the scenario of 2 users per account, ARI results are above 0.7, and metrics like AMI are above 0.8. Similar behavior is found for accounts that have 3 and 4 users.

6.3 Analysis of (Weighted) User Separation

Table 5 shows the results of the Globo.com data set, the results of the weighted average. Most of the clusters achieved a user separation above 0.60, which clusters as high as 0.84 considering simple average and 0.91 considering weighted

Table 3. Evaluation of clustering metrics for accounts with 2, 3, and 4 users, Globo.com

C	ARI	NMI	AMI	FMI	C	ARI	NMI	AMI	FMI	C	ARI	NMI	AMI	FMI
11	0.92	0.93	0.87	0.96	5	0.92	0.86	0.82	0.95	8	0.88	0.90	0.86	0.91
5	0.75	0.84	0.77	0.81	4	0.92	0.92	0.91	0.94	4	0.77	0.73	0.68	0.85
2	0.80	0.75	0.72	0.90	6	0.88	0.90	0.86	0.90	10	0.96	0.98	0.96	0.97
5	0.58	0.75	0.59	0.69	7	0.77	0.83	0.77	0.81	9	0.78	0.89	0.83	0.81

Table 4. Evaluation of clustering metrics for accounts with 2, 3, and 4 users, Last.fm

C	ARI	NMI	AMI	FMI	C	ARI	NMI	AMI	FMI	C	ARI	NMI	AMI	FMI
178	0.94	0.98	0.96	0.94	421	0.74	0.95	0.84	0.75	591	0.80	0.96	0.88	0.81
421	0.78	0.96	0.87	0.79	587	0.89	0.98	0.94	0.90	618	0.84	0.97	0.90	0.84
391	0.88	0.98	0.92	0.88	274	0.76	0.96	0.86	0.77	687	0.92	0.99	0.96	0.92

average. There were a few cases in which user separation was as low as 0.4, mainly because of the difficulty of the clustering phase in separating users with minimal user or item metadata available. Table 6 depicts the user separation results obtained for Last.fm. One may observe that results were as high as 0.9, indicating a very high effectiveness in revealing the users behind the "fake" shared accounts built for the evaluation. All in all, these results provide evidence of the effectiveness of user item similarity to cut off online sessions and clustering to group these sessions into user profiles.

Table 5. User separation analysis for shared accounts from Globo.com

Account #	N. of users	Average	Std. Dev.	Weighted Avg.
1	2	1	0	1
2	2	0.67	0.04	0.67
3	2	0.46	0.02	0.59
4	2	0.84	0.21	0.66
1	3	0.84	0.21	0.91
2	3	0.67	0.11	0.63
3	3	0.65	0.16	0.64
4	3	0.59	0.16	0.65
1	4	0.71	0.26	0.82
2	4	0.63	0.32	0.48
3	4	0.64	0.15	0.63
4	4	0.62	0.19	0.64

Table 6. User separation analysis for shared accounts from Last.fm

Account #	N. of users	Average	Std. Dev.	Weighted Avg.
1	2	0.91	0.13	0.9
2	2	0.96	0.08	0.96
3	2	0.97	0.08	0.98
1	3	0.93	0.13	0.94
2	3	0.94	0.11	0.95
3	3	0.93	0.12	0.94
1	4	0.9	0.15	0.91
2	4	0.93	0.13	0.93
3	4	0.91	0.14	0.91

6.4   Discussion

We chose Euclidean distance for session splitting as it represents the dissimilarity between the points. For the future, we intend to perform multiple cuts in each session. As for the construction of the word embedding file, noted that the metadata available on each dataset influences the result, the final assessment for the Last.fm data set is an example, and may have influenced the final result of this session breaking stage. About the clustering step, the chosen algorithm provided satisfactory results for the study, although it presents limitations related to the performance, such as choice of the initial parameters.

7   Final Considerations

There has been substantial research on methods for unveiling user profiles behind shared accounts. Existing methods, however, rely on massive amount of (user or item) metadata and homogeneous items (i.e., items of a single type) to perform satisfactorily. In this work, we take a different approach to reveal user profiles behind shared accounts for environments hosting heterogeneous items of which no item metadata can be made available. FIP-SHA applies a similarity technique to data points, t-SNE to reduce the dimensionality, and clustering.

Our experiments provided evidence of the feasibility of the two stages presented in FIP-SHA. The second stage presented more satisfactory results, as the data present in each data set influences the final result since when using a data set with good characteristics the result was better. One of the advantages of using Affinity Propagation is the replacement of the centroid by the exemplar, as they are representatives of the groupings and are still considered relevant data. As future work, we intend to assess the effectiveness of other options for clustering, and analyze the effectiveness of the proposal with other datasets.

Acknowledgements. This work was financed in part by the Coordenação de Aperfeiçoamento de Pessoal de Nível Superior - Brasil (CAPES) - Finance Code 001.

References

1. Bajaj, P., Shekhar, S.: Experience individualization on online TV platforms through persona-based account decomposition. In: 24th ACM International Conference on Multimedia, MM 2016, pp. 252–256. ACM, New York (2016)
2. Bobadilla, J., Ortega, F., Hernando, A., Gutiérrez, A.: Recommender systems survey. Knowl. Based Syst. **46**, 109–132 (2013)
3. Dueck, D.: Affinity propagation: clustering data by passing messages. Ph.D. thesis, January 2009
4. Frey, B.J., Dueck, D.: Clustering by passing messages between data points. Science **315**, 2007 (2007)
5. Hinton, G., Roweis, S.: Stochastic neighbor embedding. Adv. Neural. Inf. Process. Syst. **15**, 833–840 (2003)
6. Jiang, J.Y., Li, C.T., Chen, Y., Wang, W.: Identifying users behind shared accounts in online streaming services. In: ACM SIGIR Conference on Research & Development in Information Retrieval, SIGIR 2018, pp. 65–74. ACM (2018)
7. Kaufman, L., Rousseeuw, P.J.: Finding Groups in Data: An Introduction to Cluster Analysis. Wiley, New York (1990)
8. Li, Y., Xu, L., Tian, F., Jiang, L., Zhong, X., Chen, E.: Word embedding revisited: a new representation learning and explicit matrix factorization perspective. In: International Conference on Artificial Intelligence, pp. 3650–3656. AAAI Press (2015)
9. Ma, M., Ren, P., Lin, Y., Chen, Z., Ma, J., Rijke, M.d.: π-net: a parallel information-sharing network for shared-account cross-domain sequential recommendations. In: ACM SIGIR Conference on Research and Development in Information Retrieval, SIGIR 2019, pp. 685–694. ACM, New York (2019)
10. van der Maaten, L., Hinton, G.: Visualizing data using t-SNE. J. Mach. Learn. Res. **9**, 2579–2605 (2008)
11. Musa, J.M., Zhihong, X.: Item based collaborative filtering approach in movie recommendation system using different similarity measures. In: 2020 6th International Conference on Computer and Technology Applications, ICCTA 2020, pp. 31–34. ACM, New York (2020)
12. Sakib, S., Bakr Siddique, M.A., Rahman, M.A.: Performance evaluation of T-SNE and MDS dimensionality reduction techniques with KNN, ENN and SVM classifiers. In: 2020 IEEE Region 10 Symposium (TENSYMP) (2020)
13. Sembium, V., Rastogi, R., Tekumalla, L., Saroop, A.: Bayesian models for product size recommendations. In: 2018 World Wide Web Conference, WWW 2018, pp. 679–687. International World Wide Web Conferences Steering Committee, Republic and Canton of Geneva, CHE (2018)
14. Sottocornola, G., Symeonidis, P., Zanker, M.: Session-based news recommendations. In: The Web Conference 2018, WWW 2018, pp. 1395–1399 (2018)
15. Verstrepen, K., Goethals, B.: Top-n recommendation for shared accounts. In: 9th ACM Conference on Recommender Systems, RecSys 2015, pp. 59–66. ACM, New York (2015)
16. Yang, Y., Hu, Q., He, L., Ni, M., Wang, Z.: Adaptive temporal model for IPTV recommendation, pp. 260–271, June 2015
17. Zhang, A., Fawaz, N., Ioannidis, S., Montanari, A.: Guess who rated this movie: Identifying users through subspace clustering. In: 28th Conference on Uncertainty in Artificial Intelligence, UAI 2012, pp. 944–953. AUAI Press (2012)

A Tag-Based Transformer Community Question Answering Learning-to-Rank Model in the Home Improvement Domain

Macedo Maia[1][(✉)], Siegfried Handschuh[2], and Markus Endres[1]

[1] University of Passau, Passau, Germany
{Macedo.Sousamaia,Markus.Endres}@uni-passau.de
[2] University of St. Gallen, St. Gallen, Switzerland
sigfried.handschuh@unisg.ch

Abstract. Community Question Answering (CQA) is an Information Retrieval (IR) task that allows matching complex subjective questions and candidate answers based on user posts in community web forums. User questions and comment-based answers deal with many problems, such as redundancy or ambiguity of linguistic information. In this paper, we propose a pairwise learning-to-rank model community QA model in the home improvement domain. For a user question, this model must rank candidate answers in order of relevance. Our main contribution consists of transformer-based language models using user tags to accurate the model generalisation. To train our model, we also propose a proper CQA dataset in home improvement domain that consists of information extracted from community forums. We evaluate our approach by comparing the performance based on analysis with the state-of-the-art method on text or document similarity.

Keywords: Information retrieval · Community question answering · Neural networks

1 Introduction

Web platforms allow web users to share opinions without disclosing too much identity. It encourages users to post genuine comments about something on web forums [7]. Some situations in the real world make some web users access the web to ask about daily problems. For instance, users who want to renovate their houses can ask for tips or opinions about improvements (e.g., buying new appropriate furniture or repairing some broken installation) in some specialized community forums.

The complexity of answering user questions is in the fact of lack of further explicit information to help models to find most relevant answer. We show in Fig. 1 an example of a subjective question and two user comments as accepted

© Springer Nature Switzerland AG 2021
C. Strauss et al. (Eds.): DEXA 2021, LNCS 12924, pp. 127–138, 2021.
https://doi.org/10.1007/978-3-030-86475-0_13

answers obtained from a community QA forum[1]. This example shows a web user asking some tips to measure the height of a tree. This Questioner aims to install an antenna for internet service that needs to clear some tree because it requires a clear line of sight to work. The expected answers for this question are distinct because there are many ways that we must consider to measure the height of something. Answerer 1 suggests to find an approximated tree size by using some steps considering a pencil, moving some meters away from the tree, extend your arm, and hold the pencil so that you can measure the height of the tree on the pencil with your thumb. Meanwhile, Answerer 2 suggests to find the approximated tree height by using a formula based on the questioner's shadow, questioner's and the tree's shadow measures. We conclude that subjective questions have different answers based on the experience of different users. A possible solution to match question and correct answer is to find an association between "ways to measure" words and often terms that appear in texts that describe approach to measure something. A way to help to generalise pairwise CQA model inferences is to use further explicit information (e.g., tags) provided by users.

Question: *Is there an easy way to measure the height of a tree?*
Questioner's nickname: *JohnFx*

Answer 1: *Take a pencil, move some meters away from the tree. Outstrech your arm and hold the pencil so that you can measure the height of the tree on the pencil with your thumb. Then turn the pencil at the bottom of the tree by 90 degrees. Note where the distance measured by thumb hits the earth and measure the way from this point to the tree. This is the height of the tree.*
Answerer's nickname 1: *bennymo*

Answer 2: *Use shadows... 1-Measure your shadow; 2-Measure yourself; 3-Measure the tree's shadow. Calculate (tree's shadow * your height) / your shadow = Tree Height. You'll have to do this on a sunny day (you might also need an assistant), and the ground will have to be relatively flat (a slope will throw off the measurement).*
Answerer's nickname 2: *Tester101*

Fig. 1. Examples of questions and their respective answers

This work proposes a tag-based pairwise community QA transformer LeToR model in the home improvement domain. As a first contribution, we propose a tag-based transformer model that improves the CQA learning-to-rank model generalisation. For that, we also propose an approach to encode multiple tags as a new sentence. As a second contribution, we build and validate a CQA dataset based on users questions and their respective candidate answers to test our approach. Finally, as a third contribution, we compare our approach with different state-of-the-art baselines in pairwise matching and pairwise learning-to-rank models using

---

[1] https://diy.stackexchange.com/questions/7100/is-there-an-easy-way-to-measure-the-height-of-a-tree.

the rank-aware evaluation measures. The experiment aims to evidence the impact of multiple user tags during the model training.

2 Related Work

Some CQA challenge series inspire our task [9,10] that aim to promote related research on that area. They provided datasets, annotated data, and developed robust evaluation procedures to establish a common ground for comparing and evaluating different CQA approaches. Despite these CQA challenge series contains user questions, they have no extra explicit information like user tags.

Advances in word embedding models have improved neural network for generalising pairwise models. Word embedding refers to a group of machine learning algorithms that learn contextual high-dimensional dense word vector representations (e.g., Glove [11], Word2Vec [8], and ELMo [12]). There are some models based on Long Short-Term Memory (LSTM) [4] and the Gated Recurrent Unit (GRU) [1] to learn sequential information on words. Bilateral Multi-Perspective Matching Model (BiMPM) [15], for instance, proposes a model for text similarity. Given two sentences, P and Q, it matches the two encoded sentences in two directions, P against Q and Q against P. In each matching direction, each time step of one sentence is matched against all the other sentences' timesteps from multiple perspectives. Then, another BiLSTM layer is utilized to aggregate the matching results into a fixed-length matching vector. Finally, based on the matching vector, a decision is made through a fully connected layer. Multi-Perspective Sentence Similarity Modeling with Convolutional Neural Networks (MPCNN) [3] proposes a model for comparing sentences that use diverse perspectives. Firstly, MPCNN models each sentence using a convolutional neural network (CNN). It extracts features at multiple levels of granularity and uses multiple types of pooling. After, It compares the sentence representations at several granularities using multiple similarity metrics. We consider these last two approaches in our comparative experiments.

Unlike recurrent neural network (RNN) and its extensions (e.g., LSTM and GRU) that process each sequence element, in turn, transformer-based language models [14] process all parts concurrently, forming direct connections between individual ones through attention mechanisms.

Bidirectional Encoder Representations from Transformers (BERT) [2] is the main transformer-based language model. It consists of pretraining deep bidirectional representations of unlabeled text by jointly conditioning both left and right context in all layers. Initially, that model beats other neural network approaches based on RNN, CNN and attentive mechanisms for eleven different NLP tasks.

DistilBERT [13] is based on BERT architecture with some optimisation procedures to improve the inference speed by removing some parameters to reduce the training speed and train a transformer model by considering a lower amount of computational resources.

With the capability of modeling bidirectional contexts, denoising autoencoding based pretraining like BERT achieves better performance than pretraining approaches based on autoregressive language modeling. However, relying on corrupting the input with masks, BERT neglects dependency between the masked positions and suffers from a pretrain-finetune discrepancy. XLNet [16], a generalized autoregressive pretraining method that (1) enables learning bidirectional contexts by maximizing the expected likelihood over all permutations of the factorization order and (2) overcomes the limitations of BERT thanks to its autoregressive formulation. Furthermore, XLNet integrates ideas from Transformer-XL, the state-of-the-art autoregressive model, into pretraining. Our experiments includes all the previously described transformer models as baselines.

3   Task Definition

Given a question and a set of comments as candidate answers in natural language, a CQA LeToR model sort comments as candidate answer list to the question in order of ascending relevance score.

In Fig. 2 we represent a generic QA pairwise LeToR model architecture. For a question Q and a set of n candidate answers $[C_1, C_2, C_3, ..., C_n]$ we must build question and comment pairs $QCP = \{<Q, C_1>, <Q, C_2>, <Q, C_3>, ...,$ $<Q, C_n>\}$, where The model output is represented by an n-dimensional vector $O = \{<C_1, O_1>, <C_2, O_2>, ..., <C_n, O_n>\}$, where the pair $<C_j, O_j>$ represents the comment C_j and the relevance score O_j of $<Q, C_j>$.

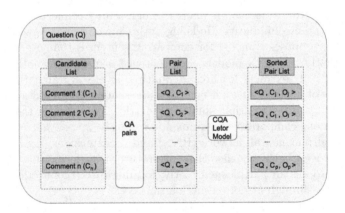

Fig. 2. CQA learning-to-rank pairwise model

4   Our Approach

Our CQA pairwise learning-to-rank transformer model is based on a question answering pairwise approach, including multiple user tags as further information of questions.

4.1   Transformer Models

Although the RNN or CNN models usually provide accurate performance in some text classification tasks, they need to use some special procedure like attention mechanism to detect the most relevant parts (for prediction purposes) in texts. In order to address this issue, transformer-based models consist of multiple layers where each one contains multiple attention heads. An attention head takes as input a sequence of vectors $h = [h_1, ..., h_n]$ corresponding to the n tokens to the input sentence. Each vector h_i is transformed into query, key, and value vectors q_i, k_i, v_i through separate linear transformations. The head computes attention weights α between all pairs of words as softmax-normalised dot products between the query and key vectors. The output o_i is a weighted sum of the value vectors (Eq. 1).

$$\alpha_{ij} = \frac{exp(q_i^T k_j)}{\sum_{l=1}^{n} exp(q_i^T k_l)}; \quad o_i = \sum_{j=1}^{n} \alpha_{ij} v_j, \tag{1}$$

BERT, for instance, is the most-known transformer model. It is pre-trained on raw corpus to perform two tasks: (1) Masked Language Modeling predicts the identities of words that have been masked out of the input text; (2) Next Sentence Prediction predicts whether the second part of the input follows the first part or is a random text.

Since the pre-trained model uses a large corpus, the vocabulary size is fixed, containing thousands or millions of words. However, when we apply a pre-trained model to some different input texts, it can detect some unknown tokens that is out of the original pre-trained model vocabulary (out-of-vocabulary (OOV) tokens). Converting all unknown token to a default representation partially solves the word embedding problem because the model loses relevant contextual information from the input data. Hence, transformer models use an algorithm that breaks an out-of-vocabulary token into subwords, such that the model represents commonly seen subwords.

4.2   Input and Tag Representation

The input in our proposed model is a user question $Q = \{w_1, w_2, ..., w_i, ..., w_m\}$, a candidate answer based on comment $C = \{v_1, v_2, ..., v_i, ..., v_n\}$, where w_i is the i-th token in Q and v_j is the j-th token in C. In order to include tags information in our model, we also consider as a third input the set of alphabetically ordered user tags $T = \{t_1, t_2, ..., t_o\}$, where $|T| = o$ and t_l is the l-th tag.

To represent multiple tags as new information, we define a new input sentence adding the coordinating conjunction term "and" to connect all tags in alphabetical order (see Fig. 3). Conjunction term helps the model identify different tags and avoid confusing them with compound names (e.g., electrical panel). We represent multiple tags in alphabetical order because it helps the model know which tags must appear before or after others. Question Q and tag-based sentence T are concatenated into a new input $\hat{Q}$ where Q comes before T (see Eq. 2).

$$\hat{Q} \leftarrow [Q, T] \tag{2}$$

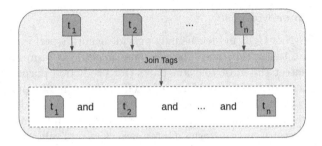

Fig. 3. Procedure to join tags and create a new sentence.

4.3   CQA Pair Matching Model

We use transformer-based language models to represent each QA pairwise representation using sequential contextual information among user question, tag-based sentence, and candidate answers.

For IR tasks like CQA, we use pre-trained transformer models and run the fine-tuning step to fit the model parameters to our data. Firstly, we convert raw input question and answer pair into a proper unified input U. Before using a transformer tokeniser, we must indicate where the input starts using a special token "$<CLT>$". We also need to inform the model where the first sentence ends and where the second sentence begins by using a special token "$<SEP>$". The tokens $<CLT>$ and $<SEP>$ are predefined for the transformer models as default. After defining U, we use the Transformer Tokeniser ($T\_Tokeniser$) (Eq. 3) to separate it into tokens.

$$U = [<CLT>, \hat{Q}, <SEP>, C, <SEP>]; \quad \hat{U} = T\_Tokeniser(U), \quad (3)$$

When we tokenise the inputs using Transformer Tokeniser, each token receive an ID. Hence, when we want to use a pre-trained transformer model, we need to convert each token in $\hat{U}$ into its corresponding unique IDs. The "attention mask" tells the model which tokens should be attended to and which must not. $Tokens\_id$ and $attention\_mask$ feed the input into the transformer model (Eq. 4).

$$tokens\_id, att\_mask = \hat{U}.convert\_tokens\_to\_ids(), \hat{U}.get\_att\_mask() \quad (4)$$

After defining the word encodings and the attention mask, we must pass these as input parameters to the transformer model (Eq. 5). The transformer output is the addictive pairwise encoding by considering whether an answer comes after a question. In Sect. 4.1 we describe information about transformer models. Finally, We feed the result to a *sigmoid* activation function in the output layer (Eq. 5) to return the pairwise matching relevance score.

$$M = Transformer(tokens\_id, att\_mask); \quad o_{rel} = \sigma(W_h M + b) \quad (5)$$

where $W_h \in \mathbb{R}^{|M|}$ and $b \in \mathbb{R}$ are learning parameters and σ represents a sigmoid function. The variable o_{rel} represents the relevance score of U where $\{o_{rel} \in \mathbb{R} \mid 0 \le o_{rel} \le 1\}$.

4.4   Model Optimisation

We use the samples in the training set described in Sect. 5 for training the model
by minimising the binary cross-entropy loss:

$$Loss = -\frac{1}{|S|}\sum_{s=1}^{|S|} y_s \log \hat{y}_s + (1 - y_s)log(1 - \hat{y}_s), \tag{6}$$

where $|S|$ is the size of the training dataset. y_s and $\hat{y}_s$ are the true and the
predicted CQA pairwise relevance, respectively. The model parameters are optimised through the AdamW optimizer [6].

4.5   Candidate Answers Ranking

Our pairwise LeToR model must rank all the QA pairs for a question Q in terms
of relevance. Let L the list of all QA pairs for q, adding the all pair relevance
scores. The result is shown in Eq. 7:

$$L = [<q, c_1, o_1>, <q, c_2, o_2>, ..., <q, c_n, o_n>]; \quad \hat{L} = desc\_sort(L), \tag{7}$$

where $|L| = n$, where n is the number of candidate answers to q and o_j is
the relevance score of the pair $<q, c_j>$. To ranking our QA pairs, we sort L in
relevance score descending order through the function $desc\_sort$, where the final
output is a sorted list $\hat{L}$ and the most relevant pair is in the first position.

5   Dataset Building and Validation

In this section, we describe the procedure of building our CQA dataset and its
validation.

5.1   Subjective CQA

We built our dataset composed of subjective texts harvested from a specialized
Q&A forum called Home Improvement (HI)[2]. This website has 77k thousands
of subscribed users with some background in this domains. They share questions and opinions about different subjects associated with home improvements.
Some moderators review new postings, correct mistakes directly and suggest new
changes for original authors. They use facts and references to support and help
the user to fix some incoherence or misunderstanding information.

Postings in these websites also contain some explicit information provided by
users. We use the number of upvotes as information to validate our gold standard dataset. Upvotes represents the number of positive votes that the message
receive by other users. we build our dataset considering the two information.

---

[2] https://dyi.stackexchange.com.

We scrapped the question comments from both home improvements forums in a dataset.

We describe quantitative information about questions and their associated tags in Table 1. Following the information in these Tables, we state that 11935 questions about home improvement topics where each question is associated with one tag, at least. There are questions associated with five different user tags.

Table 1. Distribution of number of tags by number of questions in either training, dev, and test sets in home improvement dataset.

Datasets	#1_tags	#2_tags	#3_tags	#4_tags	#5_tags	#Questions
Training	2308	2681	1673	627	187	7476
Validation	250	546	563	248	125	1732
Test	196	757	961	563	250	2727
Total	2754	3984	3197	1438	562	11935

Questions are associated with user tags. We show in Table 2 the most frequent user tags in our dataset. Precisely, we have 771 different home improvement tags in our dataset.

Table 2. Tag examples and their frequency for our home improvements dataset.

Tag	Frequency
Electrical	2941
Wiring	1136
Plumbing	827
Lighting	423
Drywall	392
Bathroom	358

5.2   Gold Standard Definition

We distributed all the questions into training, dev and test datasets in terms of question amount. Initially, we distributed all the questions using an overall distribution of about 75/25% for the training and test dataset. Afterwards, we split the original training set into training and development sets with a proportion of 80/20% approximately. In Table 3 we show the distribution of the number of questions, number of accepted answers, number of non-related answers, and QA pairs.

Table 3. Distribution of the number of questions by labels in either training, dev and test datasets for home improvements.

Datasets	#Question	#Acc_Answers	#NonRelev_Answers	#QA_pairs
Training	7476	10451	29542	39993
Validation	1732	2408	6879	9287
Test	2727	3878	10856	14734
Total	11935	16737	47277	64014

6 Evaluation

We evaluated the ranking performance of our approach with different state-of-the-art learning-to-rank models on our gold standard datasets described in Sect. 5.

6.1 Experiment Setup

For recurrent neural networks, we performed our experiments with the 300-dimension word vectors by using the pre-trained GloVE [11] word embedding model generated from a 6-billion-token corpus[3] (extracted from Wikipedia 2014 + Gigaword 5). Regarding transformer-based language models baselines, we used different configurations for each one. In Table 4 we describe the pretraining model configuration for different transformer-based algorithms. The RNN- and CNN-based models' parameters optimized through the Adam optimizer [5] and transformer-based language models parameter optimized through the AdamW optimizer [6].

Table 4. Overview of transformer-based language models configuration.

Model	Pre-trained	#layer	#hidden	#head	#param	#token
BERT	bert-base-uncased	12	768	12	110M	512
RoBERTa	roberta-base	12	768	12	125M	512
DistilBERT	distilbert-base-uncased	6	768	12	66M	512
XLNet	xlnet-base-cased	12	768	12	110M	512

We ran our experiments in a dedicated server with Intel(R) Xeon(R) CPU E5-2643 v3 with 24 cores, 128 GB DIMM DDR4 Memory, and an NVIDIA Tesla K40c GPU device.

---

[3] http://nlp.stanford.edu/data/glove.6B.zip.

6.2 Rank-Aware Evaluation Metrics

We define both measures as: (1) Mean Reciprocal Rank (MRR) is essentially the average of the reciprocal ranks of the first relevant answer among candidate answers for a set of community queries and (2) Mean Average Precision (MAP) is the mean of average precision across multiple community questions. For a single question, Average Precision (AP) is the average of the precision value obtained for the set of top k candidate answers existing after each relevant answer. MRR evaluates the performance of a ranking-based model to find the first relevant answer while MAP evaluates the capacity of same model to predict the whole candidate answer list correctly.

6.3 Results

In Table 5 we show the performance of different baselines considering different evaluation measures for the ranking task. Among RNN-based neural network models, BiPMP has the best performance in terms of ranking measures. BiPMP outperformed MPCNN and the simple Siamese BiLSTM for both evaluation measures. In terms of finding the best first answer among candidates, BiPMP outperformed MPCNN with 7.5% of the difference. BiPMP also outperformed MPCNN with 6.1% of the difference.

The previously discoursed approaches are because they depend on the word embedding models to encode the model. Real-world texts are a problem because this kind of text contains some new words or misspelt words classified as out-of-vocabulary word. These word groups have no distinct representation, and thus the model must lose relevant information. Unlike the previously described models, transformer-based language models had a better performance in all scenarios because they must encode all the parts of text input and use multiple attention mechanisms to find the most relevant parts in each training set sample. XLNet outperforms all the state-of-the-art approaches considered in our analysis. However, the difference in performance for each transformer-based model is slight, as shown in Table 5 for XLNet, BERT and DistilBERT. Among these baseline models, XLNet obtained the best results in our ranking analysis. In the result table, we realised how transformer models generalise better when we include user tags as extra information on questions. For DistilBERT and BERT with user tags information, for instance, outperformed these models without this information in around 2% of difference for both MRR and MAP measures. Despite XLNet with multiple user tags also have improvements by generalising ranking models regarding the same model, without this extra information, the difference is smaller than the other two transformer models. The explanation is in the way how attention mechanisms work for each model. DistillBert and BERT have the same parameters where the difference is because the first one is a distilled version of the second one (It must explain why BERT outperforms DistillBERT). Unlike BERT and DistilBERT, XLNet outperforms results of other approaches without using tags as extra information. However, Bert and DistilBERT obtained more advantage in extracting information on tags than XLNet. In general, BERT with tags representation outperformed the other approaches.

Table 5. Experiment results for Mean Average Precision (MAP) and Mean Reciprocal Rank (MRR).

	Home improv.	
	MRR	MAP
MPCNN	0.636	0.617
BiPMP	0.701	0.676
DistilBERT	0.773	0.753
DistilBERT+tags	**0.791**	**0.773**
BERT	0.799	0.778
BERT+tags	**0.821**	**0.804**
XLNet	0.813	0.795
XLNet+tags	**0.815**	**0.798**

7   Conclusion

We proposed a tag-based transformer community question answering the learning-to-rank model in the home improvement domain. As a first contribution, we built a CQA dataset based on users questions and comment-based answers. We validated that dataset by considering explicit information and feedback provided by different users on the website. We also proposed a way to encode multiple tags as part of the input of transformer models. In our experiment, we showed that multiple-tag-based input helps to increase the CQA generalisation performance on pairwise LeToR CQA task considering different transformer-based models. BERT and DistilBERT showed better performance using the tag information if we compare the difference between the performance of each approach without tags and with these. For future work we plan to analyse community questions from other domains and to determine the effects of specific-domain multiple tag information in the model generalisation. We also plan to explain the effects of tag information on the attention mechanism for transformer models.

References

1. Chung, J., Gülçehre, Ç., Cho, K., Bengio, Y.: Empirical evaluation of gated recurrent neural networks on sequence modeling. CoRR abs/1412.3555 (2014)
2. Devlin, J., Chang, M.W., Lee, K., Toutanova, K.: BERT: pre-training of deep bidirectional transformers for language understanding. In: Proceedings of the 2019 Conference of the North American Chapter of the Association for Computational Linguistics: Human Language Technologies, Minneapolis, Minnesota, vol. 1, pp. 4171–4186 (2019)
3. He, H., Gimpel, K., Lin, J.: Multi-perspective sentence similarity modeling with convolutional neural networks. In: Proceedings of the 2015 Conference on Empirical Methods in Natural Language Processing (EMNLP), pp. 1576–1586. Association for Computational Linguistics (ACL), Lisbon, September 2015

4. Hochreiter, S., Schmidhuber, J.: Long short-term memory. Neural Comput. **9**(8), 1735–1780 (1997)
5. Kingma, D.P., Ba, J.: Adam: a method for stochastic optimization. In: 3rd International Conference on Learning Representations, ICLR 2015, Conference Track Proceedings, San Diego, CA, USA, 7–9 May 2015 (2015)
6. Loshchilov, I., Hutter, F.: Fixing weight decay regularization. In: Proceedings of Seventh International Conference on Learning Representations (ICLR 2019) (2019)
7. Maia, M., Sales, J.E., Freitas, A., Handschuh, S., Endres, M.: A comparative study of deep neural network models on multi-label text classification in finance. In: 15th IEEE International Conference on Semantic Computing, ICSC 2021, Laguna Hills, CA, USA, 27–29 January 2021. IEEE Computer Society (2021)
8. Mikolov, T., Sutskever, I., Chen, K., Corrado, G., Dean, J.: Distributed representations of words and phrases and their compositionality. In: Proceedings of the 26th International Conference on Neural Information Processing Systems, NIPS 2013, USA, vol. 2, pp. 3111–3119 (2013)
9. Nakov, P., et al.: SemEval-2017 task 3: community question answering. In: Proceedings of the 11th International Workshop on Semantic Evaluation (SemEval-2017), pp. 27–48. Association for Computational Linguistics, Vancouver, August 2017
10. Nakov, P., et al.: SemEval-2016 task 3: community question answering. In: Proceedings of the 10th International Workshop on Semantic Evaluation (SemEval-2016), pp. 525–545. Association for Computational Linguistics, San Diego, June 2016
11. Pennington, J., Socher, R., Manning, C.: GloVe: global vectors for word representation. In: Proceedings of the 2014 Conference on Empirical Methods in Natural Language Processing (EMNLP), pp. 1532–1543. Association for Computational Linguistics, Doha, October 2014
12. Peters, M., et al.: Deep contextualized word representations. In: Proceedings of the 2018 Conference of the North American Chapter of the Association for Computational Linguistics: Human Language Technologies, Volume 1 (Long Papers), pp. 2227–2237. Association for Computational Linguistics, New Orleans, June 2018
13. Sanh, V., Debut, L., Chaumond, J., Wolf, T.: DistilBERT, a distilled version of BERT: smaller, faster, cheaper and lighter. In: NeurIPS EMC^2 Workshop (2019)
14. Vaswani, A., et al.: Attention is all you need. CoRR (2017)
15. Wang, Z., Hamza, W., Florian, R.: Bilateral multi-perspective matching for natural language sentences. In: Proceedings of the 26th International Joint Conference on Artificial Intelligence, IJCAI 2017, pp. 4144–4150. AAAI Press (2017)
16. Yang, Z., Dai, Z., Yang, Y., Carbonell, J., Salakhutdinov, R.R., Le, Q.V.: XLNet: generalized autoregressive pretraining for language understanding. In: Wallach, H., Larochelle, H., Beygelzimer, A., d'Alché-Buc, F., Fox, E., Garnett, R. (eds.) Advances in Neural Information Processing Systems, vol. 32. pp. 5753–5763 (2019)

An Autonomous Crowdsourcing System

Yu Suzuki[(✉)]

Gifu University, 1-1 Yanagido, Gifu 5011193, Japan
ysuzuki@gifu-u.ac.jp

Abstract. We propose an autonomous crowdsourcing system for constructing high-quality task results. To change the minds of low-quality workers while processing the tasks, we introduce implicit and explicit feedback. If task results are of low quality, the requesters should tell the workers whether the quality of the task results is high/low and how to improve their work. However, providing detailed and accurate feedback in real-time is difficult and time-consuming for requesters. Our method mitigates this issue by providing feedback for workers using the majority voting (implicit feedback) and the other workers directly (explicit feedback). To confirm the effectiveness of our method, we constructed a crowdsourcing system for processing a large-scale sentiment analysis task and experimented with evaluating the method. We found that our scalable crowdsourcing-based quality management method could improve task results by using implicit feedback and improved worker motivation by providing explicit feedback.

1 Introduction

Large-scale human-annotated data are essential for market surveys, machine learning, and information seeking. Howe [2] first introduced crowdsourcing. Since then, crowdsourcing services, such as Amazon Mechanical Turk[1] (MTurk) and CrowdFlower[2], have been gaining attention as platforms for generating vast amounts of human-annotated data at a low price.

We need an autonomous crowdsourcing system that outputs high-quality task results with a low duty of requesters. If we can automatically measure the quality of task results, we can easily construct an autonomous crowdsourcing system that automatically rejects low-quality task results. However, in many tasks, the quality of the output is difficult to define. For example, a requester needs to select high-quality texts from Twitter messages, but high quality in this sense is difficult to define. However, it costs too high to measure the quality of task results by requesters. To realize an autonomous crowdsourcing system, we should construct a mechanism to change the workers' minds and behavior to improve the quality of task results and increase the number of tasks.

Inspired by the idea of the peer-review process of academic papers and to provide social incentives on a large scale, we propose a method for improving

---

[1] https://requester.mturk.com.

[2] https://www.crowdflower.com.

© Springer Nature Switzerland AG 2021
C. Strauss et al. (Eds.): DEXA 2021, LNCS 12924, pp. 139–147, 2021.
https://doi.org/10.1007/978-3-030-86475-0_14

the worker-management process by using two feedback mechanisms, i.e., *implicit* and *explicit*, with which workers implicitly and explicitly evaluate each other. Implicit feedback is based on majority voting [1]. If a worker selects a majority option, he or she earns more wages, but if the worker selects a minority option, he or she earns less. Our method automatically provides feedback to the workers. Workers know whether the task results are acceptable but do not know the reason for the decision.

2   Related Work

One mechanism of generating feedback is self-correction. Shar et al. [4] proposed a method for applying the self-correction mechanism to a crowdsourcing system. In the first process of such tasks, workers process the task normally. In the second process, the workers browse the task results from themselves and other workers. If the workers feel that the task results are wrong, they can correct them. This mechanism is similar to implicit feedback in our method. However, in our experiment, we discovered that if crowdsourcing systems give concrete reasons why the task results are wrong, it is difficult to change the workers' minds.

Our method uses reputation; that is, workers evaluate each other. Whiting et al. [5] proposed a method in which workers explicitly evaluate each other by using a double-blind reputation assignment (the explicit feedback mechanism of our method is similar). They claimed that the quality of task results improves as a result of using reputation. However, the cost of workers making explicit evaluations was high. Moreover, they did not consider worker motivation. The difference between their research and ours is that our method considers both worker motivation and the quality of task results, whereas their method only takes into account the quality task results.

3   Crowdsourcing Task

The task for this study was a typical sentiment analysis task on a set of tweets, which is similar to a task in SemEval 2017 [3]. This section discusses a method of accomplishing this task using crowdsourcing without using implicit and explicit feedback.

3.1   Workflow

Suppose that a set of tweets $T = \{t_1.t_2, \cdots, t_N\}$ has been prepared. Each tweet t_i has one subject $s_j \in S$. Each individual task is to label a t_i with one or more sentiments selected from the sentiment set {*positive and negative, positive, negative, neutral,* and *unrelated*}. Workers W are hired to accomplish a set of these tasks on a crowdsourcing platform.

The crowdsourcing system assigns each tweet to K workers to maintain the quality of the sentiment assignment. The workflow of this task is as follows:

1. The crowdsourcing system instructs our task. This instruction includes 10 example tweets and their answers.
2. The system randomly selects a $t_i \in T$ to be processed when t_i is labeled by fewer than K workers and not labeled by the worker.
3. The system gives t_i and the subject $s_j \in S$ corresponding to t_i to a worker $w \in W$.
4. w assesses t_i by using the web interface and selects the most appropriate sentiment from the following options:
   - [*pos+neg*] The user who posted the tweet has both *positive* and *negative* feelings about the specified part of the subject.
   - [*positive*] The user who posted the tweet has *positive* feelings about the specified part of the subject.
   - [*negative*] The user who posted the tweet has *negative* feelings about the specified part of the subject.
   - [*neutral*] The user who posted the tweet feels something, but it is not positive or negative about the subject, or his or her emotions are vague.
   - [*unrelated*] The user who posted the tweet feels *no emotion* concerning the target; this tweet is an advertisement or was automatically generated.
5. According to this task result, the system automatically gives points (described below) to the worker.
6. The system gives implicit and explicit feedback to the worker. We do not explain these feedback mechanisms to the workers; several workers believe that the requesters directly evaluate all assessments and generate advice, and other workers may correctly or incorrectly guess the actual feedback mechanism.
7. Go to 2. If the worker wants to continue or quit the task.

The system finally assigns the labels by majority voting. If top labels for a tweet have the same number of votes, the system assigns more than two labels to the tweet.

4   Experimental Evaluation

In this section, we experimented to discover whether the implicit and explicit feedback is valid for the quality of task results and worker motivation.

4.1   Experimental Setup

Data. We prepared tweets from Twitter Search APIs[3]. We chose seven subjects, input keywords related to the subjects, and searched the tweets. We list the subjects and the number of tweets of each subject in Table 1. We used tweets written in Japanese. We retrieved tweets that contained more than one of the keywords; thus, some of the tweets were not relevant to the subjects. We removed tweets that mentioned other Twitter users, that included URLs, or that were re-tweets. As a result, we prepared 140, 744 tweets.

---

[3] https://developer.twitter.com/.

Table 1. Subject of tweets

Symbol	Subject	# tweets
s_1	Smartphone A	20,173
s_2	Smartphone B	20,128
s_3	Smartphone C	20,055
s_4	Robot vacuum	20,060
s_5	Printing services at convenience stores	20,111
s_6	Manufacturer of electric appliance A	20,003
s_7	Manufacturer of electric appliance B	20,214
Sum.		140,744

Table 2. Feedback usage and the number of workers. *Imp.* means implicit feedback. *FM* (Feedback Message) means that the workers receive feedback messages from the reviewers. *PC* (Point Change) means that the points are changed on the implicit feedback.

Group	Imp.	FM	PC	# workers
Baseline				124
Imp	✓			121
Imp+Exp(No)	✓	✓		119
Imp+Exp(Pos)	✓	✓	✓	129
Imp+Exp(Neg)	✓	✓		130
Imp+Exp(Pos+Neg)	✓	✓	✓	127
Sum.				750

Workers. Our method assigned more than five workers to each tweet. We determined the number of workers who selected appropriate options for more than 75% of tweets by using majority votes. In our preliminary experiments, we found that assigning five workers is sufficient to accomplish this goal.

When the workers signed in, our method randomly assigned them to the following six groups:

- **Baseline**: The workers in this group did not receive any feedback from our method. They earned one point for each task.
- **Imp**: The workers in this group received only implicit feedback. The workers earned 0.5 points for each task and earned 0.5 more points for each task if their selection was in the majority. As a result, the workers earned at least 0.5 points and at most 1 point.
- **Imp+Exp(No)**: The workers in this group received both implicit and explicit feedback. We used the implicit feedback mechanism and the feedback message type of explicit feedback. The workers earned 0.5 points for each task and earned 0.5 more points for each task if their selection was in

the majority. The workers received advice about their tasks, but they did not earn extra points for each task even if they received positive votes from implicit feedback. As a result, the workers earned at least 0.5 points and at most 1 point.

– **Imp+Exp(Pos)**: We used the implicit feedback mechanism and the feedback message and point change types of explicit feedback. This was almost the same as *Imp+Exp(No)*. Also, the workers earned extra points for each task if they received positive votes from other workers. As a result, the workers earned at least 0.5 points and at most 1 point.

– **Imp+Exp(Neg)**: We used the implicit feedback mechanism and the feedback message and lockout types of explicit feedback. This was almost the same as *Imp+Exp(No)*. Also, the workers' accounts were locked if they received negative votes five times in a row.

– **Imp+Exp(Pos+Neg)**: This group was the sum of *Imp+Exp(Pos)* and *Imp+Exp(Neg)*. We used all functions of our crowdsourcing platform.

Table 2 shows the usage of feedback for workers in each group. *Imp.* means implicit feedback, *FM* means workers sent and received advice using explicit feedback, *PC* means workers obtained extra points when they received positive votes, and *LO* means workers' accounts were locked if they received negative votes five times in a row. The workers, except those in *Baseline* and *Imp*, gave explicit feedback per 50 task results they received. The target of the explicit feedback was the latest task result done by another worker.

Workers did not know that these groups existed and also the detail of the groups. Moreover, the workers cannot communicate with each other except anonymized review messages.

4.2   Results

We explain the results of the crowdsourcing task from the viewpoints of task result quality and worker motivation. We defined the sentiment for a tweet as being correct if it was part of the majority. However, if there were less than two top and second-highest votes or less than three of any votes, we did not use them to calculate average accuracy.

Tasks. Table 3 lists the sentiment analysis results using our crowdsourcing platform. It shows that 5.1% of the tweets had at least positive or negative sentiment, whereas about 94% had no information about sentiment because there were many advertisements related to smartphones regarding s_1, s_2, and s_3 in the tweets. Moreover, the names of manufacturers in s_6 and s_7 included common nouns, and the keywords for gathering the tweets included common nouns; thus, many of the tweets were not related to the subjects. Regarding s_2, s_4 and s_5, the ratios of tweets categorized as *PN*, *positive*, or *negative* were higher than those of the other subjects.

As we stated in Sect. 3.1, our method assigns more than two labels if the top labels have the same number of votes. As shown in Table 3, 2.6% of tweets were

Table 3. Results of sentiment analysis. Values in parentheses are ratios of sentiments. PN corresponds to *positive and negative*, P. means *mostly positive*, N. means *mostly negative*, Neu. means *neutral*, U means *unrelated*, and Multi. means number of tweets having more than two labels.

	PN	P	N	Neu	U	All	Multi.
s_1	31 (0.2%)	127 (0.6%)	246 (1.2%)	3231 (16.0%)	16,871 (83.6%)	20,173	333 (1.7%)
s_2	123 (0.6%)	683 (3.4%)	936 (4.7%)	14,846 (73.8%)	4,036 (20.1%)	20,128	496 (2.5%)
s_3	20 (0.1%)	144 (0.7%)	188 (0.9%)	6,462 (32.2%)	13,591 (67.8%)	20,055	350 (1.7%)
s_4	134 (0.7%)	928 (4.6%)	1,148 (5.7%)	15,506 (77.3%)	3,107 (15.5%)	20,060	763 (3.8%)
s_5	67 (0.3%)	678 (3.4%)	598 (3.0%)	7,967 (39.6%)	11,528 (57.3%)	20,111	727 (3.6%)
s_6	48 (0.2%)	299 (1.5%)	220 (1.1%)	4,513 (22.6%)	15,375 (76.9%)	20,003	452 (2.3%)
s_7	29 (0.1%)	218 (1.1%)	324 (1.6%)	10,530 (52.1%)	9,690 (47.9%)	20,214	577 (2.9%)
All	452 (0.3%)	3,077 (2.2%)	3,660 (2.6%)	63,055 (44.8%)	74,198 (52.7%)	140,744	3,698 (2.6%)

assigned more than two labels; 31% of these tweets were labeled *neutral* and *unrelated* and 13% were labeled *negative* and *neutral*. This means that *neutral* is difficult for workers to understand; thus, many workers sometimes randomly selected *neutral* when they were confused about what option to select.

The workers in *Imp+Exp(Neg)* and *Imp+Exp(Pos+Neg)* had the risk of being locked out. A worker who received explicit negative feedback five times in a row would be locked out in our experimental setting. However, no workers were locked out. The worst workers received negative feedback at most two times in a row, and we believed that even these workers were good. Lock-out seemed to have only a psychological impact.

Implicit Feedback. Next, let us discuss whether implicit feedback was effective. Figure 1 shows the accuracy for each worker group. We can see that the number of workers who had low average accuracy decreased due to using implicit feedback. The minimum average accuracy of the workers in *Implicit* was 0.7, whereas that of those in *Baseline* was 0.48. Implicit feedback also improved the lower-quartile average accuracies.

To determine whether the workers' reactions to the implicit feedback were effective, the workers who received implicit feedback from others evaluated that feedback. In our crowdsourcing platform, when a worker received implicit feedback, he or she selected an opinion on whether the feedback was effective. Table 4 lists the results of this evaluation. The numbers in the column where the advice (adv. in this table) was negative, and the reply was positive mean the number of tasks in which the label was not in the majority, and the workers agreed with this feedback. From this table, 11.2% of the selected labels were decided as not appropriate for other workers, and the workers agreed with 69.0% of these decisions.

Explicit Feedback. Next, we determined whether the explicit feedback was effective. As we described above, the reviewers reviewed their task results. In our method, the system did not select reviewers by the quality of reviewers. The

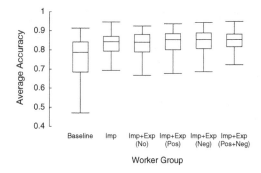

Fig. 1. Accuracy for each worker group

Table 4. Number of evaluations by workers who received *implicit* feedback: *adv.* means advice, *acc.* means accurate, and *inacc.* means inaccurate.

adv.\reply	acc.	inacc.	Sum.
acc.	311,160	65,914	377,074
inacc.	32,745	14,684	47,429
Sum.	343,905	80,598	424,503

Table 5. Number of evaluations by workers who received *explicit* feedback

adv.\reply	acc.	inacc.	Sum.
acc.	1,989	163	2,152
inacc.	77	43	120
Sum.	2,066	206	2,272

reviewers did not always review appropriately, and the messages from the reviewers were not always appropriate. Therefore, we assumed that if the workers who received feedback decided which messages were appropriate and were not then ignored the inappropriate messages, the quality of the workers who received such feedback would worsen.

We tried to determine whether explicit feedback helps improve task result quality. As we described, a worker reviews assessments of other workers and decides whether to agree or disagree with those assessments. Figure 1 shows the distribution of average accuracy for each worker group. We can see that the average accuracy of *Imp* and those of the work groups using both implicit and explicit feedback was about the same. Therefore, explicit feedback did not significantly change assessment quality.

Finally, the workers who received explicit feedback evaluated that feedback. Table 5 shows the results of this evaluation. The workers evaluated 94% of the explicit feedback as acceptable. This rate was higher than that of the implicit feedback, i.e., the workers evaluated $91\%(2,066/2,272)$ of implicit feedbacks as acceptable. Therefore, we found that explicit feedback was more acceptable than implicit feedback.

If a task result of a worker is determined as unacceptable by reviewers, that work tends to reply, stating that the decision is unacceptable because the reviewers denied that worker's task result, which he/she believed was correct. However, if the worker feels that the feedback is beneficial, the worker should reply stating that it is acceptable. In Table 5, workers who sent feedback decided that 120 labels were unacceptable, and the workers who received the feedback decided that 77 (64%) of those feedback responses were acceptable. Therefore, we found that the worker changed the task results 77 times and that such feedback will affect future labels selected by the workers.

5 Conclusion

Micro-task crowdsourcing is sufficient for generating large-scale human-annotated data, such as market surveys, machine learning, and information-seeking applications. However, workers do not always output accurate answers, and the quality of their work varies depending on their ability and motivation. To solve these issues, we construct the autonomous crowdsourcing system.

A reason for low-quality task results is the neglect of workers in performing their tasks. Requestors should accurately review all of the task results in a short time to address this issue. However, requesters are unable to do this. We proposed a method of finding which labels are in the majority, i.e., *implicit feedback*, for improving the quality of task results. We confirmed that the quality of task results improved from our experimental results, but worker motivation decreased when we used implicit feedback only.

For future work, we will develop a method for improving the quality of feedback. There is no guarantee that implicit and explicit feedback will always be accurate. In this study, we attempted to increase the quality of task results, but we did not discuss the quality of feedback. Whiting et al. [5] automatically measured rankings of workers; then the workers evaluated other workers who were lower-ranked. We argue that if a worker receives adequate and straightforward advice, the quality of his/her work will increase.

Acknowledgements. This work was partly supported by JSPS KAKENHI Great Number 19H04218, 18H03342, and 19H04221.

References

1. Chittilappilly, A.I., Chen, L., Amer-Yahia, S.: A survey of general-purpose crowd-sourcing techniques. IEEE Trans. Knowl. Data Eng. **28**(9), 2246–2266 (2016). https://doi.org/10.1109/TKDE.2016.2555805
2. Howe, J.: Crowdsourcing: Why the Power of the Crowd Is Driving the Future of Business, 1st edn. Crown Publishing Group, New York (2008)
3. Nakov, P., Ritter, A., Rosenthal, S., Stoyanov, V., Sebastiani, F.: SemEval-2016 task 4: sentiment analysis in Twitter. In: Proceedings of the 10th International Workshop on Semantic Evaluation, SemEval 2016. Association for Computational Linguistics, San Diego, June 2016

4. Shah, N.B., Zhou, D.: No oops, you won't do it again: mechanisms for self-correction in crowdsourcing. In: Proceedings of the 33rd International Conference on International Conference on Machine Learning, ICML 2016, vol. 48, pp. 1–10. JMLR.org (2016). http://dl.acm.org/citation.cfm?id=3045390.3045392
5. Whiting, M.E., Gamage, D., et al.: Crowd guilds: worker-led reputation and feedback on crowdsourcing platforms. In: Proceedings of the Computer Supported Cooperative Work and Social Computing, CSCW 2017, pp. 1902–1913 (2017). https://doi.org/10.1145/2998181.2998234

Machine Learning

The Effect of IoT Data Completeness and Correctness on Explainable Machine Learning Models

Shelernaz Azimi[(✉)] and Claus Pahl

Free University of Bozen-Bolzano, Bolzano, Italy
{shelernaz.azimi,claus.pahl}@unibz.it

Abstract. Many systems in the Edge Cloud, the Internet-of-Things or Cyber-Physical Systems are built for processing data, which is delivered from sensors and devices, transported, processed and consumed locally by actuators. This, given the regularly high volume of data, permits Artificial Intelligence (AI) strategies like Machine Learning (ML) to be used to generate the application and management functions needed. The quality of both source data and machine learning model is here unavoidably of high significance, yet has not been explored sufficiently as an explicit connection of the ML model quality that are created through ML procedures to the quality of data that the model functions consume in their construction. Here, we investigated the link between input data quality for ML function construction and the quality of these functions in data-driven software systems towards explainable model construction through an experimental approach with IoT Data using decision trees. We have 3 objectives in this research: 1. Search for indicators that influence data quality such as correctness and completeness and model construction factors on accuracy, precision and recall. 2. Estimate the impact of variations in model construction and data quality. 3. Identify change patterns that can be attributed to specific input changes.

Keywords: Explainable AI · Data quality · IoT systems · Machine learning · Data correctness · Data completeness · Decision trees

1 Introduction

There are different types of errors or faults which may occur in data sets, such as missing values or rows, invalid values or formats, or duplicated values or rows. Low quality data will result in low quality machine learning models if the model is used to learn from the data. Before using often faulty real world data and trying to find a remedial solution for observed machine learning model, we need to better understand the effects of low input data quality on the created models.

Our ultimate goal is to automate quality control of machine learning models, but to reach that the understanding the impact of a sensor producing faulty data or no data on a model trained on this data is a general requirement. The wider

© Springer Nature Switzerland AG 2021
C. Strauss et al. (Eds.): DEXA 2021, LNCS 12924, pp. 151–160, 2021.
https://doi.org/10.1007/978-3-030-86475-0_15

objective is explainable model construction. Black-box explainable AI aims at a better understanding of how ML model output depends on the model input [11]. Of particular importance is here a root cause analysis for model deficiencies. Our aim here is, based on observed model quality problems, to identify a root cause at input data level. The concrete practical benefit of this in an IoT setting for example is, that certain ML quality patterns might already point to specific problems with the data, such as outages for faulty sensors.

Therefore, we investigated different experimental scenarios with artificial and real faulty input data sets. We specifically considered 1) input data completeness and 2) input data correctness, since these are of direct relevance to IoT settings. With the experiments, we created situations with different faulty data sets and compare the results to find a connection between the type of faulty data and the ML quality assessment factors (accuracy, precision, recall). We focus here on numeric data that would for example be collected in technical or economic applications, neglecting text and image data here.

The novelty lies in the integrated investigation the quality of information that is derived from data through a machine learning approach. We proposed a quality frameworks in [1,2], but report on an in-depth experimental study here.

2   Related Work

Machine learning (ML) techniques have generated huge impacts in a wide range of applications such as computer vision, speech processing, health or IoT.

Input data quality is important. The issue of missing data is unavoidable in data collection [4,7,13,18]. Various imputation approaches, i.e., substituting missing values, have been proposed to address the issue of missing values in data mining and machine learning applications. [13] addresses missing data imputation. The authors propose a method called DIFC integrating decision tress and fuzzy clustering into an iterative learning approach in order to improve the accuracy of missing data imputation. They demonstrated DIFC robustness against different types of missing data.

Currently, missing data impacts negatively on the performance of machine learning models. Regarding concrete ML techniques, handling missing data in decision trees is a well studied problem [5]. [19] also proposed a method for dealing with missing data in decision trees. In [7], authors tackle this problem by taking a probabilistic approach. They used tractable density estimators to compute the "expected prediction" of their models. Missing data or uncertain data in general have always been a central issue in machine learning and specially classifiers. [18] focused on the accuracy of decision trees with uncertain data. The authors discovered that the accuracy of a decision tree classifier can be improved if the complete information of a data item is utilized. They extended classical decision tree algorithms to handle data tuples with uncertain data. Paper [15] describes a solution pattern that analyzed IoT sensor data and failure from multiple assets for data-driven failure analysis. The paper used univariate and multivariate change point detection models for performing analysis and adapted precision, recall and accuracy definition to incorporate the temporal window constraint. In [17], a toolkit for structured data quality learning is presented. They defined 4 core data quality constructs and their interaction to cover the majority of data quality analysis tasks.

Focusing on decision trees and missing data, we investigate the link between source data and machine learning model as a so far unexplored AI explainability concern.

3   Method

Before presenting the results of the experiments in the following section, we introduce here our methods including the description of objectives, data and implementation. In many applications, ML models are reconstructed continuously based on changing input data. We use experiments to determine the extent to which different input changes regarding data quality impact on model construction quality. In more concrete terms, the question is if changes in the data quality or the model construction have a similar impact on output quality. We consider here the following ML quality attributes. *Precision*, also known as Positive Predictive Value (PPV), answers the question of how many selected items are relevant. *Recall*, or Sensitivity, answers the question of how many relevant items were selected. *Accuracy* is the percentage of correct predictions for the test data.

For input data quality, we selected two attributes that are IoT-relevant [3]: **completeness** is the degree to which the number of data points required to reach a defined accuracy threshold has been provided and **correctness** is the degree to which data correctly reflects an object or an event described, i.e., how close a label is to the real world.

In the context of these definitions, a sample question is if minor changes in the completeness of data (as a data quality problem) or the tree depth of decision trees (as a model construction concern) have a similar impact on model accuracy. Experiments shall help to determine the scale of the impact of a given size on input variations. We use experiments to determine if certain input change patterns correlate to observable output change patterns [6]. In concrete terms, this is if minor or major changes in input and input quality result in identifiable change patterns across different output qualities (e.g., accuracy, precision, recall). The question is if observed change patterns in the ML model output can be attributed to the root cause of that change at input data level.

Our models here are decision trees – using scikit-learn[1] to both data sets for predictions. Using traffic data, we predicted the traffic volume and using weather data we predicted rain fall. The first data set was traffic data that has been taken from an application, which consisted of daily averages of traffic and number of vehicles in 72 stations around our province in a month. The total number of rows in this data set is thus 72. The second data set was weather data consisting of the

---

[1] https://scikit-learn.org/ - Machine learning library for Python.

Table 1. Incompleteness and incorrectness experiments summary

	Completeness	Correctness
Rows	In the traffic data, precision and recall behaved slightly different from accuracy but we do not see the same behavior for weather data. However, there is no significant difference	For −1000 the values fell from lower initial values than in traffic data. For −5000, accuracy, precision and recall fell but the gradient was steeper than for −1000. For −10000, all three factors fell from a lower initial value but the final values are not lower than before. Therefore, the graph gradient is slighter when in fact the higher invalid value has effected the factors correctly
Features	The stable area in the accuracy graph in the missing row does not occur in for missing features, where we see a soft fall. For the precision and recall, the sudden rise does not occur here. All factors have a steady gradient not as steep as for missing rows	Accuracy is gradually falling, but precision and recall are acting differently. There is no connection to previous cases as those were from missing rows and invalid features here. Comparing the results we can say that this results are more understandable to the lower invalid value results because like there, accuracy is showing a steep and steady fall where on the other hand precision and recall are acting differently in a more unpredictable way

minimum and maximum temperature, rainfall, wind speed, humidity, pressure, cloud and rain today as features, and the target is the possibility of rain fall the next day for 49 stations. The data from both data sets consisting of only numerical values has been processed and labeled manually.

The experimental strategy was to find the effect on accuracy, precision and recall while inducing error into the data set. We start each experiment with an initial baseline for these quality attributes. In order to check the impact of incomplete and incorrect input data on accuracy we created two different situations for each data set. For *incompleteness*, we checked the impact of *Missing Features* and *Missing Rows* on accuracy, precision and recall. For *incorrectness*, we checked the impact of *Invalid Features* and *Invalid Rows* for different invalid values on accuracy, precision and recall.

The experiments on input data completeness and incorrectness have been summarised in Table 1. For each data set in each table, we performed the experiments in two different formats, missing or invalid rows and missing or invalid features. The values were selected to reflect small, medium and large scale faulty situations. The values are in that sense meaningful in relation to the size of the data set in rows or features. For the missing or invalid rows in traffic data, we started with 2 rows and increased the number of missing rows gradually to 5, 15 and 24. For the missing or invalid features, we started with 3 features then 7 then 10 and lastly 13 features. For the missing or invalid rows in weather data

Table 2. Comparison summary (TS: Test Size, TD: Tree Depth)

Rows-TD	In traffic data, the accuracy fell with increasing the missing rows. Depths 3, 4 and sometimes 5 were the best and anything below or over were unstable. This was shown better in the weather data set. The accuracy increased until the depth of 3, 4 and sometimes 5 and then started to fall which is expected. In traffic data, the accuracy first increased with tree depth but from the depth 3 to 5 was stable and after that fluctuated irregularly. In weather data, a similar result is visible. The best accuracy was at depths 3 to 5 as well but afterwards the accuracy started to fall. The fall was more significant with higher incorrectness
Features-TD	In both data sets, accuracy started to rise until depth 4 and afterwards to fall. However, in traffic data it started to grow again after depth 8. A probable reason is over-fitting. In traffic data the accuracy rose from depth 1 to 3–4, then varies and then after reaching the depth 8 it rose again. In weather data, the accuracy rose from depth 1 to 4 and then it fell significantly. In traffic data, the first rise is expected because it's normal for accuracy to rise until the best depth but the second rise is due to a machine learning tool error or over-fitting
Row-TS	In traffic data, accuracy falls with more missing rows but improves with bigger test sizes. The best test sizes were 20% and 30%. For weather data, accuracy improved until 20% and 30% before falling again. In traffic data, accuracy gradually increased until 30% but varied afterwards. In Weather Data, the results were more clear. The accuracy first rose until the best test size and then started to fall gradually. The best test sizes were 20% and 30%
Features-TS	The best test size for both data sets were 20% and 30%. In traffic data, accuracy started to grow after 40% but according to the other experiments and weather results, probable reasons are ML errors or over-fitting. The results were similar to the previous experiments. Overall, the effect of invalid features on accuracy was less than the effect of invalid rows

we started with 6 rows and increased the number of missing rows gradually to 20, 36 and 49. For the missing or invalid features we started with 2 features then 4 then 8 and lastly 13 missing features. We observed the accuracy, precision and recall in these situations with 20% test size and tree depth of 3.

For the weather data set we tried another set of invalid values as well to test the accuracy of the machine learning tool in identifying invalid values. As we mentioned before, in the first set we tried negative values as clearly invalid, but in the second set we tried extreme positive values as potentially possible, though highly improbable rainfall values. In general, we wanted to reflect different categories of sensors values: (i) correct sensor readings within small sensor reading variation, (ii) extreme but in principle possible values, likely linked to sensor faults, and (iii) clearly incorrect reading, definitely linked to sensor faults. We are dealing with sensor data and chose invalid values that are out of the range of regular sensor readings. We generally chose 3 different incorrect settings in order to avoid unexpected behaviour from a single invalid value – typically choosing a clearly incorrect value such as –1000 and increasing this to the next order of magnitude. What we are also looking for is to find out which type of invalid values (positive or negative) can be identified better by the machine learning tool, thus allowing a better judgement of the possible root causes. The same experiments were repeated also on positive values. Compared to the negative results no significant pattern changes were identified except that the output values were less in positive values.

After observing the effect of different levels of faulty situations on accuracy, precision and recall, the next step was to try to find a concrete change pattern on each outcome factor's variation in different scenarios in order to connect those patterns to a specific scenario. To do so, we also tested the effect of different tree depths and different test sizes on normal and various faulty data sets and compare the results with each other in order to find a specific change pattern. We present the results in Table 2.

4   Observation, Analysis and Validation

The outcome of the experiments demonstrated similarity between the data sets and thus a validity of the observations as they have been confirmed in two settings. In total, we conducted more than 50 experiments that varied settings in 4 dimensions (tree depth, test size, missing/invalid features, missing/invalid rows), which cannot be presented here in full. As a summary of the findings, we can state that:

1. *Incorrectness More Significant than Incompleteness.* The incorrectness has a bigger effect on the accuracy than the incompleteness. The most probable reason for it is that in incompleteness the machine learning tool may ignore the missing rows or features and not engage them in the predictions and calculations, but regarding incorrectness the tool is forced to use all the values either correct or incorrect therefore it cannot control or minimize the damage to the accuracy.
2. *Rows More Significant than Features.* Missing or invalid rows have a stronger impact on the accuracy than missing or invalid features. Here again, the causes may be different factors, but the most probable one may be the fact that dealing with a complete missing or invalid row is more difficult than dealing with some missing or invalid features. Remedying the reduction of accuracy is more difficult with missing or invalid rows than missing or invalid features, see Fig. 1.
3. *Data Set Differences.* In the analysis of the experiments, we noted that the results of the weather data was easier to process than the traffic data. In the traffic data set, the volume of data might have been rather low.

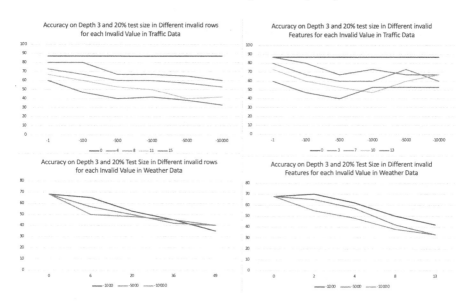

Fig. 1. Experimental results for the effect of different invalid values on both data sets for invalid rows and invalid features.

4. *Overfitting.* As a general observation, With very high results in the outcome, we tend have a machine learning tool problem like over-fitting, but when we have low results in outcomes, it means that the problem lies more likely in the data or sensors.

5. *Incorrect and Improbable Data.* Regarding positive and negative values, i.e., highly improbable vs. certainly incorrect data, we observed for weather data that the results for positive invalid values were lower than negative invalid values. This situation needs to be tested on other data sets to determine a reason. However, for weather data and with some negative values as inputs, a plausible explanation is that it is difficult to identify a real negative error, but for positive values, since the values were very high, it was easier for the algorithm to identify them.

   In conclusion, the observations are validated in both data sets and are practically applicable in machine learning quality analysis. They can be used in root cause analyses to identify possible faults in a IoT architecture such as sensor or connectivity problems. This provides a post-hoc explanation to black-box explainable model construction.

   However, a clear identification of the reason behind the observation is not always possible. The problem here is the *white-box explainability* of machine learning models. As deep learning and other highly accurate black-box models develop, the social demand or legal requirements for interpretability and explainability of machine learning models are becoming more significant [16]. Nowadays, the two terms are beginning to have different meanings, with interpretability describing the fact that the model is understandable by its nature (e.g. decision trees) and explainability corresponding to the capacity of a black-box model to be explained using external resources (e.g., visualizations). However, white-box explainability is beyond the scope of the paper here.

We used two data sets to investigate the *correctness* of the results and *applicability* for multiple domains. While the observations are generally of *practical benefit*, another important aspect is the *explainability* of the observations. Our observations apply to sensor-based IoT settings where all the data came from IoT sensors. The question is whether or not we can utilise the observations in a root cause analysis.

The missing or invalid rows situation is more likely to happen in real-life situations than missing or invalid features. Data is received from sensors. If a sensor is faulty or the data is not received due to a connection problem, all the data from that sensor is lost (and not a part of data), unless we have different sensors for different factors. In the latter case, it would be possible to have missing features. For example, if a weather sensor can calculate different factors like temperature, humidity, pressure, wind and etc., then if the sensor is faulty, we will lose all the measurement at the same time. If we have different sensors for each measurement, then if the sensor is faulty, we will lose only some at the same time, but not all of them. For invalid values, it depends on the type of sensor and factors. For instance, –50C is generally unlikely for a temperature reading, but still possible to happen; on the other hand below –100C can be assumed incorrect. These observation can be used to deduce probable root causes in sensor-based IoT environments such as faulty sensors or incorrect data processing.

5 Conclusion

More and more software applications are based on functions generated using ML from larger volumes of data available in contexts such as the Internet-of-Things (IoT) instead of being manually programmed [14]. With less human involvement in the construction process of the software, quality assurance becomes more important.

We focused on the link between input data quality for ML function construction and the quality of these functions in data-driven software applications. An important observation is the range of quality concerns that apply. For input data, we considered correctness and completeness as data quality concerns. For ML model construction, the usual accuracy, precision and recall were considered. We organized our work in three steps. In first step, we determined a framework of indicators that influence data quality such as correctness and completeness and model construction factors on accuracy, precision and recall as described above. Then, we experimentally analysed the impact of variations in model construction and data quality on ML model quality and in the final step, we aimed to identify change patterns that can be attributed to specific input changes caused by for instance faults in the environment in the context of a root cause analysis. This provides a post-hoc explanation for a black box explainability setting.

The observations were validated in two data sets and are practically applicable in machine learning quality analysis and root cause analysis. However, a clear identification of the reason behind the observation is not always possible. More work on the white-box explainability of results is needed. Other application domains could here be considered, such as mobile learning that includes the usage of multimedia content being delivered to mobile learners and their devices [9,12]. A further direction is the implementation of self-adaptive ML quality management in an IoT-edge continuum [8,10].

Acknowledgments. This work has been performed partly within a Ph.D. Programme funded through a bursary by the Südtiroler Informatik AG (SIAG).

References

1. Azimi, S., Pahl, C.: A layered quality framework in machine learning driven data and information models. In: ICEIS (2020)
2. Azimi, S., Pahl, C.: Root cause analysis and remediation for quality and value improvement in machine learning driven information models. In: ICEIS (2020)
3. Azimi, S., Pahl, C.: Continuous data quality management for machine learning based data-as-a-service architectures. In: CLOSER (2021)
4. Ehrlinger, L., Haunschmid, V., Palazzini, D., Lettner, C.: A DaQL to monitor data quality in machine learning applications. In: Hartmann, S., Küng, J., Chakravarthy, S., Anderst-Kotsis, G., Tjoa, A.M., Khalil, I. (eds.) DEXA 2019. LNCS, vol. 11706, pp. 227–237. Springer, Cham (2019). https://doi.org/10.1007/978-3-030-27615-7_17
5. Harp, S., Goldman, R., Samad, T.: Imputation of missing data using machine learning techniques. pp. 140–145 (1996)
6. Javed, M., Abgaz, Y.M., Pahl, C.: Ontology change management and identification of change patterns. J. Data Semant. **2**(2–3), 119–143 (2013)
7. Khosravi, P., Vergari, A., Choi, Y., Liang, Y., Broeck, G.: Handling missing data in decision trees: a probabilistic approach (2020)
8. von Leon, D., Miori, L., Sanin, J., Ioini, N.E., Helmer, S., Pahl, C.: A lightweight container middleware for edge cloud architectures. In: Buyya, R., Srirama, S.N. (eds.) Fog and Edge Computing, pp. 145–170. Wiley, Chichester (2019)
9. Melia, M., Pahl, C.: Constraint-based validation of adaptive e-learning courseware. IEEE Trans. Learn. Technol. **2**(1), 37–49 (2009)
10. Mendonça, N.C., Jamshidi, P., Garlan, D., Pahl, C.: Developing self-adaptive microservice systems: challenges and directions. IEEE Softw. **38**(2), 70–79 (2021)
11. Mittelstadt, B.D., Russell, C., Wachter, S.: Explaining explanations in AI. In: Proceedings of the Conference on Fairness, Accountability, and Transparency (FAT 2019). ACM (2019)
12. Murray, S., Ryan, J., Pahl, C.: Tool-mediated cognitive apprenticeship approach for a computer engineering course. In: International Conference on Advanced Learning Technologies (ICALT), pp. 2–6. IEEE Computer Society (2003)
13. Nikfalazar, S., Yeh, C.H., Bedingfield, S., Khorshidi, H.: Missing data imputation using decision trees and fuzzy clustering with iterative learning. Knowl. Inf. Syst. **62**, 2419–2437 (2020)

14. Pahl, C., Azimi, S.: Constructing dependable data-driven software with machine learning. In: IEEE Software (2021)
15. Patel, D., Nguyen, L.M., Rangamani, A., Shrivastava, S., Kalagnanam, J.: Chief: a change pattern based interpretable failure analyzer. In: International Conference on Big Data, pp. 1978–1985 (2018)
16. Roscher, R., Bohn, B., Duarte, M.F., Garcke, J.: Explainable machine learning for scientific insights and discoveries. IEEE Access **8** (2020)
17. Shrivastava, S., Patel, D., Zhou, N., Iyengar, A., Bhamidipaty, A.: Dqlearn : a toolkit for structured data quality learning. In: International Conference on Big Data, pp. 1644–1653 (2020)
18. Tsang, S., Kao, B., Yip, K., Ho, W.s., Lee, S.: Decision trees for uncertain data. In: Proceedings - International Conference on Data Engineering (2009)
19. Twala, B., Jones, M., Hand, D.: Good methods for coping with missing data in decision trees. Patt. Recogn. Lett. **29**, 950–956 (2008)

Analysis of Behavioral Facilitation Tweets for Large-Scale Natural Disasters Dataset Using Machine Learning

Yu Suzuki[1(✉)], Yoshiki Yoneda[2], and Akiyo Nadamoto[2]

[1] Gifu University, 1-1 Yanagido, Gifu 5011193, Japan
ysuzuki@gifu-u.ac.jp
[2] Konan University, 8-9-1 Okamoto, Higashi-Nada, Kobe 6588501, Japan

Abstract. In this paper, we extract behavioral facilitation tweets from a set of tweets. We analyze and compare the extracted BF tweets related to a large-scale disaster with those related to normal-scale disasters. Specifically, we did two analysis, such that 1) Compare three neural network-based classifiers such as LSTM, BiLSTM, and BERT using a set of tweets related to large-scale disasters, and propose the best method of extracting BF tweets, and 2) Analyzing characteristics of large-scale disaster and comparing them with those of normal disaster.

Keywords: Twitter · Social media · Information extraction · Deep learning · Behavioral facilitation · LSTM · BiLSTM · BERT

1 Introduction

Large-scale natural disasters, such as Typhoons, floods, and pandemic outbreaks of infectious diseases, occur frequently. When these natural disasters occur, many people suggest that people should take some action to help victims via social media such as Twitter and Facebook. For example, "Close curtains or shades to ward off flying objects from outside of your home!" and "Do not feel constraint to tidy your room" are the tweets to suggest to take some action. The goal of this paper is to extract these tweets "Behavioral Facilitation tweets (*BF tweets*)".

BF tweets are essential for reducing damages by disasters and assisting damaged areas. When the people in the damaged area catch and believe these BF tweets, they will follow the BF tweets' suggestions. However, these suggestions are not always helpful, and some suggestions are harmful because these suggestions are not always correct. The paper [2] reports that 53% of 7,177 fake tweets are behavioral facilitation tweets. From this result, we observe that the ratio of fake tweets to all BF tweets is high. Therefore, we need a method for extracting behavioral facilitation tweets from all tweets and alerts that the tweets may be fake. From these alerts, the victims do not believe in fake tweets. Therefore, it is essential for victims to extract and present BF tweets automatically in disaster situations. In the literature, some methods have been proposed to extract and

© Springer Nature Switzerland AG 2021
C. Strauss et al. (Eds.): DEXA 2021, LNCS 12924, pp. 161–169, 2021.
https://doi.org/10.1007/978-3-030-86475-0_16

present BF tweets at disaster situations such as references [3] and [9]. As a result, it has been shown that LSTM is better than rule-based classifiers, Support Vector Classifiers (SVC). They also show that the characteristics of BF tweets in large-scale disasters are different from that of normal-scale disasters.

We need the best method to extract BF tweets in the event of a large-scale disaster. Many classifiers are proposed for texts, but good classifiers should differ when the scales of disasters are different. Therefore, we find better classifiers for extracting BF tweets. Then we analyze the characteristics of BF tweets during a large-scale disaster.

In this paper, we contribute the following two points.

– Comparing two neural network-based classifiers using large-scale disasters.
  We compared the other two neural network-based classifiers, such as Bidirectional LSTM (BiLSTM) and Bidirectional Encoder Representations from Transformers (BERT).
– Analyzing characteristics of large-scale disasters based on comparing with normal disaster.
  We compare the tweets related to large-scale disasters (Typhoons in 2019) with normal-scale disasters (small Typhoons in 2018), and analyze tweets' characteristics of each disaster.

2   Related Work

Some researchers have studied analyzing or predicting user's behavior from Twitter. Silva et al. [6] studied how individual behavior data may influence relationships in a social network. Xu et al. [7] performed a comprehensive analysis of user posting behavior on a popular social media website, Twitter. Mogadala et al. [4] proposed a method to predict the mood transition of a Twitter user by regression analysis on the tweets posted over the Twitter timeline. Yamamoto et al. [8] analyzed the characteristics of Growth-type users and proposed a user growth prediction method to collect a large number of tweets strategically. These methods extract the authors' behavior and predict their future behavior. In contrast, we propose extracting behavioral facilitation tweet, which influences other people.

3   Extraction of Behavioral Facilitation Tweets

This section describes how to extract BF tweets from the tweets related to a disaster. In our system, we deal with the following procedure.

1. *Prepare twitter dataset*: Using Twitter filtering API, we obtain tweets that include the keyword "Typhoon."
2. *Data cleaning*: From the tweets, we remove URLs and user names.
3. *Data labeling*: Using crowdsourcing, we label whether the tweets are BF tweets or non-BF tweets.

4. *Building the classifier:* We randomly select 80% of all tweets as a training set and set the rest of the tweets as a test set. Using this training set, we build the classifier. We used LSTM, BiLSTM, and BERT.

5. *Validate the tweets into BF tweets and non-BF tweets:* We validate the accuracy using the test set.

In this section, we explain step 4. in detail. Primarily, we explain the settings of each machine learning model.

3.1 A Classifier Based on LSTM

LSTM [1] is a machine learning model that can consider time series data. In natural language analysis, LSTM is widely used when considering the order of terms in sentences. We used Chainer[1] to implement LSTM.

First, we convert distributed expressions for each term using fastText[2] with NWJC2Vec[3] as a dictionary. We used all terms in the sentences and set them as the term array $W = [w_1, w_2, \cdots, w_n]$; we do not select terms according to a part-of-speech. We removed terms that are not in the dictionary.

Next, we describe the hyperparameters for LSTM. The hidden layer is one, the vector size is 300, the batch size is 500, the number of the epoch is 20, the learning rate is 0.001, the dropout ratio is 0.5, and we used Adam as an optimizer. The number of units is the same as the number of terms for each tweet. The last LSTM unit outputs values to the fully connected layer and output the degree of BF tweets. Then, using the Softmax layer, the system outputs whether the tweets are BF tweets or not.

3.2 A Classifier Based on BiLSTM

As we described in Sect. 3.1, LSTM only considers the order of terms in the tweets. However, the reverse order of terms is also helpful for classifying tweets. Therefore, we used BiLSTM [5] to consider both the order and the reverse order of terms. We used Chainer to implement BiLSTM. The dictionary and the term array W are the same as that used for LSTM.

Next, we show hyperparameters for BiLSTM. The hidden layer is 2, vector size is 300, the batch size is 500, the number of the epoch is 20, the learning rate is 0.0001, the dropout ratio is 0.5, and we used Adam as an optimizer. The number of input units is the same as the number of terms for each tweet. We construct one LSTM unit for the terms and another LSTM unit for the terms' reverse order. Then, we concatenate the outputs of both LSTM units and connect them to the fully connected layer. Then, using the softmax layer, the system outputs whether the tweets are BF tweets or not.

---

[1] https://chainer.org/.

[2] https://fasttext.cc.

[3] https://pj.ninjal.ac.jp/corpus_center/nwjc/ (in Japanese).

3.3   A Classifier Based on BERT

BERT is one state-of-the-art method for classifying tweets, which can consider the order of tweets. BERT is based on Transformer, which is one implementation of transfer learning. BERT is an encoder-decoder model using only attention; this does not use RNN (recurrent neural network) and CNN (convolutional neural network). The accuracy of BERT for machine translation is exceptionally high. Therefore, we use BERT to extract BF tweets.

We decide hyperparameters for BERT using a grid search. The hidden transformer layer is 12, and the vector size is 768, the batch size is 32, the number of the epoch is 10, the number of attention head is 12, the learning rate is 0.00002, warm up ratio of learning ratio is 0.1, and dropout ratio is 0.1. The number of input units is the same as the number of terms in each tweet. The system outputs whether the tweets are BF tweets or not using the softmax layer.

4   Experiment 1: Comparison of Models for Classification Accuracy

We compared LSTM, BiLSTM, and BERT accuracy for extracting BF tweets to find the most accurate classifier in this experiment.

4.1   Data

We set a Typhoon Faxai[4] of 2019 to the target of a large-scale disaster in this experiment. The Typhoon landed in next Tokyo on September 15, 2019, and causes significant damage to the east part of Japan. We collect 10,000 tweets that contain the term "Typhoon" and which are post between September 6 and September 16, 2019.

Next, we decide which tweets are BF tweets or not using crowdsourcing. We assign five workers for each tweet, and the workers decide whether the tweets are BF tweets or not. Then, we decide that the tweets are BF tweets when more than three workers decide that the tweets are BF tweets, and the tweets are not BF tweets when all workers decide that the tweets are not BF tweets. We do not use the tweets that are decided as BF tweets by one or two workers. As a result, we used 7,201 tweets as a dataset, 2,406 tweets are BF tweets, and 4,795 tweets are not BF tweets.

However, the numbers of BF tweets and not BF tweets are imbalanced; they are not suitable for constructing machine learning-based classifiers. To avoid this problem, we used under-sampling. We randomly select 2,406 tweets from the set of not BF tweets. As a result, the dataset we use includes 4,812 tweets, and the number of both BF tweets and not BF tweets are 2,406, respectively.

---

[4] https://en.wikipedia.org/wiki/Typhoon_Faxai.

Table 1. Result of experiment 1.

Method	Prec.	Recall	F1	Acc.	AUC
LSTM	**0.942**	0.873	0.906	0.910	0.967
BiLSTM	0.926	0.905	0.915	0.916	0.964
BERT	0.904	**0.944**	**0.938**	**0.937**	**0.979**

Table 2. Processing time (sec.)

Classifier	Learn	Val.	Total
LSTM	23.75	**5.84**	29.59
BiLSTM	34.27	6.56	40.83
BERT	**11.46**	8.11	**19.60**

4.2 Method

First, we randomly select 80% of all BF and non BF tweets as a training dataset. We extract BF tweets using three classifiers introduced in Sect. 3 and construct three classifiers. Then, we validate the classifiers using the rest 20% of BF and non BF tweets as a test dataset. Finally, we calculate the precision and recall ratio, F1-score, accuracy, and Area Under the Curve (AUC).

4.3 Result

Table 1 shows the precision and recall ratio, F1-scores, accuracy, and AUC. From this result, the accuracies of these three classifiers are practical. The precision ratio of LSTM is better than that of BiLSTM and BERT. The recall ratio and F1-score are better than those of LSTM and BiLSTM. Therefore, we discover that BERT is the best method for extracting BF tweets, but LSTM and BiLSTM are also accurate.

Next, we analyze these results in detail. Table 3 shows several tweets as an example. In this table, *Ans.* means whether the tweet is a BF tweet or not. We check the columns if a tweet is a BF tweet. LSTM, BiLSTM, and BERT indicate whether each model decides the tweet as BF tweets, respectively.

#1 is accurately extracted tweets using LSTM, BiLSTM, and BERT. In this target disaster (Typhoon in 2019), the massive damaged area is tiny. Therefore, many tweets are related to the specific area; then, the tweets include the area's name. In our experiment, almost all of the tweets that include the area's name are considered BF tweets. For example, tweets #1 and #2 include the area's name, such as *Kujyukuri city* and *Hachi-jo town*. All classifiers we used can extract these BF tweets.

#2 is the example of a tweet that are extracted using only BERT. In this tweet, the user who post this tweet does not facilitate behaviors to the victims; he/she only post her opinions. It is hard to identify opinions and behavioral facilitation because the terms' usages are almost the same.

#3 is not BF tweet, and only BERT does not extract this tweet. When we use the classifiers of LSTM and BiLSTM, the classifiers do not recognize the target of the facilitations. For example, the user only said that "Typhoon has gone." The user does not facilitate behaviors to the victims. If the user said that "You should go to the shelter because the Typhoon will come.," the tweets should be considered as BF tweets. Occurrences of terms for each tweet are similar to each other. Therefore, if classifiers accurately considers the order of terms, the accuracy of the classifiers will improve.

Finally, we compared the processing time of each classifier. We use Geforce GTX 1080Ti as GPU for the experiment. We classified 1,000 tweets using each classifier and calculated the averaging elapsed time. Table 2 shows the results. From this result, we found that BERT is the fastest for learning, and LSTM is the fastest for validation. The processing time of BERT is 1.39 times slower than that of LSTM. When we have a large amount of tweets to classify, BERT should be the slowest but accurate classifier. However, BERT is still a reasonable choice for extracting BF tweets because it takes only 8 s for classifying about 1,000 tweets. From these results, we conclude that BERT is the most appropriate classifier for extracting BF tweets because BERT is the most accurate, and the processing time is short enough.

In this section, we used a large-scale disaster dataset. However, we assume that the characteristics of BF tweets of large-scale disasters and normal-scale disasters should be different. Therefore, in the next section, we analyze and compare the BF tweets of large-scale disasters with normal-scale disasters using BERT.

5   Experiment 2: Analysis Characteristics of BF-Tweets in a Large-Scale Disaster Situation

The damages of large-scale disasters are different from normal disasters. Therefore, we can predict that the content of BF tweets is different. Yoneda et al. [9] said that the difference between the two cases appears clearly. Many BF tweets in large-scale disasters and its rumor are more significant than the number of normal disaster situations [2]. We consider that if we extract characteristics of a large-scale disaster of BF tweets extracted by our proposed BERT model, we can help extract and present BF tweets in such situations. Then, we compare the results of tweet BF tweets in a large-scale disaster that we extract using BERT with the normal disaster. We analyze the characteristics of the method of extracting BF tweets in a large-scale disaster.

5.1   Experimental Conditions

In this experiment, we used two datasets; *Large*: Large-scale disaster dataset: a set of tweets related to Typhoon in 2019, which we used for experiment 1 described in Sect. 4.1. *Normal*: Normal-scale disaster dataset: a set of tweets related to typhoons in 2018 except large-scale typhoons. We extract 566 BF tweets from 3000 from *Large* dataset and *Normal* dataset, respectively, using crowdsourcing.

We used crowdsourcing for labeling tweets. We assigned ten workers for each tweet, and we identify the tweets as the BF tweets if the tweets are judged as the BF tweets by six or more workers. We formulate hypotheses and analyze the *Large* dataset with *Normal* dataset.

Table 3. BF tweets extracted by three classifiers

ID	Ans.	LSTM	BiLSTM	BERT	Tweet
#1	✓	✓	✓	✓	For victims in Kujyukuri city areas that are damaged by Typhoon, The following public baths are open for free. Please bring a towel and shampoo
#2	✓			✓	Last year, the roof of my house was brown away during the typhoon. From this experience, I can say that you should cover one more sheet to the roof because the sheets that are provided by the government to fix the roof is too thin
#3		✓	✓		Let me see. I will sleep earlier than usual, because I do not feel asleep well. Typhoon has gone

Table 4. Examples of tweets in *Large*-scale disaster dataset (Typhoon in 2019) and in *Normal*-scale disaster dataset (Typhoon in 2018)

#	Dataset	Tweet
#1	*Large*	There was a Typhoon. Are you safe? Many trees knocked down at Ueno park. Some fallen trees have been removed, but some trees remain. It would help if you took care of the rest of the trees
#2	*Large*	From Sammu city fire department: It's ready to provide drinking water again. Please bring empty bottles to Sammu city hall
#3	*Large*	I experienced a heavy Typhoon last year. There is a large scale power outage in Osaka. From my experience, you should charge portable batteries and Nintendo Switch. You should keep some food and water, and you should reduce foods in your refrigerator. When I faced the large scale power outage two days, foods in my refrigerator went bad
#4	*Normal*	It's cloudy because of typhoon. We should be careful of heat stroke. Let's survive this heat!
#5	*Normal*	Typhoon Number 26 formed. Pay attention!

5.2   Results

Table 4 shows the tweets which are decided as BF tweets by BERT. *Large* is a set of tweets of the large-scale disaster situation, and *Normal* is a set of the normal-scale disaster situation.

From this result, we observed that the names of damaged places appear more frequently in *Large* than *Normal*. Because a seriously damaged area by large-scale disaster is relatively small, and people try to express the damages clearly. For example, there is the place name "Ueno Park" in #1, and "Yamatake city fire department" and "Yamatake city hall" in #2. Primarily, we found many tweets which are related to a small area. For example, Yamatake city appeared in #2 is a small city. Therefore, this information is not broadcasted via TVs or newspapers. However, there is no place name in #4 and #5.

From these observations, we create three hypotheses and confirm whether the hypotheses are true or not.

5.3 Discussion

We formulate the following three hypotheses as follows:

Hypothesis 1: The number of place names is more significant in BF tweets of a large-scale disaster than that of a normal-scale disaster.

We consider that when the scale of the disaster is large, the victims would actively help each other. Eventually, the number of place names is large in *Large*. Our analysis results show that 10.4% of BF tweets of large-scale disasters include place names. On the other hand, 2.2% of BF tweets of normal disasters include local place names. Five times BF tweets of large-scale disasters include local place names. From the results, we conclude that hypothesis 1 is "True."

Hypothesis 2: Many Large-scale BF tweets include user's experiment.

We consider that everybody knows a large-scale disaster causes severe damage, and the victims have a great sense of crisis. People want to tell their experience of similar disaster situations and advice the victims. We assume that BF tweets of large-scale disasters are based on users' experiences. The results are that 6.8% of BF tweets of large-scale disasters include users' experience. On the other hand, 0.8% of BF tweets of normal disaster include users' experience. About 8% times BF tweets of large-scale disasters include local users' experience. From the results, we conclude that hypothesis 2 is "True."

Hypothesis 3: BF tweets of large-scale disasters include a firm tone.
We consider that people that are both victims and non-victims have a great sense of crisis. People want to pay attention to the BF-Tweets, and they become strong tone. We assume that BF tweets of large-scale disasters include a firm tone. The results are that 42.0% of BF tweets of large-scale disasters include a firm tone. On the other hand, 49.8% of BF tweets of normal disasters include a firm tone. From the results, hypothesis 3 is "False."

We found the characteristics of BF tweets of large-scale disasters include more local place names and users' experiments from our experiments. The tone of BF tweets is not so different from BF tweets of normal disasters. Then we should consider these two points to extract BF tweets of a large-scale disaster.

6 Conclusion

We extract the behavioral facilitation tweets for large-scale disaster data using LSTM, BiLSTM, and BERT. From experiment 1., we found that BERT is the best classifier for extracting BF tweets. Therefore, we analyzed the results of the extracted tweets by BERT.

This research aims to discover the differences between the BF tweets of large-scale disasters and normal-scale disasters. We formulated three hypotheses 1) The number of place names increases in BF tweets of a large-scale disaster, 2) BF tweets of large-scale disaster based on user's experience, and 3) BF tweets of large-scale disaster include strong tone. In our experiment, we discovered whether these hypotheses are true or false, respectively. We found that hypotheses 1) and 2) are true from our experimental result, and 3) are false.

Fake detection is a future work. In this paper, we extract BF tweets. We found that several BF tweets are fake, and we should alert these fake BF tweets to the users. There is much research on fake news detection. Therefore, we should integrate these methods with our proposed method and find which tweets should be alerted.

Acknowledgements. This work was partly supported by Research Institute of Konan University, and by JSPS KAKENHI Great Number 19H04218, 19H04221, and 20K12085.

References

1. Hochreiter, S., Schmidhuber, J.: Long short-term memory. Neural Comput. **9**(8), 1735–1780 (1997)
2. Miyabe, M., Nadamoto, A., Aramaki., E.: How do rumors spread during a crisis?: analysis of rumor expansion and dis-affirmation on twitter after 3.11 in Japan. Int. J. Web Inf. Syst. **10**, 394–412 (2014)
3. Mizuka, K., Suzuki, Y., Nadamoto, A.: Extraction of commentary tweets about news articles. In: Proceedings of the 19th International Conference on Information Integration and Web-based Applications and Services, pp. 188–192 (2017)
4. Mogadala, A., Varma, V.: Twitter user behavior understanding with mood transition prediction. In: Proceedings of the 2012 Workshop on Data-Driven User Behavioral Modelling and Mining from Social Media (DUBMMSM 2012), pp. 31–34 (2012)
5. Schuster, M., Paliwal, K.K.: Bidirectional recurrent neural networks. IEEE Trans. Signal Process. **45**(11), 2673–2681 (1997)
6. Silva, A., Valiatinn, H., Guimarães, S., Jr., W.M.: From individual behavior to influence networks: a case study on twitter. In: Proceedings of the 17th Brazilian Symposium on Multimedia and the Web, p. 18 (2011)
7. Xu, Z., Zhang, Y., Wu, Y., Yang, Q.: Modeling user posting behavior on social media. In: Proceedings of the 35th international ACM SIGIR Conference on Research and Development in Information Retrieval (SIGIR 2012), pp. 545–554 (2012)
8. Yamaoto, S., Wakabayashi, K., Satoh, T.Y., Nozaki, N.K.: Twitter user growth analysis based on diversities in posting activities. Int. J. Web Inf. Syst. **13**, 370–386 (2017)
9. Yoneda, Y., Suzuki, Y., Nadamoto, A.: Detection of behavioral facilitation information in disaster situation. In: Proceedings of the 21st International Conference on Information Integration and Web-based Applications and Services, pp. 255–259 (2019)

Using Cross Lingual Learning
for Detecting Hate Speech in Portuguese

Anderson Almeida Firmino[1]([✉])[iD], Cláudio Souza de Baptista[1][iD],
and Anselmo Cardoso de Paiva[2]

[1] Information Systems Laboratory, Federal University of Campina Grande,
Campina Grande, Paraiba, Brazil
andersonalmeida@copin.ufcg.edu.br, baptista@computacao.ufcg.edu.br
[2] Applied Computing Center, Federal University of Maranhão, Maranhão, Brazil
paiva@nca.ufma.br

Abstract. Social media growth all around the world brought benefits
and challenges to society. One problem that must be highlighted is hate
speech proliferation on the Internet. This article proposes a technique
to hate speech detection in texts, which uses a Cross-Lingual Learning
classifier. In our experiments, we used a public dataset in Portuguese
and achieved one of the greatest F1 Scores within our state-of-the-art.
Besides, this work is the first one to perform a cross-lingual learning task
for hate speech detection using a corpus in Portuguese.

Keywords: Hate speech detection · Cross-Lingual Learning · Deep
learning · Social media

1 Introduction

Mobile technology growth has impacted social media. According to a recent
survey[1], people prefer to use their smartphones and social media for news consumption instead of printed newspapers and television. People usually prefer
platforms such as Facebook and Twitter for gathering information and news.

Besides the benefits and convenience that social media has provided, the
anonymity provided by such means may be harmful to society, keeping in mind
that people tend to have a more aggressive behavior while using their social
networks [1]. An example of this is the growing proliferation of hate speech
on the Internet. Fortuna and Nunes [1] define hate speech as a language that
attacks and incites violence against certain groups of people based on their specific characteristics such as physical appearance, religion, lineage, nationality or
ethnic origin and gender [2].

---

[1] https://www.journalism.org/2021/01/12/news-use-across-social-media-platforms-
in-2020.

This research was partially funded by Brazilian Coordination for the Improvement of
Higher Education Personnel (CAPES) and the Brazilian Research and Development
Council (CNPq).

© Springer Nature Switzerland AG 2021
C. Strauss et al. (Eds.): DEXA 2021, LNCS 12924, pp. 170–175, 2021.
https://doi.org/10.1007/978-3-030-86475-0_17

As discussed by Pikuliak et al. [3], most languages do not have enough data available to create state-of-the-art models. Therefore, the ability to create intelligent systems for these languages is restricted. The importance of Natural Language Processing (NLP) tasks for languages with fewer resources has emerged recently during various crises in regions of the World where people speak languages that are not commonly dealt with in the NLP community, such as the Ebola outbreaks in West Africa (e.g. Niger-Congo languages).

This work proposes to perform hate speech detection in texts using a multilingual approach. The languages used in this work are Italian and Portuguese since both originate from the same mother language - Latin. Pelle and Moreira [4] published the Portuguese dataset we used, and the Italian dataset was made available in Evalita 2018, in the task Hate Speech Detection - Bosco et al. [5].

The contributions of this work lie because this is the first work to use CLL with hate speech data in Portuguese, besides having achieved one of the best results in the literature with the OffComBr2 database - provided by Pelle and Moreira [4].

The remainder of this paper is structured as follows. Section 2 focuses on reviewing briefly the state-of-the-art on hate speech detection in texts using CLL. Section 3 addresses the methodology and the dataset used by this research work. Section 4 presents the results and evaluation method. Finally, Sect. 5 highlights the conclusion and further work to be undertaken.

2   Related Work

Silva et al. [6] developed a novel approach to detect hate speech in Portuguese, which comprises a CNN model and a psycho-linguistic dictionary, the Linguistic Inquiry and Word Count (LIWC), with Logistic Regression (LR+LIWC). They used three Brazilian datasets; OffComBr2 and OffComBr3 [4] and HSD [7], and compared the baseline results with theirs. Using a CNN along with a 300-dimensional word embedding achieved the best F1 score.

The idea of Pagmunkas and Pati [9] was to develop a multilingual hate speech detection approach, in which the concept of transfer learning from a more resource-rich language to another with fewer resources is used. The languages studied were English, Spanish, Italian, and German. The best results were achieved using Hurtlex and multilingual embeddings as features and an LSTM architecture, and the best approach used Joint Learning and Multilingual Embeddings.

Ranasinghe and Zampieri [10] used multilingual word embeddings to detect hate speech. Besides carrying out experiments with different languages, different domains were also tested. Data were obtained in English, Spanish, Hindi, and Bengali. The XLM-R framework was used to perform the classification. The idea of using a multilingual approach is to train the model in a richer language and test it in another language with fewer resources. Ranasinghe and Zampieri trained the model in English and tested it in the other three languages researched. The results showed went beyond the state-of-the art of each dataset and language.

3 Methodology

This section describes the Portuguese and Italian datasets used. Then, we show the methodology used and how we apply cross-lingual learning to detect hate speech in Portuguese texts.

We used two datasets in this research: one comprising data in the Portuguese language (OffComBr-2 - [4]), with comments containing hate speech collected from the Brazilian news site g1.globo.com. Pelle and Moreira [4] gathered comments from pages on politics and sports. They selected a sample of 1,250 random comments and two judges noted each comment.

The other dataset comprises Facebook posts in Italian, made publicly available at the Evalita 2018 conference [5], in the Hate Speech Detection task. The dataset was developed by a research group in Pisa, created in 2016 [11], and contains about 17,000 Facebook comments, extracted from 99 posts from selected pages.

The main idea of using Cross Lingual Learning is to use one language with more resources to train a model and then use this model in a language with fewer resources. So, we chose Italian as the source language (the language with more resources) and Portuguese as the target language. We used the XLM-RoBERTa framework - called XLM-R [12] because it has presented good results in Cross-Lingual Learning tasks, achieving an accuracy 23% higher than that of BERT in using low-resource languages [13].

According to Pikuliak et al. [3], in Cross-Lingual Learning tasks, the transfer learning from the language with the most resources to the one with the least resources can be done in three ways: Zero Shot Transfer, Joint Learning and Cascade Learning. The first occurs when no data from the target language is used in the training. The joint learning approach consists of using both languages at the same time in the training; and cascade learning occurs when there is a pre-training with the source language and then the model is fine tuned with the target language.

Thus, we have conducted experiments following these three approaches. In all cases, the source language was Italian and the target language was Portuguese. We used the base and the large versions of XLM-R. The base version contains approximately 270M parameters, with 12 layers, 768 hidden states, 3072 feed-forward hidden states, and 8-heads; and the large version contains 550M parameters, with 24 layers, 1024 hidden states, 4096 feed-forward hidden-state and 16-heads [12].

We have performed some experiments initially using the first two approaches listed above (zero-shot transfer and joint learning). In these experiments, we only trained the XLM-R model using the Italian data (adding some Portuguese data sometimes) and we tested it with the Portuguese data. In the second round of experiments, we performed a fine-tuning on the model. We trained it with Italian data; after that, we did another training - this time with Portuguese data - and then we tested it with Portuguese data. This process is shown in Fig. 1.

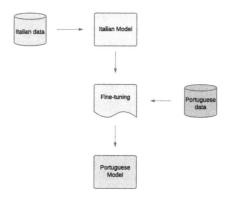

Fig. 1. The fine-tuning process on cross-lingual learning.

4 Evaluation and Results

This section presents the experiments used to evaluate our proposed model to identify hate speech in Portuguese in tweets. Moreover, we discuss the results of the performed experiments.

In our experiments, we used a Nvidia Tesla K80 GPU to train the models used. We obtained $1e^{-5}$ as the best value for the learning rate and we adopted 3 as the best value for the number of epochs. First of all, we have performed some experiments using the zero-shot transfer and joint learning approaches.

In these experiments, we performed the training with the Italian dataset with 90% used for training and the 10% remaining used for validation. In the test step, we used the Portuguese dataset (zero-shot) or splitted the Portuguese dataset in two parts (Table 1 shows the amount of Portuguese data used in training; the remaining data was used for testing). The experiment results are displayed in Table 1. As we can see, the results from the large model are greater than the base one. Also, using the joint learning approach, when we increased the amount of target language in the training process, the F1 score was increased. But we observed if we added data from the target language above a threshold, the F1 score decreased. This threshold was about 70%, and this result is listed in Table 1.

We also conducted experiments using the cascade learning approach. Here, we also trained the model with the Italian dataset, with 90% used for training and the 10% remaining for validation. In zero-shot experiments, we used 70% of data for training, 10% for validation, and 20% for test. In joint learning experiments, we also set apart some Portuguese data for the training with Italian data (as we can see in Table 1).

Table 1. Experiment results.

Model	Approach	Fine-Tuning	F1 score
XLM-R (base)	zero-shot	No	58%
XLM-R (base)	joint learning (20%)	No	69%
XLM-R (base)	joint learning (40%)	No	73%
XLM-R (large)	zero-shot	No	71%
XLM-R (large)	joint learning (20%)	No	75%
XLM-R (large)	joint learning (40%)	No	77%
XLM-R (large)	joint learning (70%)	No	74%
XLM-R (large)	zero-shot	Yes	80%
XLM-R (large)	joint learning (30%)	Yes	86%
XLM-R (large)	joint learning (50%)	Yes	74%

We trained the XLM-R with the Italian dataset and then fine-tuned it with the Portuguese dataset. We also used the zero-shot and joint learning within this approach (adding or no data from Portuguese in the training step). The experiments are listed in Table 1. We notice that when we used the joint learning approach; we obtained the best result in our experiments. But we observed we cannot add too much data from Portuguese in the training step. The F1 score decreased sharply from 86% to 74% - when the amount of Portuguese data was increased from 30% to 50% in the training step.

Silva et al. [6], Lima and Bianco [14], and Soto et al. [8] also used the Off-ComBr2 database [4]. In Table 2, a comparison with the F1-Score of these works is displayed. It is worth noticing that our work is the only one that used cross-lingual learning and it achieved one of the best results using the OffComBr2 database.

Table 2. Comparison of related work results.

Work	F1 score
Baseline (Pelle e Moreira, [4])	77%
Silva et al. [6]	89%
Lima and Bianco [14]	72%
Soto et al. [8]	86%
Our approach	86%

5   Final Remarks

In this paper, we presented an approach for hate speech detection in texts using a cross-lingual learning approach. Among other works that used the same dataset, we had one of the best F1 scores so far.

As further work, we point to the usage of other languages to perform the training in the model (such as English) to verify an improvement when classifying the texts. Besides, we suggest the use of more iterations on the fine-tuning step.

References

1. Fortuna, P., Nunes, S.: A survey on automatic detection of hate speech in text. ACM Comput. Surv. (CSUR) **51**(4), 1–30 (2018)
2. Bourgonje, P., Moreno-Schneider, J., Srivastava, A., Rehm, G.: Automatic classification of abusive language and personal attacks in various forms of online communication. In: Rehm, G., Declerck, T. (eds.) GSCL 2017. LNCS (LNAI), vol. 10713, pp. 180–191. Springer, Cham (2018). https://doi.org/10.1007/978-3-319-73706-5_15
3. Pikuliak, M., Šimko, M., Bielikova, M.: Cross-lingual learning for text processing: a survey. Expert Syst. Appl. **165**, 113765 (2021)
4. Pelle, R.P., Moreira, V.P.: Offensive comments in the Brazilian Web: a dataset and baseline results. In: Anais do VI Brazilian Workshop on Social Network Analysis and Mining. SBC, July 2017
5. Bosco, C., Felice, D. O., Poletto, F., Sanguinetti, M., Maurizio, T.: Overview of the EVALITA 2018 hate speech detection task. In: EVALITA 2018-Sixth Evaluation Campaign of Natural Language Processing and Speech Tools for Italian, vol. 2263, pp. 1–9. CEUR (2018)
6. Silva, S.C., Serapião, A.B., Paraboni, I.: Hate-speech detection in Portuguese using CNN and psycho-linguistic dictionary. J. Inf. Data Manage. **5**, 1–12 (2019)
7. Fortuna, P.C.T.: Automatic detection of hate speech in text: an overview of the topic and dataset annotation with hierarchical classes (2017)
8. Soto, C., Nunes, G., Gomes, J.: Avaliação de técnicas de word embedding na tarefa de detecção de discurso de ódio. In: Anais do XVI Encontro Nacional de Inteligência Artificial e Computacional (pp. 1020–1031). SBC, October 2019
9. Pamungkas, E.W., Patti, V.: Cross-domain and cross-lingual abusive language detection: a hybrid approach with deep learning and a multilingual lexicon. In: Proceedings of the 57th Annual Meeting of the Association For Computational Linguistics: Student Research Workshop, pp. 363–370, July 2019
10. Ranasinghe, T., Zampieri, M.: Multilingual offensive language identification with cross-lingual embeddings. arXiv preprint arXiv:2010.05324 (2020)
11. Del Vigna, F., Cimino, A., Dell'Orletta, F., Petrocchi, M., Tesconi, M.: Hate me, hate me not: Hate speech detection on facebook. In Proceedings of the First Italian Conference on Cybersecurity (ITASEC17), pp. 86–95 (2017)
12. Conneau, A., et al.: Unsupervised cross-lingual representation learning at scale. arXiv preprint arXiv:1911.02116 (2019)
13. Devlin, J., Chang, M. W., Lee, K., Toutanova, K.: BERT: pre-training of deep bidirectional transformers for language understanding. arXiv preprint arXiv:1810.04805
14. Lima, C., Dal Bianco, G.: Extração de característica para identificação de discurso de ódio em documentos. In: Anais da XV Escola Regional de Banco de Dados, pp. 61–70. SBC, April 2019

MMEnsemble: Imbalanced Classification Framework Using Metric Learning and Multi-sampling Ratio Ensemble

Takahiro Komamizu[✉]

Nagoya University, Nagoya, Japan
taka-coma@acm.org

Abstract. In classification, class imbalance is a factor that degrades the classification performance of many classification methods. Resampling is one widely accepted approach to the class imbalance; however, it still suffers from an insufficient data space, which also degrades performance. To overcome this, in this paper, an undersampling-based imbalanced classification framework, MMEnsemble, is proposed that incorporates metric learning into a multi-ratio undersampling-based ensemble. This framework also overcomes a problem with determining the appropriate sampling ratio in the multi-ratio ensemble method. It was evaluated by using 12 real-world datasets. It outperformed the state-of-the-art approaches of metric learning, undersampling, and oversampling in recall and ROC-AUC, and it performed comparably with them in terms of Gmean and F-measure metrics.

Keywords: Class imbalance · Ensemble · Metric learning · Muti-ratio undersampling

1 Introduction

Class imbalance [6] is a universal phenomenon for labeled datasets and causes classification performance to degrade. Class imbalance refers to the characteristics of datasets in which the number of instances in a class is much bigger than that in other classes. The large gap in the numbers tempts classifiers to be biased toward the majority class. This problem has been observed and dealt with in various domains, such as the clinical domain [2], and economic domain [12].

Resampling is an effective preprocessing for dealing with class imbalance. It draws instances in order to balance the numbers of instances in different classes. Resampling can be categorized into oversampling and undersampling. Oversampling generates instances of the minority class (e.g., SMOTE [3]), and undersampling eliminates instances of the majority class (e.g., EasyEnsemble [10]). Though resampling is a successful method, classifiers still suffer from insufficient data spaces, which is not changeable by resampling.

This paper proposes a classification framework, **MMEnsemble**, consisting of *metric learning*, *multi-ratio ensemble*, and *asset-based weighting*.

© Springer Nature Switzerland AG 2021
C. Strauss et al. (Eds.): DEXA 2021, LNCS 12924, pp. 176–188, 2021.
https://doi.org/10.1007/978-3-030-86475-0_18

- **Metric Learning**: Metric learning methods such as LMNN [19] learn a data transformation so that instances in different classes can be distinguishable. Recent metric learning approaches [18] have shown that selecting subsets of training instances for metric learning improves the classification performance of an imbalanced classification. On the basis of this idea, in MMEnsemble, metric learning is incorporated into an undersampling-based ensemble.
- **Multi-ratio Ensemble**: When applying undersampling, the sampling ratio is an important parameter. It determines the number of drawn majority instances. A recent study [8] has shown that incorporating multiple sampling ratios in an ensemble manner improves the classification performance.
- **Asset-based Weighting**: When applying a multi-ratio undersampling-based ensemble, weak classifiers for different sampling ratios have different assets. A classifier with a large sampling ratio may correctly classify the majority class, and another classifier with a small sampling ratio may correctly classify the minority class. To capture the assets, MMEnsemble introduces a weighting scheme that weighs on classifiers that can correctly classify instances that are hard for the other classifiers to classify.

The contributions of this paper are summarized as follows.

- **MMEnsemble – a novel framework**: MMEnsemble is a framework composed of metric learning, multi-ratio undersampling-based ensemble, and asset-based weighting. This framework overcomes the weakness of metric learning regarding the class imbalance by applying undersampling beforehand, and it releases users from the burden of choosing sampling ratios in undersampling by using ensemble of base classifiers in various sampling ratios and automatic weighting schemes.
- **Superior Classification Performance**: In an experiment using 12 real-world datasets, MMEnsemble outperforms the state-of-the-art approaches, especially for recall and ROC-AUC metrics, and it performs comparably to them on Gmean and F-measure metrics. This experiment indicates that this approach can achieve higher recall scores, which would be useful for many real-world applications.

The rest of this paper is organized as follows. Section 2 introduces the related work on resampling-based approaches. Section 3 explains MMEnsemble in detail, and Sect. 4 shows the experimental evaluation using 12 real-world datasets. Finally, Sect. 5 concludes this paper.

2   Related Work: Resampling Approaches

To deal with class imbalance, there are basically three groups of approaches, namely, resampling, cost-adjustment [4], and algorithm modification [17]. Resampling is commonly used because it has shown robust performance and applicability to any classifiers. Resampling approaches can be roughly classified into two categories, namely, oversampling and undersampling.

2.1 Oversampling

A simple oversampling approach is to randomly copy minority instances so that the numbers of minority and majority instances become the same. This approach easily causes overfitting. To cope with the overfitting problem, oversampling approaches generate synthetic minority instances that are close to the minority. SMOTE [3] is the most popular synthetic oversampling method. It generates synthetic minority instances on the basis of the nearest neighbor technique. Since SMOTE does not take majority instances into consideration, the generated instances can easily overlap with majority instances. This degrades the classification performance. To overcome this weakness, various extended approaches have been proposed. Comprehensive experiments by [9] were conducted to investigate a large number of SMOTE variants and compares these variants with diverse kinds of datasets.

2.2 Undersampling

Undersampling-based approaches can be classified into two categories: instance selection and ensemble. Instance selection is to choose majority instances that are expected to contribute to better classification. Major approaches choose majority instances hardly distinguishable from minority instances. NearMiss [11] samples majority instances close to the minority instances. Instance hardness [15] is a property that indicates the likelihood that an instance will be misclassified.

The ensemble approach is to combine multiple weak classifiers each of which is learned on individual pieces of undersampled training data. A comprehensive experiment on ensemble approaches can be found in [5], Basically, there are two types of ensemble approaches.

- **Boosting:** BalanceCascade [10] is an iterative method that removes correctly classified majority instances. RUSBoost [14] is a weighted random undersampling approach for removing majority instances that are likely to be classified correctly.
- **Bagging:** Ensemble of Undersampling [7] is one of the earlier ensemble methods using undersampled training data. EasyEnsemble [10] is an ensemble-of-ensemble approach that ensembles AdaBoost classifiers [13] for each piece of undersampled training data. MUEnsemble [8] extends EasyEnsemble to incorporate multiple sampling ratios. DDAE [20] is the state-of-the-art bagging-based approach composed of metric learning and cost-sensitive learning.

The proposed method, MMEnsemble, is classified in the bagging category. A major distinction of MMEnsemble from the others (except DDAE) is that it incorporates metric learning to overcome the issue of insufficient data spaces in resampling methods. There are two major differences between DDAE and MMEnsemble. One is the control of undersampling (called *data block*); MMEnsemble undersamples data with respect to the sampling ratio, while DDAE undersamples on the basis of the number of blocks, which is not dependent on the imbalance ratio of datasets. The other is the choice of weak

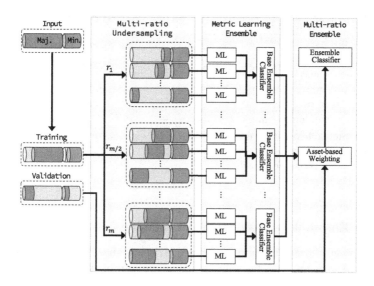

Fig. 1. MMEnsemble – proposed framework

classifier; DDAE uses the nearest neighbor classifier, which is considered to fit metric learning, while MMEnsemble uses the AdaBoost classifier. Additionally, but importantly, DDAE has (at least) three hyper-parameters that need to be tuned, while MMEnsemble has only one, which is a much smaller parameter space.

In terms of the multi-ratio ensemble, MMEnsemble uses assets of weak classifiers that are obtained from the process of validating weak classifiers, while MUEnsemble uses a heuristic weighting (i.e., Gaussian function-based weighting). Therefore, to capture the characteristics of weak classifiers, a comprehensive hyper-parameter tuning is required. The experimental evaluation in this paper (Sect. 4) shows the superiority of the asset-based weighting scheme over the heuristic weighting in MUEnsemble.

3   MMEnsemble

Figure 1 shows an overview of MMEnsemble, which consists of three phases: the multi-ratio undersampling phase, metric learning phase, and multi-ratio ensemble phase. The first phase is imitated from MUEnsemble [8], that is, for each sampling ratio $r_i \in R$ (R is a predefined set of sampling ratios), multiple undersampled sets of instances with r_i are drawn from the training data. In the second phase, for each drawn set, metric learning is performed and a base ensemble classifier, called *MLEnsemble*, is trained using this drawn set that is transformed by the metric learning. In the last phase, given $|R|$ base classifiers from the previous phase, the final ensemble classifier is constructed by the asset-based weighting. In the following sections, the technical details of MLEnsemble and the ensemble with the asset-based weighting are introduced.

Algorithm 1. MLEnsemble

Input: Training data $D^{(train)} = (D_{maj}, D_{min})$, sampling ratio r, the number of weak
    classifiers n
Output: Base   ensemble   classifier   with   metric   learner C_r $=$
    $((c_1, m_1), (c_2, m_2), \ldots, (c_n, m_n))$
1: **for** $i = 1$ to n **do**
2:      $D'_{maj} \leftarrow$ Randomly sample D_{maj} s.t. $\frac{|D'_{maj}|}{|D_{min}|} = r$
3:      Train metric learner m_i using (D'_{maj}, D_{min})
4:      $D' \leftarrow$ Transform (D'_{maj}, D_{min}) using m_i
5:      Train weak classifier c_i using D'
6: **end for**

3.1 Base Ensemble Classifier – MLEnsemble

MLEnsemble is a bagging classifier with metric learning. Its procedure is summarized in Algorithm 1. The training data are sampled multiple times with replacement to obtain particular sets of instances (Line 2). For each set, a metric learner is trained by using the set so that it transforms the set into a sufficient data space for distinguishing instances of different classes (Lines 3–4). Using the transformed set, a weak classifier is trained (Line 5).

3.2 Ensemble Using Asset-Based Weighting

Typical ensemble methods use the weighted voting strategy. These methods often use equal weights for all base classifiers, and they are not aware of class imbalance. In contrast, for the case of an ensemble of base classifiers in different sampling ratios, the weights of base classifiers are more sensitive, and thus need to be carefully designed. [8] showed that a heuristic weighting using a Gaussian function is superior to the equal weighting. The Gaussian-based weighting is calculated as follows.

$$W_{gauss}(r) = \frac{1}{\sum_{r \in R} W_{gauss}(r)} \cdot \exp\left(-\frac{(r - \mu)^2}{2\sigma^2}\right), \tag{1}$$

where μ and σ^2 are tunable parameters. When $\mu = 1.0$, most of the weight is on the base classifier trained using the balanced data, and the weights gradually decrease as r increases and decrease from μ.

    The heuristic weighting approach does not take the classification performances of base classifiers into consideration. There are typically some instances that can be correctly classified by only a few base classifiers. To improve the classification performance with the ensemble mechanism, base classifiers classifying such instances correctly are important. Also, these base classifiers are expected to not incorrectly classify instances that are correctly classified by the other base classifiers. In this paper, this is called an *asset* of a base classifier. Formally, given set $C = \{C_j\}_{j=1}^s$ of base classifiers with size s and validation set $D^{(val)}$, for each instance $(d_i, \ell_i) \in D^{(val)}$, where d_i is a feature vector, and ℓ_i is a class label of the i-th instance, the number T_i of base classifiers that correctly classify the i-th instance is obtained. That is, $T_i = |\{C_j | C_j \in C, C_j.predict(d_i) = \ell_i\}|$. This number indicates how hard (or easy) an instance is to classify. The intuition behind using this number for weighting base classifiers is that the lower the number, the more weights on a base classifier if it correctly classifies the instance. This intuition is formalized by the following formula.

$$W_{asset}(r) = \frac{1}{\sum_{r \in R} W_{asset}(r)} \cdot \sum_{(d_i, \ell_i) \in D^{(val)}} \delta\left(C_r.predict(d_i), \ell_i\right) \cdot T_i^{-k}, \quad (2)$$

where k is a tunable parameter for emphasizing the importance of the classifiers that correctly classify instances that other classifiers cannot, δ function is the Kronecker delta (i.e., 1 if the two arguments are equal, 0 otherwise).

4   Experimental Evaluation

In this experiment, MMEnsemble was evaluated to answer the questions below.

Q1 Does MMEnsemble outperform the state-of-the-art imbalanced classification methods of metric learning, oversampling and undersampling?

Q2 Is the combination of metric learning and multi-ratio ensemble effective?

Q3 Does the asset-based weighting help improve the classification performance? and what is the effect of choice of its hyper-parameter k (Eq. 2)?

4.1   Settings

Datasets: The datasets for the experiment were obtained from the OpenML dataset [16] and KEEL repository [1]. Table 1 shows the total number of records (#records), the number of minority instances (#minor), dimensionality (#dim), and the imbalance ratio (IR), which is $\frac{\#major}{\#minor}$. D1-D6 were obtained from the OpenML dataset, and the rest were obtained from the KEEL repository.

Evaluation: The evaluation metrics were *Recall, Gmean, F_2,* and *AUC*. Let *TP, FN, TN,* and *FP* be the true positives, false negatives, true negatives, and false positives. $Recall = \frac{TP}{TP+FN}$ measures how many positive (minority) instances are correctly classified. $Gmean = \sqrt{Recall \cdot TNR}$ is the geometric mean of the recalls of both classes, where $TNR = \frac{TN}{TN+FP}$. $F_\beta = \frac{(1+\beta^2) Recall \cdot Precision}{Recall + \beta^2 Precision}$ is the harmonic mean of the recall and precision, where $Precision = \frac{TP}{TP+FP}$, and β determines the weight on the recall. In this experiment, β was set to 2 because the higher recalls are preferred in many real-world applications. *AUC* is the area under the receiver operation characteristic curve.

To accurately estimate these evaluation metric values, the experimental process was repeated 50 times. In the process, a dataset was randomly separated into 70% for training and 30% for testing, and the classifiers were trained on the training set and evaluated using the test set. The overall metric scores were the macro average of the 50 trials.

Table 1. Datasets

ID	Name	#records	#minor	#dim	IR
D1	cm1	498	49	21	9.2
D2	kc3	458	43	39	9.7
D3	mw1	403	31	37	12.0
D4	pc1	1,109	77	21	13.4
D5	pc3	1,563	160	37	8.8
D6	pc4	1,458	178	37	7.2
D7	yeast1-7	459	30	7	14.3
D8	abalone9-18	731	42	8	16.4
D9	yeast6	1,484	35	8	41.4
D10	abalone19	4,174	32	8	129.4
D11	wine3-5	691	10	11	68.1
D12	abalone20	1,916	26	8	72.7

Baseline Methods: MMEnsemble was compared with the state-of-the-art methods of metric learning, and the ensemble approach. **IML** [18] is a state-of-the-art approach of metric learning and copes with the class imbalance. IML incorporates LMNN [19] and iteratively selects training samples to improve the data transformation. For the undersampling and ensemble method, **DDAE** [20] is the state-of-the-art and also includes metric learning. Since the experiment setting is the same as [20], the results of IML and DDAE were copied from it.

To answer Q2, MMEnsemble was compared with **EasyEnsemble** [10], **MUEnsemble** [8], and MLEnsemble (in this paper), which are an undersampling-based ensemble, a multi-ratio undersampling-based ensemble, and a metric learning incorporating EasyEnsemble, respectively. The difference between EasyEnsemble and MMEnsemble shows the benefit of integrating both the metric learning and the multi-ratio ensemble. Similarly, the difference between MMEnsemble and MUEnsemble shows the benefit of metric learning to improve the performance for the imbalanced classification.

Parameters: The parameters of EasyEnsemble, MLEnsemble, and MUEnsemble were set as follows. The sampling ratio in EasyEnsemble and MLEnsemble was set to 1.0. The metric learning method was LMNN with the k parameter of kNN set to 3. For MUEnsemble, the predefined set R of sampling ratios is set to $\{0.2, 0.4, \ldots, 2.0\}$, and Gaussian weighting was used with parameters, μ and σ^2, of 1.0 and 0.2, where μ was fixed to 1.0 have the parameter be the same as the former methods, and the best σ^2 was experimentally explored from $\{0.1, 0.2, \ldots, 1.0\}$. For MMEnsemble, the base classifier, MLEnsemble, was set the same as above, R is the same as MUEnsemble, and k of the asset-based weighting was chosen from $\{0.1, 0.2, \ldots, 5.0\}$.

Table 2. Comparison with State-of-the-art Methods of Metric Learning (IML), and Ensemble (DDAE) – [†]means that scores were copied from [20].

Data	IML[†]				DDAE[†]				MMEnsemble			
	Rec	Gm	F_2	AUC	Rec	Gm	F_2	AUC	Rec	Gm	F_2	AUC
D1	.313	.520	.287	.589	.813	**.775**	**.580**	.776	**.863**	.756	.546	**.819**
D2	.692	.805	**.652**	.814	.846	**.823**	.625	.823	**.952**	.750	.534	**.868**
D3	.500	.635	.345	.653	.750	**.815**	**.588**	.817	**.793**	.772	.528	**.866**
D4	.852	.657	.408	.679	**.963**	**.819**	**.573**	.830	.944	**.819**	.548	**.895**
D5	.510	.578	.342	.582	.735	.743	.536	.744	**.867**	**.794**	**.598**	**.854**
D6	.814	.725	.574	.730	.932	.804	.676	.813	**.963**	**.873**	**.748**	**.934**
D7	.667	.716	.471	.718	.833	**.841**	**.649**	.841	**.933**	.808	.512	**.883**
D8	.600	.709	.375	.719	.700	.814	.603	.824	**.886**	**.877**	**.650**	**.941**
D9	.700	.798	.407	.805	.900	.883	.421	.883	**.931**	**.920**	**.585**	**.976**
D10	.667	.626	.037	.628	**1.000**	**.839**	.075	.852	.935	.835	**.128**	**.876**
D11	.000	.000	NA	.500	.333	.550	.156	.620	**.894**	**.842**	**.188**	**.939**
D12	.800	.802	.252	.802	**1.000**	**.964**	**.556**	.965	.992	.943	.451	**.982**

4.2 Results

To answer the questions, the experimental results are shown from three perspectives: an overall comparison (corr. Q1), the ablation study (corr. Q2), and a comparison over the k parameter and other weighting schemes (corr. Q3).

Overall Comparison: Table 2 showcases the metric scores of MMEnsemble with the state-of-the-art methods. In the table, the highest scores in a row are boldfaced. MMEnsemble totally outperformed IML and ProWSyn, and it outperformed DDAE in recall and AUC, and it was comparable with DDAE in terms of Gmean and F_2 metrics. It is noteworthy that MMEnsemble totally outperformed the others on the AUC metric, and it achieved almost the best performance on the recall metric. For the real-world applications, a high recall is preferable; therefore, this superiority of MMEnsemble is practically useful. On the contrary, the Gmean and F_2 scores were comparable with DDAE. On datasets, D5, D6, D8, D9, and D11, MMEnsemble clearly outperformed DDAE, however, on the other datasets, MMEnsemble was inferior to DDAE or comparable. This was caused by the low TNR and precision scores for MMEnsemble, coming from the weighting scheme design (i.e., the asset-based weighting). Asset-based weighting is designed to emphasize the base classifiers that correctly classify instances that others cannot. This increases the chance of increasing the number of false positives.

Impact of the Combination: Table 3 shows a comparison of MMEnsemble and its basic approaches. The comparisons of EasyEnsemble to MLEnsemble and MUEnsemble show that the classification performance could be slightly increased

Table 3. Ablation Study – EE, ML, and MR stand for EasyEnsemble, metric learning, and multi-ratio ensemble, respectively.

Data	MLEnsemble (EE + ML)				MUEnsemble (EE + MR)				MMEnsemble (EE + ML + MR)			
	Rec	Gm	F_2	AUC	Rec	Gm	F_2	AUC	Rec	Gm	F_2	AUC
D1	.751	.695	.475	.754	.812	.698	**.484**	**.783**	**.820**	**.699**	.483	**.783**
D2	.854	**.742**	**.518**	.831	.821	.718	.490	.826	**.891**	.731	.509	**.862**
D3	.790	.720	.461	.817	.761	.700	.439	.820	**.864**	**.761**	**.506**	**.860**
D4	.875	.804	.533	.871	**.880**	.788	.509	.860	.873	**.816**	**.548**	**.885**
D5	.821	.760	.554	.821	.828	.753	.546	.828	**.844**	**.781**	**.581**	**.837**
D6	.921	.844	.707	.907	.946	**.883**	**.764**	**.934**	**.971**	.873	.747	.921
D7	.787	.746	**.444**	.830	.792	.743	.438	.818	**.860**	**.749**	**.444**	**.859**
D8	.835	.822	**.537**	.913	.769	.757	.440	.840	**.911**	**.835**	.531	**.959**
D9	**.893**	.874	.438	.951	.850	.857	.427	.935	.885	**.890**	**.508**	**.973**
D10	.835	.762	.101	.828	.911	.770	.096	.834	**.999**	**.828**	**.112**	**.887**
D11	.735	.697	.144	.797	**.785**	**.753**	**.178**	**.841**	.765	.724	.160	.795
D12	.882	.875	.330	.951	.870	.840	.248	.931	**.987**	**.923**	**.363**	**.985**

by adding either the metric learning or the multi-ratio ensemble. The architectural difference between MUEnsemble and MMEnsemble is whether metric learning is involved; therefore, to observe the performance improvement caused by the difference, MMEnsemble was incorporated with the Gaussian weighting (Eq. 1). On the basis of this comparison, MMEnsemble showed its superiority to MUEnsemble, that is, the metric learning successfully improved the data space in the sets of data for each sampling ratio. In addition, as it can be seen by comparing the columns of MMEnsemble in Table 2 and Table 3, MMEnsemble with the asset-based weighting was superior to that with the Gaussian weighting; therefore, MMEnsemble clearly outperformed MUEnsemble.

Effect of the Asset-Based Weighting: Figure 2 shows the effect of the hyper-parameter k on the asset-based weighting. A basic finding is that the recall scores dropped as k increased. This is because the higher the k, the more weights are given to the base classifiers that can correctly classify instances that are incorrectly classified by the other base classifiers. This leads to a higher TNR and precision; therefore, as k increases, the Gmean and F_2 scores increase, and similarly, AUC scores gradually increase.

Table 4 shows a comparison of the asset-based weighting with uniform weighting. Uniform weighting gave equal weights for all base classifiers. MMEnsemble with uniform weighting tended to achieve high recall scores, but low scores for the other metrics. This indicates that taking the average performance among the base classifiers trained using datasets of different sampling ratios increases the number of instances classified to the minority class.

Although the details are omitted due to space limitations, it is noteworthy that MLEnsemble with the asset-based weighting showed a similar classification per-

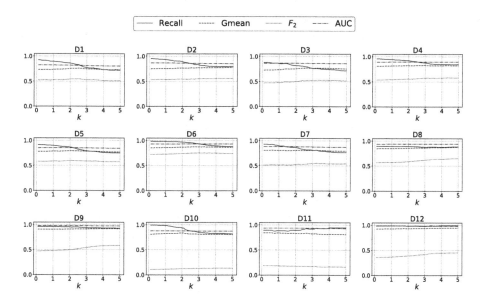

Fig. 2. Effect of k in Asset-based Weighting – As k increases, recall scores decrease, and the Gmean, F_2, and AUC scores increase.

formance to that with the uniform weighting. This is because the base classifiers in MLEnsemble are close in terms of classification tendency, that is, correctly classified instances are almost common among these classifiers. Thus, the weights on these classifiers calculated by Eq. 2 become similar values. This fact indicates that the asset-based weighting is effective for ensemble classifiers of which the classification tendencies of the base classifiers differ from each other. MMEnsemble is this kind of ensemble classifier, that is, base classifiers are trained for different sampling ratios; thus, the tendencies of these base classifiers differ from each other.

4.3   Lessons Learned

Q1: Does MMEnsemble outperform the state-of-the-art imbalanced classification methods of metric learning, oversampling and undersampling?—In terms of recall and AUC, MMEnsemble achieved the state-of-the-art, while MMEnsemble was comparable with DDAE in Gmean and F_2. This indicates that MMEnsemble can achieve a higher recall, but its performance in TNR and precision is limited. Though many real-world applications expect a higher recall, a high TNR and precision with high recall is ideal; therefore, improving MMEnsemble for these metrics without sacrificing high recall is a promising next direction.

Q2: Is the combination of metric learning and multi-ratio ensemble effective?—Yes, the combination contributes to improving the classification performance for all metrics. The comparison between MMEnsemble and MLEnsemble revealed that the multi-ratio ensemble improved the performance,

Table 4. Comparison between Uniform and Asset-based Weighting

Data	Uniform				Gauss				Asset			
	Rec	Gm	F_2	AUC	Rec	Gm	F_2	AUC	Rec	Gm	F_2	AUC
D1	**.893**	.637	.456	.781	.820	.699	.483	.783	.863	**.756**	**.546**	**.819**
D2	.950	.711	.502	.818	.891	.731	.509	.862	**.952**	**.750**	**.534**	**.868**
D3	.813	.692	.435	.815	**.864**	.761	.506	.860	.793	**.772**	**.528**	**.866**
D4	**.954**	.788	.505	.891	.873	.816	**.548**	.885	.944	**.819**	.548	**.895**
D5	**.923**	.748	.550	.840	.844	.781	.581	.837	.867	**.794**	**.598**	**.854**
D6	**.972**	.846	.710	.925	.971	**.873**	.747	.921	.963	**.873**	**.748**	**.934**
D7	.915	.742	.432	.882	.860	.749	.444	.859	**.933**	**.808**	**.512**	**.883**
D8	.900	.817	.509	.931	**.911**	.835	.531	**.959**	.886	**.877**	**.650**	.941
D9	.910	.872	.413	.954	.885	.890	.508	.973	**.931**	**.920**	**.585**	**.976**
D10	.924	.758	.091	.837	**.999**	.828	.112	**.887**	.935	**.835**	**.128**	.876
D11	.633	.666	.152	.810	.765	.724	.160	.795	**.894**	**.842**	**.188**	**.939**
D12	.873	.858	.303	.953	.987	.923	.363	**.985**	**.992**	**.943**	**.451**	.982

and that between MMEnsemble and MUEnsemble revealed that the metric learning improved the performance.

Q3: Does the asset-based weighting help improve the classification performance? and what is the effect of choice of its hyper-parameter k (Eq. 2)?—Asset-based weighting improved classification performance compared with the two weighting schemes (uniform and Gaussian) for all metrics; however, the hyper-parameter k must be carefully determined because it is sensitive to recall. As k increases, recall decreases, while the Gmean and F_2 increase. This indicates that a higher k improves classifiers in terms of the TNR and precision. Thus, k can be tuned in terms of users' preferences on recall or precision.

5   Conclusion

In this paper, a novel undersampling-based ensemble framework, MMEnsemble was proposed. MMEnsemble integrates three techniques, metric learning, multi-ratio ensemble, and asset-based weighting to overcome the insufficient data space issue in the previous undersampling-based ensemble approaches. An experimental evaluation revealed the superiority of MMEnsemble to the state-of-the-art methods, especially for the recall and AUC metrics. The major limitation of MMEnsemble (also in the other methods) is that it can achieve higher recall scores but sacrifices precision.

Acknowledgments. This work was partly supported by JSPS KAKENHI JP21H0355 and the Kayamori Foundation of Informational Science Advancement.

References

1. Alcalá-Fdez, J., Fernández, A., Luengo, J., Derrac, J., García, S.: KEEL datamining software tool: data set repository, integration of algorithms and experimental analysis framework. J. Multiple Valued Log. Soft Comput. **17**(2-3), 255–287 (2011)
2. Bhattacharya, S., Rajan, V., Shrivastava, H.: ICU mortality prediction: a classification algorithm for imbalanced datasets. In: AAAI, vol. 2017, pp. 1288–1294 (2017)
3. Chawla, N.V., Bowyer, K.W., Hall, L.O., Kegelmeyer, W.P.: SMOTE: synthetic minority over-sampling technique. J. Artif. Intell. Res. **16**, 321–357 (2002)
4. Elkan, C.: The foundations of cost-sensitive learning. IJCAI **2001**, 973–978 (2001)
5. Galar, M., Fernández, A., Tartas, E.B., Sola, H.B., Herrera, F.: A review on ensembles for the class imbalance problem: bagging-, boosting-, and hybrid-based approaches. IEEE Trans. Syst. Man Cybern. Part C **42**(4), 463–484 (2012)
6. He, H., Garcia, E.A.: Learning from imbalanced data. IEEE Trans. Knowl. Data Eng. **21**(9), 1263–1284 (2009)
7. Kang, P., Cho, S.: EUS SVMs: ensemble of under-sampled SVMs for data imbalance problems. In: King, I., Wang, J., Chan, L.-W., Wang, D.L. (eds.) ICONIP 2006. LNCS, vol. 4232, pp. 837–846. Springer, Heidelberg (2006). https://doi.org/10.1007/11893028_93
8. Komamizu, T., Uehara, R., Ogawa, Y., Toyama, K.: MUEnsemble: multi-ratio undersampling-based ensemble framework for imbalanced data. In: Hartmann, S., Küng, J., Kotsis, G., Tjoa, A.M., Khalil, I. (eds.) DEXA 2020. LNCS, vol. 12392, pp. 213–228. Springer, Cham (2020). https://doi.org/10.1007/978-3-030-59051-2_14
9. Kovács, G.: An empirical comparison and evaluation of minority oversampling techniques on a large number of imbalanced datasets. Appl. Soft Comput. **83**, 105662 (2019), (IF-2019 = 4.873)
10. Liu, X., Wu, J., Zhou, Z.: Exploratory undersampling for class-imbalance learning. IEEE Trans. Syst. Man Cybern. Part B **39**(2), 539–550 (2009)
11. Mani, I., Zhang, I.: kNN approach to unbalanced data distributions: a case study involving information extraction. In: ICML 2003 Workshop on Learning from Imbalanced Datasets, vol. 126 (2003)
12. Pozzolo, A.D., Caelen, O., Johnson, R.A., Bontempi, G.: Calibrating probability with undersampling for unbalanced classification. In: SSCI, vol. 2015, pp. 159–166 (2015)
13. Schapire, R.E.: A brief introduction to boosting. IJCAI **1999**, 1401–1406 (1999)
14. Seiffert, C., Khoshgoftaar, T.M., Hulse, J.V., Napolitano, A.: RUSBoost: a hybrid approach to alleviating class imbalance. IEEE Trans. Syst. Man Cybern. Part A **40**(1), 185–197 (2010)
15. Smith, M.R., Martinez, T., Giraud-Carrier, C.: An instance level analysis of data complexity. Mach. Learn. **95**(2), 225–256 (2013). https://doi.org/10.1007/s10994-013-5422-z
16. Vanschoren, J., van Rijn, J.N., Bischl, B., Torgo, L.: OpenML: networked science in machine learning. SIGKDD Explor. **15**(2), 49–60 (2013)
17. Wang, H., Gao, Y., Shi, Y., Wang, H.: A fast distributed classification algorithm for large-scale imbalanced data. In: ICDM, vol. 2016, pp. 1251–1256 (2016)
18. Wang, N., Zhao, X., Jiang, Y., Gao, Y.: Iterative metric learning for imbalance data classification. IJCAI **2018**, 2805–2811 (2018)

19. Weinberger, K.Q., Blitzer, J., Saul, L.K.: Distance metric learning for large margin nearest neighbor classification. In: NIPS, vol. 2005, pp. 1473–1480 (2005)
20. Yin, J., Gan, C., Zhao, K., Lin, X., Quan, Z., Wang, Z.: A novel model for imbalanced data classification. In: AAAI, vol. 2020, pp. 6680–6687 (2020)

Evaluate the Contribution of Multiple Participants in Federated Learning

Zhaoyang You⬛, Xinya Wu⬛, Kexuan Chen⬛, Xinyi Liu⬛,
and Chao Wu$^{(\boxtimes)}$⬛

Zhejiang University, Hangzhou, China
{zhaoyangyou,xinya.wu,kexuan.chen,xinyi.liu,chao.wu}@zju.edu.cn

Abstract. To address the challenge of distributed data source, Federated Learning meets with great demand of algorithmic predictions and decisions without taking a risk of privacy leakage, leaving data valuation a consequent task. Establishing an effective profit distribution model enables multiple participants to get involved in a fair incentive. Shapley Value serves as an excellent measure for calculating the contribution of the model since it provides a fair dividend of payoffs. However, it fails under the existence of data replication or dataset partition. In this work, we design a function to recalculate Shapley Value overcoming the failure mentioned before. The testing experiments have proved that new calculation improves the SV performance for about 50% compared with the original index such as the model accuracy or total loss.

Keywords: Data pricing · Federated learning · Shapley value

1 Introduction

Federated Averaging algorithm (5), asks participants to submit their models to get an aggregated one through training and being averaged by central node, allowing model parameters to be transferred alone. However, the validity of valuation in federated learning is questioned by specific attributes of data. In face of these challenges, this paper formulates a fair approach for data valuation in federated learning focusd on the model rather than dataset. It adopts Shapley Value to valuate the contribution of participants after applying a fair index, which has been proved to acquire some excellent properties.

Choosing a proper index to calculate Shapley Value warrants discussion. Accuracy and loss function are two classic methods to calculate Shapley Value (2), but these two methods neglect the circumstances mentioned before, which means double rewards will be paid if a participant replicates or divides the dataset and trains two models simultaneously to submit.

To build a relationship between a model and a unknown dataset, we propose a function to map the model to the expected number of i.i.d. data ensuring the same accuracy. It also provides us a good index to calculate Shapley Value which can solve the problems of replicating and dividing cases we mention above.

C. Strauss et al. (Eds.): DEXA 2021, LNCS 12924, pp. 189–194, 2021.
https://doi.org/10.1007/978-3-030-86475-0_19

2   Method

2.1   Shapley Value for Models

Supposing we have N models $\{M_i\}_{i=1}^{N}$ from multiple participants, Shapley Value for model M_i can be calculated as follows:

$$SV(M_j) = \frac{1}{N!} \sum_{\pi \in \prod(D)} U(\pi[1:i]) - U(\pi[1:i-1]) \tag{1}$$

where U denotes the utility function, i.e., the index to calculate the Shapley Value. $\pi \in \prod(M)$ is a permutation of models M, $\pi[1:i]$ is the first i-elements of the permutation. So $U(\pi[1:i])$ denotes the utility of the model aggregated from model M_1 to M_i.

Shapley Value serves as a good way to calculate the marginal contribution of each model submitted by participants. Furthermore, Shapley Value is equipped with some good properties:

1. (Symmetry) If $U(S \cup M_i) = U(S \cup M_j)$ for any $S \subseteq M - \{M_i, M_j\}$, then $SV(M_i) = SV(D_j)$.
2. (Dummy model) If $U(S \cup \{M_i\}) = U(S)$ for all $S \subseteq M - \{M_i\}$, then $SV(M_i) = 0$.
3. (Additivity) $SV(M_i, U_1) + SV(M_i, U_2) = SV(M_i, U_1 + U_2)$

For property 1, if a participant replicates the data and trains another model to submit, the new model will get exactly the same Shapley Value. For property 2, if a participant submits a useless model, its Shapley Value will be null. Property 1 and 2 show that Shapley Value is fair enough to conquer some dishonesty from several participants to some extent. For property 3, it means the Shapley Value can be used in multiple task valuation.

2.2   Invalid Shapley Value

Though Shapley Value has many good properties, there are still some problems left behind. (3) has shown that Shapley Value is inadequate for freely replicable goods. That is to say, if a participant replicates his data and trains a new model, the sum of Shapley Value of theses two models are not equal to the original one.

To specifically discuss an invalid case, let's suppose that we only have two participants who have two identical datasets A and B. The Shapely Value of A and B are both $\frac{1}{2}$. If the first participant replicates the dataset A and get A', the Shapely Value of A, A' and B all change into $\frac{1}{3}$ of the original. The total Shapley Value of the first and the second participants get to $\frac{2}{3}$ and $\frac{1}{3}$, which means the former participant can earn more profit.

2.3 Method

To solve the problems mentioned above, we first design a map function from the model to expected number of i.i.d data of the dataset. In the next section, we will prove some properties of the expected number of i.i.d data, which enables each data point in dataset has the same value and makes it possible to calculate the Shapley Value of the model aggregated by different smaller models. To simplify the analysis, we take the classification model into consideration.

Proposition: For a n_c classification model, N_e is the expected number of i.i.d. data required to make the model achieve accuracy α, then

$$U = C \ln \frac{1 - 1/n_c}{1 - \alpha}, \tag{2}$$

where C is a positive real number that depends on the sample distribution.

2.4 Properties

In this section, we will prove favorable properties of the method proposed in 4.1 when we map the model to the expected i.i.d. number of data points and use this number as the index to calculate Shapley Value, and illustrate how it works in the case of traditional calculation of Shapley Value ceasing to be effective.

To begin with, two basic propositions are come up.

Proposition 1: Suppose $\{M_i\}_{i=1}^N$ are N models trained on N dataset $\{D_i\}_{i=1}^N$. M is the aggregating model of $M_1, M_2, ..., M_n$. M^* is the model trained on $D_1 \cup D_2 \cup ... \cup D_n$. Then the accuracy of M is equal to M^*.

Proposition 2: The value of model is directly proportional to the expected number of i.i.d. data points to trained it.

Proposition 3: Consider we gather n different models $M_1, M_2, ..., M_n$ without overlapping. Suppose M is aggregated by M_1 and M_2. $SV(M_1)$ is the Shapley Value for M_1 in n models. $SV(M_2)$ is the Shapley Value for M_2 in n models. $SV(M)$ is the Shapley Value for M in $n-1$ models. When we define the utility function as $U(M) = |M|$, where $|\cdot|$ denotes the expected number of i.i.d datapoint to train the model, then $SV(M) = SV(M_1) + SV(M_2)$.

The purpose of federated learning is to collect models instead of the dataset to protect the data privacy. The first proposition stands for a perfect condition that we can aggregate models without any loss, which guarantees that our aggregated models have the same performance as those datasets we randomly collect to train. To fully understand the second proposition, we can consider an ideal i.i.d. dataset. Each data point is equivalent because they are from the same distribution and can their sample order be changed freely. Thus, each data point in this dataset has the same value. For proposition 3, we can see our function guarantees a perfect outcome.

3   Experiment

In this part, we use our theoretical formula to get the utility function and then calculate Shapley Value based on it. After being compared with other methods, we have proved that the Shapley Value calculated in this method reaches the closest approximation to the expected value.

3.1   Utility Function

In this section we will contrast our theoretical formula with experimental curves.

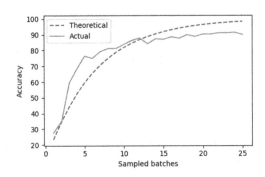

Fig. 1. Curve of sample number and accuracy. Actual curve is the average results from two experiments in mnist. Theoretical curve is formula 2 where C = 0.16.

Figure 1 shows the experimental results of accuracy batch-numbers curves. Batch size is set to 16, and model is Lenet (4). When we add parameters to estimate the curve, utility function is no longer a fixed expression and cannot be used for calculation. So unfortunately it is impractical to fit more parameters. These deviation would also become smaller and acceptable with the increasing feature space complexity and the growing model capacity, so it is a good idea to apply this function in calculation of Shapley Value.

Here we apply the definition Shapley Value formula. Specifically, we randomly sampled 3 datasets with different sample numbers from Mnist dataset, then we copy one of these datasets as an independent participant. We use LeNet5 and ResNet18 as the trained model. The results of experiments are as followed.

Minst and LeNet5
Let A, B, C stand for these 3 datasets respectively and we have the size: $|A| = 128, |B| = 256, |C| = 256$. Let's consider 2 conditions. The first case is B is divided into B and D with $|B| = 128$ and $|D| = 128$, while in the other case we set $|B| = 128$ and set $D = B$, which means D is a replica of B.

In the first case, the new price ratio is expected to be 1:1:2:1.

In the second case, it's easy to find out that the value of all datasets will be proportional to their size without D. Since dataset D shares the value of B equally, their true price ratio would be 2:1:4:1 on account of our principles.

Table 1. SV for 3 datasets, A, B, C randomly sampled from mnist dataset.

Dataset	A	B	C	MSE
Sample num	128	256	256	
Expect value	0.2	0.4	0.4	
SV-ACC	0.309	0.346	0.344	0.60%
SV-L	0.326	0.338	0.336	0.80%
OURS	0.273	0.364	0.363	**0.26%**

Table 2. SV for 4 datasets, A, B, C randomly sampled from mnist dataset and original B split into B and D.

A	B	C	D	MSE
128	128	256	128	
0.2	0.2	0.4	0.2	
0.245	0.233	0.279	0.244	0.49%
0.248	0.243	0.257	0.252	0.68%
0.236	0.207	0.322	0.236	**0.22%**

To evaluate the model performance, we compare our method to calculate data Shapley Value with another two classic methods, which are loss function on test set used in (2), and testing accuracy used in (1). SV-L stands for using loss $U = -\int_{k \in testset} l_k$ as utility function and SL-ACC stands for using test accuracy. Given the fact that different utility functions will bring about different total value, we prefer use percentages to contrast discrepancy.

The results shown in Table 1 conform to our expectation. Through MSE, the mean squared error between calculated value and expected value also proved that our model is the closest to the predicted value, while other methods have larger mean squared error. The MSE of our methods is 0.0026 in Table 1, and 0.0022 in Table 2, which is much smaller than that of other methods.

3.2 Noisy Labels

In this experiment, we consider the case of wrong samples in the dataset. If a node has a negative contribution, the accuracy of the whole model will be damaged, so we shall not add it in the end. At this point, the sharing degree of the node to the task is 0, and the reward obtained in the whole reward allocation should also be 0. We hope that the SV calculated by the utility function can have a reflection on it. In other words, when the contribution of nodes equals to 0, SV also equals to 0. And if the nodes have negative impact on the models, SV should be a negative value.

We use the settings of experiments with mnist and LeNet5 from part 1 to conduct this experiment. We gradually increase the proportion of noise-label data points in D and calculate the marginal utility contribution and Shapley Value of D.

As what the figure above indicates, with the increasing proportion of false labels in dataset D, the accuracy rate of dataset D for the final aggregation model decreases. After adding to 20% error samples, the accuracy rate declines. Therefore, it is inadvisable to add D to the final model at this time. Meanwhile, SV calculated by the three utility functions decrease with the increase of the proportion of false tags in dataset D. When the error sample size of D is 20%, the final model will not add D, the SV calculated by our utility function is closer to 0, indicating our utility function possesses a better performance and desirable result (Fig. 2).

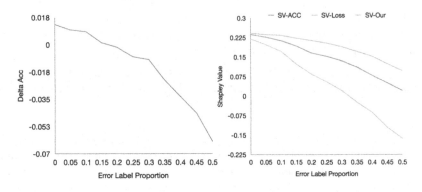

Fig. 2. Margin utility contribution of D

4   Conclusion

To solve the pricing problem with multiple participants, this paper proposes principles to reasonably allocate revenues and a novel utility function in Shapely Value is well-designed. Moreover, the fairness of this solution has been proved, which avoids fraud effectively. After applying the theory into experiments, we find out it works not only in Federal Learning environment but also in other scenarios, where the SV performance improved by 50%.

Admittedly, the utility function is proved to suit well merely for classification tasks. Expanding the application scenarios remains exploration for further research.

References

1. Agarwal, A., Dahleh, M., Sarkar, T.: A marketplace for data: an algorithmic solution. In: The 2019 ACM Conference (2019)
2. Ghorbani, A., Zou, J.: Data Shapley: equitable valuation of data for machine learning. arXiv preprint arXiv:1904.02868 (2019)
3. Jia, R., et al.: Towards efficient data valuation based on the Shapley value. In: Proceedings of the 22nd International Conference on Artificial Intelligence and Statistics (2019)
4. LeCun, Y., Bottou, L., Bengio, Y., Haffner, P., et al.: Gradient-based learning applied to document recognition. Proc. IEEE **86**(11), 2278–2324 (1998)
5. McMahan, B., Moore, E., Ramage, D., Hampson, S., Arcas, A.: Communication-efficient learning of deep networks from decentralized data. In: Artificial Intelligence and Statistics, pp. 1273–1282 (2017)

DFL-Net: Effective Object Detection via Distinguishable Feature Learning

Jia Xie[1,2], Shouhong Wan[1,2], and Peiquan Jin[1,2(✉)]

[1] University of Science and Technology of China, Hefei, China
geroci@mail.ustc.edu.cn, {wansh,jpq}@ustc.edu.cn
[2] Key Laboratory of Electromagnetic Space Information, CAS, Hefei, China

Abstract. The one-stage anchor-based approach has been an efficient and effective approach for detecting objects from massive image data. However, it neglects many distinguishable features of objects, which will lower the accuracy of object detection. In this paper, we propose a new object detection approach that improves existing one-stage anchor-based methods via a Distinguishable Feature Learning Network (DFL-Net). DFL-Net integrates distinguishable features into the learning process to improve the accuracy of object detection. Notably, we implement DFL-Net by a full-scale fusion module and an attention-guided module. In the full-scale fusion module, we first learn the distinguishable features at each scale (layer) and then fuse them in all layers to generate full-scale features. This differs from prior work that only considered one or limited scales and limited features. In the attention-guided module, we extract more distinguishable features based on some positive or negative samples. We conduct extensive experiments on two public datasets, including PASCAL VOC and COCO, to compare the proposed DFL-Net with several one-stage approaches. The results show that DFL-Net achieves a high mAP of 83.1% and outperforms all its competitors. We also compare DFL-Net with three two-stage algorithms, and the results also suggest the superiority of DFL-Net.

Keywords: Object detection · Feature learning · Neural network · Feature fusion · Attention mechanism

1 Introduction

With the advance of ubiquitous intelligence, more and more smart devices and systems have been deployed in buildings or outdoor environments. A fundamental issue in intelligence systems and services is to detect various kinds of objects effectively [1–6]. For example, a robot for home services needs to identify humans, dogs, cats, cups, and many other types of objects. So far, the techniques for object detection can be divided into three categories, namely two-stage detectors [1, 2], one-stage anchor-based detectors [3–7], and anchor-free detectors [8–10]. This study is within the scope of one-stage anchor-based detectors, but with a particular focus on improving the performance of existing one-stage detectors.

© Springer Nature Switzerland AG 2021
C. Strauss et al. (Eds.): DEXA 2021, LNCS 12924, pp. 195–206, 2021.
https://doi.org/10.1007/978-3-030-86475-0_20

Briefly, two-stage detectors first extract regions of interest and then predict the confidence and coordinate of objects. However, they require additional operation, such as RoI Pooling [1], which results in complex network structures and long inference time. Anchor-free detectors predict the confidence and the coordinates of the objects without generating anchors, but they can hardly get stable results because they have to perform additional operations to detect critical points such as Corner Pooling [8]. One-stage anchor-based detectors predict confidence and offsets of anchors. Compared to two-stage detectors, one-stage anchor-based detectors have a better trade-off between network complexity and inference time performance and have received much attention in recent years.

One-stage anchor-based detectors generate some anchors during initialization. In this paper, the anchors after regression are marked as samples. For a given threshold θ, samples are divided into positive samples (P) and negative samples (N). The previous work [4] mainly focused on difficult negative samples. In this paper, we introduce two new kinds of negative samples and consequently classify negative samples into three types, including difficult negative samples (DN), simple negative samples (SN), and unrelated negative samples (UN), which will be explained in Sect. 3. Our work will ignore UN later to prevent overfitting.

So far, it is still a challenging issue to determine which features the network should be fused to detect objects at different scales better. In addition, it is also not clear how to mine the features of different samples, i.e., P, DN, SN, and UN, to distinguish the foreground from the background better while avoiding overfitting.

Aiming to address the first challenge, most of the existing methods use the top-down feature fusion method [11] or the bi-directional feature fusion method [7] to enrich the distinguishable features of different layers. However, because they do not fuse the features of non-adjacent layers, this may leave out some distinguishable features that the current layer cannot capture. For the second challenge, the existing one-stage anchorbased detectors proposed two mining sample strategies. The first strategy proposes to mine the features of P and DN while ignoring the features of SN and UN, which leads to insufficient learning of background features. The second strategy proposes to mine all types of features, which will increase the training time and cause overfitting, i.e., the number of N is much larger than the number of P, which will induce the network to predict the object as the background.

In this paper, we propose a new object detection approach that improves existing one-stage anchor-based methods via a Distinguishable Feature Learning Network (DFL-Net). DFL-Net integrates distinguishable features into the learning process to improve the accuracy of object detection. This differs from prior work that only considered one or limited scales and limited features. Briefly, we make the following contributions in this paper:

(1) We propose a new distinguishable feature learning network called DFL-Net. Differing from prior work, DFL-Net considers distinguishable features when learning the features for image-based object detection. To the best of our knowledge, this is the first study that integrates distinguishable features in object detection.

(2) We present a full-scale fusion method in DFL-Net to learn distinguishable features from different scales (layers). In particular, we propose first to learn distinguishable

features from each layer and then to fuse the features of all layers to output full-scale features.

(3) We further propose an attention-guided method in DFL-Net to refine the features with high-response foregrounds. Our experiments show that this design can save 38.7% of GPU memory usage and improve the accuracy of object detection. Memory usage refers to the GPU memory consumed by the network after loading the specified batch image into the GPU.

(4) We conduct extensive experiments on two public datasets, including PASCAL VOC and COCO, to compare the proposed DFL-Net with three one-stage detectors. The results show that DFL-Net achieves a high ratio of 83.1% and outperforms existing one-stage detectors. Moreover, it delivers higher accuracy than two-stage detectors like Faster R-CNN and R-FCN.

2 Related Work

So far, object detection based on deep learning has been a research focus in recent years due to the high effectiveness of deep learning models. Generally, the deep learning-based approaches for object detection can be classified into two categories, namely anchor-based detection and anchor-free detection. The anchor-based method can be further divided into two types, i.e., one-stage object detection and two-stage object detection. Note that all these approaches have their unique advantages and limits. As a result, there is no evidence that one method could always outperform other ones.

The one-stage approach automatically generates dense anchors on feature maps during initialization. Then, the offsets and confidence of those anchors are predicted. The representatives of the one-stage method include SSD [3], DSSD [4], and FSSD [14]. Single Shot Multibox Detector (SSD) extracts several layers of feature maps through the Feature Pyramid Network (FPN) [11]. FPN downsamples the feature maps derived by the convolutional neural network (CNN) to obtain multiple feature maps with a decreasing scale. Deconvolutional Single Shot Detector (DSSD) is based on SSD. It adds an auxiliary layer after FPN, i.e., obtain some feature maps with gradually increasing scale through the deconvolution module, and then merge with the feature maps corresponding to the scale of FPN, and then make a prediction through a prediction module. Feature Fusion Single Shot Multibox Detector (FSSD) adds a lightweight feature fusion module based on SSD. It first fuses several feature maps of different scales in the shallow layer of FPN into a feature map f and then obtains some feature maps with decreasing scale through FPN based on f. Based on the FSSD fusion method, Attentive Single Shot Multibox Detector (ASSD) [15] first obtains the feature map through FPN, and then uses the self-attention mechanism to capture the high-response region of the feature map. Some other models focus on improving the one-stage approach. For example, Retinanet [5] proposed a focal loss function to deal with the sample imbalance in one-stage object detection and performed deconvolution and fusion operations on FPN. RefineDet [6] proposed a cascade structure to refine anchors by filtering some simple negative anchors and designed the transfer connection module to perform top-down [5] fusion on FPN. Based on RefineDet, DAFS [7] proposed to dynamically update the corresponding features of each anchor after the anchor refinement.

The two-stage approach first detects the regions of interest (ROI), which are also called proposals, from images and then extracts the features in ROIs for further classification and regression. The Faster R-CNN [1] proposed a region proposal network (RPN) to generate proposals. Then, it used the ROI Pooling to collect the input feature maps and proposals and integrated them into the proposed feature map. Finally, it used the subsequent fully connected layer to perform regression and classification. Based on the Faster R-CNN, R-FCN [2] replaces the fully connected layer in the neural network with a convolutional layer, yielding a faster training process. The Cascade R-CNN [16] devised multiple cascaded detectors to extract the proposals with different IoUs in each detector. Although the two-stage approach has higher accuracy than the one-stage approach, it has worse time efficiency than the one-stage method because it has to employ additional algorithms such as RPN and RoI Pooling [1] to generate proposals. Thus, the two-stage method is not suitable for timely object detection.

Recently, some researchers proposed the anchor-free approach for object detection, which can directly predict the coordinates and the confidence of objects. Cornernet [8] predicts several key points and then divides these points into pairs, which are the coordinates of the upper-left and lower-right of objects. Centernet [17] predicts center points and scales of objects. FoveaBox [10] is based on Retinanet. It directly predicts the bounding box of objects. The anchor-free approach eliminates the need to generate anchors, which further streamlines the structure of the network. However, the anchor-free method usually has unstable predictions due to the lack of anchors.

3 Design of DFL-Net

3.1 High-Level Idea of DFL-Net

The general idea of DFL-Net is to integrate distinguishable features into the learning of features for one-stage anchor-based object detection and help the network distinguish objects and backgrounds better.

$$
S = \begin{cases}
Positive\ Samples(P) & IoU \in (\theta, 1) \\
Difficult\ negative\ samples(DN) & IoU \in (\theta, \theta - \epsilon) \\
Unrelated\ negative\ samples(UN) & IoU \in (\theta - \epsilon, \epsilon) \\
Simple\ negative\ samples(SN) & IoU \in (0, \epsilon)
\end{cases}
\tag{1}
$$

The definitions of the four types of samples (S), namely P, DN, SN, and UN, are given in Eq. (1). Here, P, DN, SN, and UN represent positive samples, difficult negative samples, simple negative samples, and unrelated negative samples. The parameter ϵ is a manually designed threshold that can help us divide the negative samples (N) into DN, UN, SN. The proportion of DN and SN in the negative samples is relatively small and balanced, which is helpful for the network to distinguish between backgrounds and objects. Moreover, we ignore UN, which has a large proportion in the negative samples, to avoid the overfitting of the network.

Since SSD [3] is a classic one-stage anchor-based object detector containing feature pyramids, DFL-Net is an optimization of SSD. The architecture of DFL-Net is shown in Fig. 1. DFL-Net consists of a full-scale fusion module and an attention-guided module.

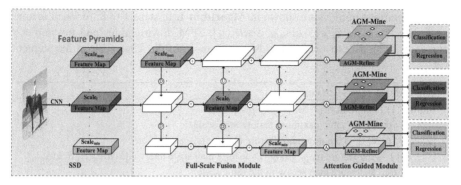

Fig. 1. The architecture of DFL-Net.

3.2 Full-Scale Fusion

The full-scale fusion module (SFM) fuses the features of each layer with the features of the remaining layers so that each layer of distinguishable features can well represent objects of corresponding scales. The full-scale fusion module is shown in Fig. 1, DFL-Net based on ResNet will generate seven layers of feature maps of different scales, which is the same as baseline SSD. For convenience, we only show the feature maps of blue, red, and green, which correspond to the $Scale_{max}$, $Scale_i$, and $Scale_{min}$ in DFL-Net, respectively.

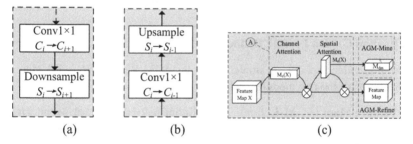

Fig. 2. The downsampling, upsampling and AGM operations in DFL-Net. Here, S_i and C_i in (a) and (b) denote the feature maps with **$Scale_i$** and **$Channel_i$** in DFL-Net, respectively

In the SFM module in Fig. 1, for the red feature map with medium-scale $Scale_i$ $\in[Scale_{min}, Scale_{max}]$, we downsample it in Fig. 2(a) and upsample it in Fig. 2(b) recur-sively. The feature maps of other scales also recur-sively upsample and downsample to obtain the feature maps of all remaining scales. Finally, we operate on the feature maps of the same scale. Figure 2(a) consists of the operations of $conv1 \times 1$ and *downsample*, and Fig. 2(b) includes the operations of $conv1 \times 1$ and *upsample*. The $conv1 \times 1$ is to change the current channel C_i to the upper channel C_{i+1} and the below channel C_{i+1}. *upsample* and *downsample* are to change the current scale S_i to the upper large-scale S_{i-1} and the below small-scale S_{i+1}.

The operations in SFM are shown in **Algorithm 1**. In steps 1 to 8, for each feature map f with the scale $S_i \in [Scale_{min}, Scale_{max}]$, SFM downsamples and upsamples f recursively. In steps 9 to 11, SFM adds feature maps of the same scale at the element level.

Algorithm 1 *Full-Scale Fusion*

Input: (1) *FeatureMaps*[7] are *feature pyramids*.
    (2) $Scale_{max}$ is the maximum feature map scale.
    (3) $Scale_{min}$ is the minimum feature map scale.
    (4) f is the feature map. (5) ϕ is a feature map scale interval $[Scale_{max}, Scale_{min}]$.
    (6) f_{sets} is feature map sets with the same scale.
Output: *FeatureMaps*[7]
1: **for each** f with scale $S_i \in \phi$ **do**
2:    **for each** $Scale_d \in [S_i, Scale_{min}]$ **do**
3:       Recursively *downsample* f to $Scale_d$
4:    **end for**
5:    **for each** $Scale_u \in [S_i, Scale_{max}]$ **do**
6:       Recursively *upsample* f to $Scale_u$
7:    **end for**
8: **end for**
9: **for each** f_{sets} with scale $S_i \in \phi$ **do**
10:    $FeatureMaps[S_i] \leftarrow$ Element-wise addition of f_{sets}
11: **end for**
12: **return** *FeatureMaps*[7]

3.3 Attention Guided Feature Refinement

The attention-guided module is implemented by AGM-Refine and AGM-Mine. The implementation of the two modules is described below.

Refine Feature Maps (AGM-Refine). AGM-Refine helps the network better detect the foreground by refining the features of the high-response foregrounds. First, inspired by CBAM [12], but differing from CBAM embedding channel and spatial modules into each block of ResNet, we propose a method named AGM-Refine, which put channel and spatial attention modules behind the feature pyramid instead of embedding them into backbones. Compared with CBAM, our approach saves 38.7% memory usage and improves mAP by 2%. As shown the Fig. 2(c), for a feature map $X \in \mathbb{R}^{C \times H \times W}$ from SFM, the process of refining feature maps is the same as that in CBAM where C, H, and W refer to the channel, height, and width of an input feature map. The channel attention module generates $M_c(X) \in \mathbb{R}^{C \times 1 \times 1}$. The spatial attention module generates $M_s(X) \in \mathbb{R}^{1 \times H \times W}$. The symbol denotes multiplication along with the channel level and the spatial level, i.e., $X = X \otimes M_c(X)$ and $X = X \otimes M_s(X)$.

Mine Distinguishable Features of Specified Samples (AGM-Mine). AGM-Mine helps the network filter out a large number of backgrounds by mining the features of SN with high contrast to the foregrounds and DN very similar to the foregrounds. As shown in Fig. 2(c), for a given feature map $M_s(X) \in \mathbb{R}^{1 \times H \times W}$ generated by the spatial attention module, AGM-Mine in Fig. 2(c) can be described by Eqs. (2)–(4):

$$M_a\left[1, i, j, k\right]_{k=1}^{Num} = M_s(X)\left[1, i, j\right] \tag{2}$$

$$M_{sn}\left[1, i, j, k\right]_{k=1}^{Num} = \begin{cases} 1 M_a\left[1, i, j, k\right] \leq \theta_{sn} \\ 0 M_a\left[1, i, j, k\right] \geq \theta_{sn} \end{cases} \tag{3}$$

$$M_{dm} = M_{sn} + M_p + M_{dn} \tag{4}$$

Here, i ranges from 0 to $H - 1$, j ranges from 0 to $W - 1$, and k ranges from 1 to *Num*. As each spatial location $[i, j]$ of the feature map usually generates Num anchors, we assign the value of $M_s(X)\left[1, i, j\right]$ to *Num* anchors which attached to the spatial location $[i, j]$. $M_a \in \mathbb{R}^{1 \times H \times W \times Num}$. The smaller the value of M_a, the more likely these features can well represent SN. We only need to design a threshold to mine features of a specified number of SN. The symbol M_{sn} denotes a mask of SN features. The threshold θ_{sn} is set to the $(N_{pos} \times Ratio_{sn})$th smallest value in M_a. $Ratio_{sn}$ indicates the times for AGM-Mine to mine SN of P. N_{pos} denotes the number of P. Besides, M_p and M_{dn} denote the mask of features of P and DN, which are the same as those in SSD. M_{dm} denotes the distinguishable mask, which can mine features of P, DN, and SN.

4 Performance Evaluation

In this section, we report the experimental results of DFL-Net as well as its competitors, including several one-stage anchor-based detectors and two-stage detectors.

4.1 Settings

Datasets. We conducted experiments on two datasets, including PASCAL VOC [13] and COCO [19]. Both datasets are widely used as public datasets in object detection. For PASCAL VOC, we use PASCAL VOC 2007 *Trainval* and PASCAL VOC 2012 *Trainval* as the training datasets and PASCAL VOC 2007 *Test* as the test dataset. For COCO, we use COCO Trainval35k as the training dataset and COCO Test-dev as the test dataset.

Compared Methods. Since our baseline is SSD, we focus on one-stage anchor-based SSD, DSSD, and FSSD. Also, we compared our proposal with two-stage approaches, including Faster R-CNN and R-FCN.

Evaluation Metrics. We adopt mAP as the evaluation metric, which is commonly used in object detection. AP is the PR (Precision-Recall) curve area of the specified class, and mAP is the average of the APs of all categories.

Training Details on COCO and VOC. We employ PyTorch to conduct experiments on COCO and PASCAL VOC 2007 with an input size of 513×513 and 321×321. In the following experiments, we denote an input size of 513×513 as the large input size and 321×321 as the small input size. The experiments on COCO and VOC were run on four and one NVIDIA GeForce 2080ti, respectively. The batch size of each GPU is set to 18 and 6, respectively. We train 30epochs with a 1^{-3} learning rate, then 20 epochs with a 1^{-4} learning rate, and finally 20 epochs with a 1^{-5} learning rate.

Table 1. The results on PASCAL VOC 2007 TEST.

Method	Datasets	Backbone	Input size	FPS	Map (%)
Faster R-CNN [1]	07+12	VGG-16	~1000 × 600	7	73.2
Faster R-CNN [1]	07+12	ResNet-101	~1000 × 600	2.4	76.4
OHEM [22]	07+12	VGG-16	~1000 × 600	7	78.9
R-FCN [2]	07+12	ResNet-101	~1000 × 600	9	80.5
SSD300 [3]	07+12	VGG-16	300 × 300	46	77.5
FSSD300 [14]	07+12	VGG-16	300 × 300	65.8	78.8
SSD321 [3]	07+12	ResNet-101	321 × 321	11.2	77.1
DSSD321 [4]	07+12	ResNet-101	321 × 321	9.5	78.6
ASSD321 [15]	07+12	ResNet-101	321 × 321	27.5	79.5
DFL-Net321	07+12	ResNet-101	321 × 321	**58**	**79.6**
RefineDet320 [6]	07+12	VGG-16	320 × 320	40.3	80
YOLO [23]	07+12	GoogleNet	448 × 448	45	63.4
YOLOv2 [24]	07+12	Darknet-19	544 × 544	40	78.6
SSD512 [3]	07+12	VGG-16	512 × 512	19	79.5
FSSD512 [14]	07+12	VGG-16	512 × 512	35.7	80.9
SSD513 [3]	07+12	ResNet-101	513 × 513	6.8	80.6
DSSD513 [4]	07+12	ResNet-101	513 × 513	5.5	81.5
RefineDet512 [6]	07+12	VGG-16	512 × 512	24.1	81.8
ASSD513 [15]	07+12	ResNet-101	513 × 513	16	83
DFL-Net513	07+12	ResNet-101	513 × 513	**50**	**83.1**

4.2 Results

Most previous one-stage anchor-based methods perform experiments on PASCAL VOC 2007 and COCO with the input size of 320 × 320, 321 × 321, 512 × 512, and 513 × 513. Thus, we keep the same conditions to compare with those methods.

The Results on the PASCAL VOC 2007 Test. The results on the PASCAL VOC 2007 TEST are shown in Table 1, where *07+12* refers to the training datasets PASCAL VOC 2007 and PASCAL VOC 2012 *Trainval* datasets. The three digits attached to the method name means the image size. For example, DFL-Net513 means the DFL-Net running on the images with the size of 513 × 513. The first part shows the results of the advanced two-stage detectors; the second part shows the results of the advanced one-stage detectors with small input size, and the third part shows the results of the advanced one-stage detectors with a large inputs size. Compared to recently presented one-stage and two-stage approaches, DFL-Net513 achieves the highest mAP of 83.1%. Compared with the baseline SSD513 and SSD321, DFL-Net513 and DFL-Net321 have improved mAP by 2.5% and 2.5%, respectively. FPS refers to the number of images supported by the network, mAP refers to evaluation metrics.

The Results on the COCO Test-dev. The results on the COCO Test-dev are shown in Table 2. COCO *Trainval dataset* refers to the 135k training images on COCO. Compared with SSD513 and SSD321, DFL-Net513 and DFL-Net321 have improved mAP by

Table 2. The results on the COCO Test-dev.

Method	Backbone	AP	AP_{50}	AP_{75}	AP_S	AP_M	AP_L
Two-Stage							
Faster R-CNN [1]	VGG-16	21.9	42.7	–	–	–	–
R-FCN [2]	ResNet-101	29.9	51.9	–	10.8	32.8	45.0
CoupleNet [20]	ResNet-101	34.4	54.8	37.2	13.4	38.1	52
One-Stage							
SSD300 [3]	VGG-16	25.1	43.1	25.8	6.6	25.9	41.4
SSD321 [3]	ResNet-101	28	45.4	29.3	6.2	28.3	49.3
FSSD300 [14]	VGG-16	27.1	47.7	27.8	8.7	29.2	42.2
DSSD321 [4]	ResNet-101	28	46.1	29.2	7.4	28.1	47.6
ASSD321 [15]	ResNet-101	29.2	47.8	30.9	6.9	33.3	47.9
RefineDet320 [6]	VGG-16	29.4	49.2	31.3	10	32	44.4
DFL-Net321	ResNet-101	**30**	49.1	**31.6**	7.2	**34.9**	47.6
SSD512 [3]	VGG-16	28.8	48.5	30.3	10.9	31.8	43.5
SSD513 [3]	ResNet-101	31.2	50.4	33.3	10.2	34.5	49.8
FSSD512 [14]	VGG-16	31.8	52.8	33.5	14.2	35.1	45
DSSD513 [4]	ResNet-101	33.2	53.3	35.2	13	35.4	51.1
RetinaNet500 [5]	ResNet-101	34.4	53.1	36.8	14.7	38.5	49.1
ASSD513 [15]	ResNet-101	34.5	55.5	36.6	15.4	39.2	51
RefineDet512 [6]	VGG-16	33	54.5	35.5	16.3	36.3	44.3
YOLOv3 [21]	Darknet-53	33	57.9	34.4	18.3	35.4	41.9
DFL-Net513	ResNet-101	**35.1**	**55.5**	**37.2**	14.6	**39.2**	**52.1**

3.9% and 2.0%, respectively. DFL-Net513 surpasses most of the one-stage anchor-based models using SSD as the baseline, such as ASSD513, DSSD513, and FSSD512. At the same time, compared with the two-stage approaches R-FCN and CoupleNet [20], DFL-Net513 also improves mAP by 5.9% and 0.9%, respectively.

4.3 Ablation Study

All ablation studies were conducted on the PASCAL VOC 2007 Test by ResNet-101 using the large input size of 513×513. In addition to the modules of DFL-Net, all other settings are the same as those in SSD. Below, we discuss the ablation study of SFM, AGM-Refine, and AGM-Mine.

Ablation Study on Cascade Use of SFM, AGM-Refine, and AGM-Mine. As shown in Table 3, mAP in the baseline SSD is 80.6%. When using SSD and SFM, mAP is improved by 1.2%. When using SSD, SFM, and AGM-Refine, mAP was further improved by 2.1%. When using SSD, SFM, AGM-Refine, and AGM-Mine, DFL-Net improved mAP by 2.5% and achieved a superior mAP of 83.1%.

Ablation Study on Cascade Use of SFM, AGM-Refine, and AGM-Mine. As shown in Table 4, AGM-*Refine$_{embed}$* and AGM-*Refine$_{behind}$* denotes channel and spatial attention

Table 3. The Results of cascade use of SSD, SFM, AGM-Refine, and AGM-Mine on PASCAL VOC 2007 TEST.

mAP(%)	SSD	SFM	AGM-Refine	AGM-Mine
80.6	√			
81.8	√	√		
82.7	√	√	√	
83.1	√	√	√	√

Table 4. Comparison between AGM-Refine$_{embed}$ and AGM-Refine$_{behind}$

mAP (%)	Memory usage (M)	Method
81.8	975.5	SSD+SFM
80.7	1590	SSD+SFM+ AGM-*Refine*$_{embed}$
82.7	975.8	SSD+SFM+ AGM-*Refine*$_{behind}$

modules are embedded in a block of ResNet or put behind SFM, respectively. Compared with SSD + SFM, attention module embedded (AGM-*Refine*$_{embed}$ in ResNet leads to 1.1% reduction of mAP, but our approach (AGM-*Refine*$_{behind}$) improves mAP by 0.9%. Compared with SSD+SFM+ AGM-*Refine*$_{embed}$, our approach improves mAP by 2% and saves 38.7% of memory usage. Compared to DFL-Net with no attention module, our dual-channel module only increased memory usage by 0.3M. In this paper, our AGM-Refine refers to the AGM-*Refine*$_{behind}$ module.

5 Conclusion and Future Work

Object detection has been a fundamental technology for intelligent systems and services. In this paper, we propose a distinguishable feature learning network called DFL-Net based on SSD to improve the effectiveness of object detection. DFL-Net can better distinguish background and objects while avoiding misclassification. The results show that DFL-Net achieves a superior mAP of 83.1%, which outperforms most compared SSD-like detectors and three two-stage detectors, including Faster R-CNN and R-FCN. The results also suggest the superiority of DFL-Net.

The two modules in DFL-Net, namely the Full-Scale Fusion Module (SFM) and the Attention Guided Module (AGM), are both plug-and-play. Therefore, in future work, we will extend the two modules into other one-stage object detectors. In addition, we will consider improving other optimization models for transformer and feature fusion [26–30].

Acknowledgments. This paper is supported by the National Science Foundation of China (grant no. 62072419).

References

1. Ren, S., He, K., Girshick, R., Sun, J.: Faster R-CNN: towards real-time object detection with region proposal networks. In: NIPS, pp. 91–99 (2015)
2. Dai, J., Li, Y., He, K., Sun, J.: R-FCN: object detection via region-based fully convolutional networks. In: NIPS, pp. 379–387 (2016)
3. Liu, W., et al.: SSD: single shot multibox detector. In: Leibe, B., Matas, J., Sebe, N., Welling, M. (eds.) ECCV 2016. LNCS, vol. 9905, pp. 21–37. Springer, Cham (2016). https://doi.org/10.1007/978-3-319-46448-0_2
4. Fu, C.Y., Liu, W., Ranga, A., Tyagi, A., Berg, A.C.: DSSD: deconvolutional single shot detector. arXiv preprint arXiv:1701.06659 (2017)
5. Lin, T.Y., Goyal, P., Girshick, R., He, K., Dollár, P.: Focal loss for dense object detection. In: ICCV, pp. 2980–2988 (2017)
6. Zhang, S., Wen, L., Bian, X., Lei, Z., Li, S.Z.: Single-shot refinement neural network for object detection. In: CVPR, pp. 4203–4212 (2018)
7. Li, S., Yang, L., Huang, J., Hua, X.S., Zhang, L.: Dynamic anchor feature selection for single-shot object detection. In: ICCV, pp. 6609–6618 (2019)
8. Law, H., Deng, J.: Cornernet: detecting objects as paired keypoints. In: Ferrari, V., Hebert, M., Sminchisescu, C., Weiss, Y. (eds.) Computer Vision – ECCV 2018. LNCS, vol. 11218, pp. 765–781. Springer, Cham (2018). https://doi.org/10.1007/978-3-030-01264-9_45
9. Zhou, X., Wang, D., Krähenbühl, P.: Objects as points. arXiv preprint arXiv:1904.07850 (2019)
10. Kong, T., Sun, F., Liu, H., Jiang, Y., Shi, J.: FoveaBox: beyond anchor-based object detector. arXiv preprint arXiv:1904.03797 (2019)
11. Lin, T.Y., Dollár, P., Girshick, R., He, K., Hariharan, B., Belongie, S.: Feature pyramid networks for object detection. In: CVPR, pp. 2117–2125 (2017)
12. Woo, S., Park, J., Lee, J.-Y., Kweon, I.S.: CBAM: convolutional block attention module. In: Ferrari, V., Hebert, M., Sminchisescu, C., Weiss, Y. (eds.) ECCV 2018. LNCS, vol. 11211, pp. 3–19. Springer, Cham (2018). https://doi.org/10.1007/978-3-030-01234-2_1
13. Everingham, M., Van, L., Williams, C.K., Winn, J., Zisserman, A.: The PASCAL visual object classes challenge 2007 (2007). http://host.robots.ox.ac.uk/pascal/VOC/voc2007
14. Li, Z., Zhou, F.: FSSD: feature fusion single shot multibox detector. arXiv preprint arXiv:1712.00960 (2017)
15. Yi, J., Wu, P., Metaxas, D.N.: ASSD: attentive single shot multibox detector. In: CVIU, vol. 189 (2019)
16. Cai, Z., Vasconcelos, N.: Cascade R-CNN: delving into high quality object detection. In CVPR, pp. 6154–6162 (2018)
17. Duan, K. Bai, S., Xie, L., Qi, H., Huang, Q., Tian, Q.: CenterNet: keypoint triplets for object detection. In: ICCV, pp. 6569–6578 (2019)
18. He, K., Zhang, X., Ren, S., Sun, J.: Deep residual learning for image detection. In: CVPR, pp. 770–778 (2016)
19. Lin, T.-Y., et al.: Microsoft COCO: common objects in context. In: Fleet, D., Pajdla, T., Schiele, B., Tuytelaars, T. (eds.) ECCV 2014. LNCS, vol. 8693, pp. 740–755. Springer, Cham (2014). https://doi.org/10.1007/978-3-319-10602-1_48
20. Zhu Y., Zhao C., Wang J., Zhao X., Wu Y., Lu H.: CoupleNet: coupling global structure with local parts for object detection. In: ICCV, pp. 4126–4134 (2017)
21. Redmon, J., Farhadi, A.: YOLOv3: an incremental improvement. In: CoRR, abs/1804.02767 (2018)
22. Shrivastava, A., Gupta, A., Girshick, R.: Training region-based object detectors with online hard example mining. In CVPR, pp. 761–769 (2016)

23. Redmon, J., Divvala, S., Girshick, R., Farhadi, A.: You only look once: unified, real-time object detection. In: CVPR, pp. 779–788 (2016)
24. Redmon, J., Farhadi, A.: YOLO9000: better, faster, stronger. In: CVPR, pp. 6517–6525 (2017)
25. Simonyan, K., Zisserman, A.: Very deep convolutional networks for large-scale image detection. arXiv preprint arXiv:1409.1556 (2014)
26. Yang, X., Wan, S., Jin, P., Zou, C., Li, X.: MHEF-TripNet: mixed triplet loss with hard example feedback network for image retrieval. In: Zhao, Y., Barnes, N., Chen, B., Westermann, R., Kong, X., Lin, C. (eds.) ICIG 2019. LNCS, vol. 11903, pp. 35–46. Springer, Cham (2019). https://doi.org/10.1007/978-3-030-34113-8_4
27. Sun, Z., Cao, S., Yang, Y., Kris, K.: Rethinking transformer-based set prediction for object detection. arXiv preprint arXiv:2011.10881 (2020)
28. Tian, Q., Wan, S., Jin, P., Xu, J., Zou, C., Li, X.: A novel feature fusion with self-adaptive weight method based on deep learning for image classification. In: Hong, R., Cheng, W.-H., Yamasaki, T., Wang, M., Ngo, C.-W. (eds.) PCM 2018. LNCS, vol. 11164, pp. 426–436. Springer, Cham (2018). https://doi.org/10.1007/978-3-030-00776-8_39
29. Yang, X., Wan, S., Jin, P.: Domain-invariant region proposal network for cross-domain detection. In: ICME, pp. 1–6 (2020)
30. Ma, J., Chen, B.: Dual refinement feature pyramid networks for object detection. arXiv preprint arXiv:2012.01733 (2020)

Transfer Learning for Larger, Broader, and Deeper Neural-Network Quantum States

Remmy Zen[(✉)] and Stéphane Bressan

National University of Singapore, Singapore, Singapore
remmy@u.nus.edu, steph@nus.edu.sg

Abstract. Neural-network quantum states are a family of unsupervised neural network models simulating quantum many-body systems. We investigate the efficiency and effectiveness of neural-network quantum states with deep restricted Boltzmann machine with different sizes, breadths, and depths. We propose and evaluate several transfer learning protocols for the improvement of scalability, effectiveness, and efficiency of neural-network quantum states with different numbers of visible nodes, hidden nodes per layer, and hidden layers. The results of a comparative empirical performance evaluation confirm the advantages of deep neural-network quantum states and of the proposed transfer learning protocols.

Keywords: Neural-network quantum states · Transfer learning · Restricted Boltzmann machine

1 Introduction

There is an intrinsic and original relationship between energy-based neural networks [24] and quantum many-body systems [42], which is a system that consists of many interacting particles. Carleo and Troyer [3] put this relationship under the spotlight when they proposed a family of unsupervised neural network models capable of finding the ground state energy of a quantum many-body system described by its Hamiltonian. The authors called this family of neural network models *neural-network quantum states*. However, the number of parameters of a neural-network quantum states model grows with the size of the systems. Indeed, the number of particles in the many-body quantum system is the number of visible nodes of the neural network. This makes the optimisation problem harder for larger systems, thus requiring **larger networks**. In view of this challenge, Zen et al. [48] devised several transfer learning protocols to improve the scalability, efficiency, and effectiveness of neural-network quantum states.

One of the reasons that neural network models succeed is due to their universal function approximation ability. The universal approximation theorem [20,23] states that a neural network with one hidden layer with enough hidden nodes can model any smooth function.

© Springer Nature Switzerland AG 2021
C. Strauss et al. (Eds.): DEXA 2021, LNCS 12924, pp. 207–219, 2021.
https://doi.org/10.1007/978-3-030-86475-0_21

Are **broader networks**, i.e. networks with more hidden nodes in some hidden layers, and **deeper networks**, i.e. networks with more hidden layers, better? The representational capability of a network increases with the breadth of its hidden layers. However, this may result in a network that is too large to train efficiently. Several studies [30,34,36] argue theoretically and empirically that a deeper network can be more effective. However, some also argue that deeper networks are harder to train [10,31].

In this paper, we first investigate the efficiency (speed) and effectiveness (accuracy) of neural-network quantum states with restricted Boltzmann machine with different breadths and depths. Then, with the objective of studying large many-body systems, we propose and evaluate the potential of existing [48] and new transfer learning protocols for the improvement of scalability, effectiveness, and efficiency of neural-network quantum states with different numbers of visible nodes, hidden nodes per layer, and hidden layers. In a slight abuse of terminology, we refer to the corresponding neural networks as "*deep restricted Boltzmann machines*". We comparatively and empirically evaluate the different existing and proposed models, as well as the reference matrix product states method [32], for the case of a one-dimensional quantum Ising model in the ferromagnetic phase.

The remainder of this paper is structured as follows. Section 2 synthesises related works. Section 3 presents the necessary background on neural-network quantum states. Section 4 presents and discusses the different transfer learning protocols. Section 5 presents and analyses the results of a comparative performance evaluation of the different methods. Section 6 summarises the key findings and highlights the possible extensions for this work.

2   Related Work

Quantum many-body systems are difficult to simulate and study because the state space of the system grows exponentially as the size of the systems increases. Meanwhile, a simulation with very large systems is needed to fully understand the behaviour of different materials. To overcome these challenges, numerical methods, such as quantum Monte Carlo methods [14] and tensor networks [32], have been widely used by adding several constraints to the simulation of quantum many-body systems. For example, quantum Monte Carlo methods only work on systems with specific characteristics, while tensor networks struggle to simulate systems beyond one dimension. Recently, still in the family of Monte Carlo, Carleo and Troyer [3] proposed neural-network quantum states that represent a quantum many-body system with a neural network, specifically a restricted Boltzmann machine.

Many attempts have been made to evaluate the effectiveness of neural-network quantum states to represent different quantum many-body systems [8, 12,27,49]. Several studies have investigated the effectiveness of neural-network quantum states by looking at different properties, such as the physical properties [9] and the representational power [21]. Neural-network quantum states have also been shown to be applicable for the simulation of a quantum computer [22] and the reconstruction of an unknown quantum state [43].

Several authors have also studied different neural network architectures, such as multilayer perceptrons [2,6,38], convolutional neural networks [7,25], and recurrent neural networks [16] to improve the effectiveness of neural-network quantum states. Neural-network quantum states could also be trained in a supervised manner with a multilayer perceptron [2]. It is also possible to convert restricted Boltzmann machine to multilayer perceptron with specific weights and activation functions that have been studied for neural-network quantum states in [6]. In this paper, we focus on neural-network quantum state with restricted Boltzmann machine architecture. The deep restricted Boltzmann machine that we consider and describe below are different from deep Boltzmann machine [39] and deep belief network [19].

One of the key successes of neural networks can be credited to their being a universal function approximator. The universal approximation theorem states that a neural network with one hidden layer and enough hidden units can model any smooth function to any desired level of accuracy [20,23]. However, this may result in a network that is too large to train efficiently and does not generalise well because it memorises all of its input [13]. Several works [30,34,36] both empirical and theoretical have shown that deeper networks are more efficient and effective than shallow network because it can learn hierarchical features of its input. However, the same problem appears because deeper networks are also harder to train [10,31]. The author of [15] proposed residual network to make training of deeper network easier. However, to improve one percent of accuracy, one needs to approximately double the number of layers [46]. Other works [1,26,37] showed that shallow but broad neural networks can learn as effectively as or even more effective than deep neural networks. However, it seems that the optimal accuracy is only achieved by balancing the width and depth of the network [40,41,46]. Thus, making the depth and the width of the network an hyper-parameter that needs to be fine tuned.

One finds diverse results in the neural-network quantum states literature about how many hidden layers and hidden nodes are needed. Saito and Kato [38] mentioned that a single hidden layer neural network converges faster than a deeper one for a Bose-Hubbard model. The authors mentioned that a deeper network may improve the convergence but it may also be counterproductive because of the increased complexity. Cai and Liu [2] mentioned that a two-layer network is more effective than a one-layer network for a one-dimensional Heisenberg model. However, further increasing the number of layers does not significantly improve the performance and there is no established result about the number of hidden nodes. Choo et al. [6] compared restricted Boltzmann machines and multilayer perceptrons to find the excited states of a one-dimensional Heisenberg model. They mentioned that single hidden layer multilayer perceptron has comparable performance to the restricted Boltzmann machine. Their results showed that multilayer perceptron with two hidden layers is more effective than restricted Boltzmann machine. However, increasing the number of hidden units of restricted Boltzmann machine improves the effectiveness. For a Bose-Hubbard model, they found that a multilayer perceptron with two hidden layers is more effective than a multilayer perceptron with one hidden layer even if the latter has a large number of hidden units.

In this paper, we propose to use transfer learning to improve the efficiency and effectiveness for larger, broader, and deeper neural-network quantum states. In common machine learning settings, transfer learning refers to the reuse of a model trained for a particular task to help with the training of another related task. This is usually done by transferring the weights of the initial network [29,35]. Transfer learning has been successfully used to improve the efficiency and effectiveness of various machine learning tasks and architectures [33,44]. In quantum computing, transfer learning has been studied for hybrid classical-quantum neural networks [5,28]. For neural-network quantum states, transfer learning has been used to improve scalability of Ising and Heisenberg model [48] and to find quantum critical points of a system [47].

3   Background

3.1   Quantum Many-Body Systems

Quantum many-body system [42] consists of particles interacting with each other and with external fields in a discrete or continuous multi-dimensional space.

We study the quantum Ising model [4] where each particle is pinned on a lattice. The model is fully characterised by its position and its binary discrete spin, which can be in an up $(+1)$, or in a down (-1) state. A configuration of the system is given by the value for each spin. Since the particle is a binary discrete spin, there are 2^N possible configurations. For example, $x = (+1, +1, -1, -1)$ is one of the possible configuration for a system in the Ising model with four particles.

A state of a quantum system is described by the wave function ψ. The normalised modulus square $|\psi(x)|^2 / \sum_x |\psi(x)|^2$ gives the probability of the configuration x in the state.

Each particle in the Ising model interacts with its nearest neighbours with magnitude J and with an external field with magnitude h. The dynamics of a system of N particles with amplitude values J and h is described the Hamiltonian matrix H of size $2^N \times 2^N$ defined in Eq. 1, where $neighbour(\cdot)$ is a function that returns the neighbour particles of a particle and σ_i^α are Pauli matrices where $\alpha = x, z$ and i indicates the position of the spin it acts upon.

$$H = -h \sum_i^N \sigma_i^x - J \sum_i^N \sum_{j \,\in\, neighbour(i)} \sigma_i^z \sigma_j^z. \tag{1}$$

The energy of a state is the expected value of the local energy of the configurations. The local energy function of a configuration x, $E_{loc}(x)$, is defined in Eq. 2, where $H_{x,x'}$ is the entry of the Hamiltonian matrix for the configurations x and x'.

$$E_{loc}(x) = \sum_{x'} H_{x,x'} \frac{\psi(x')}{\psi(x)} \tag{2}$$

By diagonalising the Hamiltonian matrix, one obtains the eigenvalues and the corresponding eigenvectors that correspond to the possible energy levels and the corresponding possible states of the system, respectively, after a measurement of its energy has been performed. In particular, the energy of the state of the system with the lowest possible energy, the ground state, is the lowest eigenvalue of the Hamiltonian matrix, which is called the ground state energy.

It is computationally very expensive to diagonalise this matrix for large N. The reader recalls that the Hamiltonian matrix has a size exponential in the number of particles. In this context, a system with 32 particles can already be considered large as it has a state-space made of more than four billion configurations and a Hamiltonian of size more than sixteen quintillions. Fortunately, the variational principle applies and one can look for a surrogate function that approximates the wave function and minimises the local energies and, find, in this manner, the ground state energy. The zero-variance property also guarantees that the value of the local energies converges. In addition, the Hamiltonian for Ising systems is generally sparse.

3.2   Deep Neural-Network Quantum States

Carleo and Troyer [3] proposed to use a restricted Boltzmann machine [24] as a surrogate of the wave function $\psi(x)$ to find the ground state energy of a system. A typical restricted Boltzmann machine consists of a visible layer x and a hidden layer h. It is possible to add more hidden layers for the purpose of neural-network quantum states, thus making it similar to the multilayer perceptron [6].

We consider deep restricted Boltzmann machines similar to multilayer perceptrons [6]. They consist of one visible layer and L hidden layers $h^{(l)}$, where l is the index of the layer. The visible layer x consists of N visible nodes. The number of nodes in a hidden layer $h^{(l)}$ is a multiple, α, of the number of visible nodes N. Each hidden layer is connected to the previous layer by the weight matrix $W^{(l)}$. For simplicity, we omit the biases in the presentation but use them in the implementation. The activation function is cosh. Each of the N visible nodes of the deep restricted Boltzmann machine of a neural-network quantum state represents one of the N particles of the system and its value represents the value of the binary spin of that particle, therefore, taking one of the values -1 and $+1$.

Training the neural-network quantum state follows a Monte Carlo variational approach that trains the deep restricted Boltzmann machine in an unsupervised manner to become a surrogate of the wave function that minimises the local energy. The deep restricted Boltzmann machine is initialised with random weights. One iteration of the training involves sampling the configurations, with the Metropolis sampling described in [3], evaluating the expectation value of the local energy in Eq. 3 (notice that only the entries of the Hamiltonian corresponding to neighbouring configurations are needed), and calculating the gradient to update the weights of the restricted Boltzmann machine with gradient descent. We repeat the process until a predefined stopping criterion is reached. The final expectation value of the local energy represents the approximated energy at the ground state of the system.

$$E_{loc}(x) = \sum_{x'} H_{x,x'} \prod_n \frac{\cosh\left(\cdots\left(\sum_l \cosh\left(\sum_k x'_k W^{(1)}_{kl}\right)\right)\cdots W^{(L)}_{l,n}\right)}{\cosh\left(\cdots\left(\sum_l \cosh\left(\sum_j x'_j W^{(1)}_{jl}\right)\right)\cdots W^{(L)}_{l,n}\right)} \quad (3)$$

4   Methodology

The objective of the methods presented in this paper is to estimate the ground state energy of a quantum many-body system in the Ising model. We use deep restricted Boltzmann machine neural-network quantum states. We consider larger, broader, and deeper neural-network quantum state by varying the number of hidden nodes and hidden layers and consider one baseline and three transfer learning protocols illustrated in Fig. 1 and presented below.

Our baselines, referred to as *cold start* borrowing the terminology from transfer learning literature [45], are deep neural-network quantum states as described in Subsect. 3.2.

Instead of training a broad and deep network directly, we propose an iterative approach that starts from the trained neural-network quantum state solution of a small system (shown in Fig. 1a) and repeatedly uses one of the three protocols to initialise and train a larger, broader, and deeper neural-network quantum state illustrated in Fig. 1b, 1c, and 1d, respectively, until the target architecture is reached.

Transfer Learning to a Larger Network. We present a transfer learning protocol, called *transfer to a larger network*, from a trained deep restricted Boltzmann machine neural-network quantum state of size N with L hidden layers and $\alpha \times N$ hidden nodes to a deep restricted Boltzmann machine neural-network quantum state of size $2 \times N$ with L hidden layers and $\alpha \times 2 \times N$ hidden nodes. The transfer learning protocol extends the $(L, 2)$−tiling protocol proposed in [48] for neural-quantum states with one hidden layer to deep neural-network quantum with more than one hidden layer. The weight matrix of the target network is double the size of the weight matrix of the base network. The protocol transfers the weights of the base network to the first half of nodes connected to the first half of the hidden nodes of the target network and then repeat them for the other half nodes but couples them to the new hidden nodes of the target network. The rest of the weights are initialised randomly. After transferring the weights from the base network, the target network is further trained until it converges. Figure 1b shows an example of the target network after applying $(L, 2)$−tiling protocol from the base network in Fig. 1a.

Transfer Learning to a Broader Network. We present a transfer learning protocol, called *transfer to a broader network*, from a deep restricted Boltzmann machine neural-network quantum state of size N with L hidden layers and $\alpha \times N$

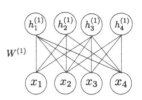

(a) Baseline cold start with $N = 2, L = 1$, and $\alpha = 1$.

(b) Transfer to larger network with $N = 4, L = 1$, and $\alpha = 1$.

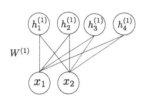

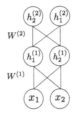

(c) Transfer to broader network with $N = 2, L = 1$, and $\alpha = 2$.

(d) Transfer to deeper network with $N = 1, L = 2$, and $\alpha = 1$.

Fig. 1. Overview of transfer baseline and three learning protocols. Blue solid lines are transferred weights from the original network. Black dashed line are weights that are initialised randomly. Biases are omitted for simplicity. (Color figure online)

hidden nodes to a deep restricted Boltzmann machine neural-network quantum state of size N with L hidden layers and $2 \times \alpha \times N$ hidden nodes. In this protocol, the base network is a trained one-layer restricted Boltzmann machine neural-network quantum state for size N with $\alpha \times N$ hidden nodes. The target network is a one-layer restricted Boltzmann machine neural-network quantum state for size N with $2 \times \alpha \times N$ hidden nodes. In this case, the number of hidden nodes in the weight matrix is doubled. We transfer the weights of the base network for both the first and second half of the hidden nodes of the target network. All weights are determined by this transfer. There is no need for random initialisation of any weights.

Transfer Learning to a Deeper Network. We present a transfer learning protocol, called *transfer to a deeper network*, from a deep restricted Boltzmann machine neural-network quantum state of size N with L hidden layers and $\alpha \times N$ hidden nodes to a deep restricted Boltzmann machine neural-network quantum state of size N with $L + 1$ hidden layers and $\alpha \times N$ hidden nodes. In this protocol, the base network is a trained restricted Boltzmann machine neural-network quantum state for size N with L hidden layers and $\alpha \times N$ hidden nodes. The target network is a one-layer restricted Boltzmann machine neural-network quantum state for size N with $L + 1$ hidden layers and $\alpha \times N$ hidden nodes. In this case, we need new weights from the new hidden layer to the previous, which are simply initialised randomly. Indeed, the proposed protocol is reminiscent of the layer-wise pre-training done for deep belief network in [19]. Note that an alternative strategy would be to initialise the new weights as a diagonal implementing the identity function. While this transfer is initially effective it only implements the composition of the base network with the identity function and does not yield any interesting benefit.

5   Performance Evaluation

We comparatively evaluate the performance of the three transfer learning protocols compared to the cold start approach. We evaluate the effectiveness by computing the relative error of the energy of the ground state compared to the matrix product states method [32]. We evaluate the efficiency by measuring the time needed to reach convergence. For the transfer learning protocols, we measure the cumulative time.

The value reported is from 20 realisations of the same calculation. Each realisation has the same random seed throughout all different experiments. We study a one-dimensional Ising model with size $N = 64$, open boundary conditions, $J = 2$, and $h = 1$. The network is initialised from a zero-mean normal distribution with 0.01 standard deviation following [18] even though there are more advanced solutions [11, 47]. At each iteration of the training, we take 1,000 samples and update weights with gradient descent algorithm RMSProp optimiser [17] with a learning rate of 0.001. Following [48], the training stops after it reaches the dynamic stopping criteria, which is set to be when the ratio of the standard deviation and the mean of the local energy is less than 0.01.

5.1   Broader Networks

We evaluate the efficiency and effectiveness of the cold start (CS) approach and two transfer learning protocols: to a larger network (TL) and to a broader network (TB). We vary the ratio between the hidden nodes and visible nodes $\alpha = \{1, 2, 4\}$, while keeping the number of hidden layer $L = 1$.

For the transfer to a larger network, we start from a cold start network of $N = 8$, use $(L, 2)$−tiling protocol iteratively to $N = 64$. For transfer learning to broader networks, we use the result of the transfer to a larger network for $N = 64$ and use the protocol to transfer to $\alpha = 2$ and subsequently to $\alpha = 4$.

Figure 2a shows box plots of the effectiveness evaluation of the cold start and two transfer learning protocols. We see that the transfer learning protocols are more effective and far more robust than cold start.

For cold start, we see that for $\alpha = 1$, none of the 20 realisations get relative energy error lower than ≈ 0.02. However, this is not the case for $\alpha = 2$ and $\alpha = 4$. This indicates that increasing α is a good strategy. We also see that the quartile bar for $\alpha = 2$ is smaller than $\alpha = 4$. Our finer-grain analysis by looking at the individual realisation shows that increasing the number of hidden nodes is a good strategy but adding too much might also be counterproductive.

Figure 2b shows the effectiveness evaluation of just the two transfer learning protocols. We see that the two transfer learning protocols are competitive.

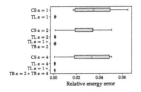 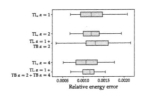 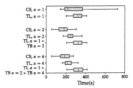

(a) Effectiveness evaluation.   (b) Effectiveness evaluation   (c) Efficiency evaluation
                                 of the transfer.

Fig. 2. Box plot of effectiveness and efficiency evaluation of the cold start (CS), transfer to a larger network (TL), and transfer learning to a broader network (TB) for Ising model size 64 using one-layer restricted Boltzmann machine neural-network quantum state with $\alpha = \{1, 2, 4\}$.

For both transfer learning protocols, we see that increasing the α marginally improves the effectiveness. Transfer to a larger network performs marginally more effective than transfer learning to broader networks.

It is interesting to see that some realisation of transfer to a broader network for $\alpha = 2$ has a lower minimum value than the transfer to a larger network of $\alpha = 1$. This means that some realisations are improved after applying the protocol. By further increasing the α to four, the maximum value of the error gets lowered but the minimum value gets increased again. Our finer-grain analysis by looking at the evolution for each realisation shows that transfer to a broader network helps to slightly improve the effectiveness.

Figure 2c shows box plots of the efficiency evaluation of the cold start and two transfer learning protocols. We see that cold start are generally more efficient but converging to a local minimum because it is not very effective. We see that, for transfer to a larger network, increasing α slightly improves the efficiency. We see that time for transfer to a broader network is very similar to transfer to a larger network of $\alpha = 1$ showing that not a lot of training is needed. However, transfer to a larger network is still more efficient than transfer to a broader network.

5.2   Deeper Networks

We evaluate the efficiency and effectiveness of the cold start (CS) approach and two transfer learning protocols: to a larger network (TL) and to a deeper network (TD). We vary the number of hidden layer $L = \{1, 2, 3\}$, while keeping $\alpha = 1$ at every layer. The training is similar to the broader networks except that we apply the transfer to a deeper network to $L = 2$ and subsequently to $L = 3$.

Figure 3a shows box plots of the effectiveness evaluation of the cold start approach and two transfer learning protocols. We see again that the transfer learning protocols are more effective and more robust than the cold start.

For the cold start, we also see that some realisations for $L = 2$ and $L = 3$ could approximate the ground state energy effectively. We see that the error for $L = 3$ is relatively lower than $L = 1$ and $L = 2$. This indicates that adding

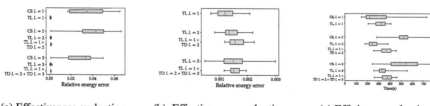

(a) Effectiveness evaluation.   (b) Effectiveness evaluation   (c) Efficiency evaluation
                                     of the transfer.

Fig. 3. Box plot of effectiveness and efficiency evaluation of the cold start (CS), transfer learning to larger networks (TL), and transfer learning to deeper systems (TD) for Ising model size 64 using restricted Boltzmann machine neural-network quantum state with number of hidden layer $L = \{1, 2, 3\}$ and $\alpha = 1$.

more layer is generally more effective for cold start. Our finer-grain analysis also confirms this hypothesis.

Figure 3b shows the evaluation of the two transfer learning protocols. We again see that the two transfer learning protocols are competitive. For both transfer learning protocols, we see that increasing L slightly make it less effective. We also see that the transfer to a deeper network is more robust than transfer to a larger network but not as effective.

Figure 3c shows box plots of the efficiency evaluation of the cold start approach and two transfer learning protocols. Similar to before, we see that cold start are efficient because it converges to a local minimum and not very effective. We see that, for transfer to a larger network, increasing the number of hidden layers to two slightly improves the efficiency but increasing it further does not improves it further. We also see that transfer to a larger network is more efficient than transfer to a deeper network.

By comparing Fig. 2 and Fig. 3, in this case, we see that adding more hidden nodes seems to be the more effective and efficient strategy.

6   Conclusion

We investigated the efficiency and effectiveness of deep neural-network quantum states and of three transfer learning protocols for the improvement of scalability, effectiveness, and efficiency of neural-network quantum states with different numbers of visible nodes, hidden nodes per layer, and hidden layers.

The results of a comparative empirical performance evaluation with systems in a ferromagnetic phase of the one-dimensional Ising model showed that broader and deeper networks are generally more efficient and effective than narrow and shallow ones, although marginally and preferably broader. More significantly, the results showed that the transfer learning from a small deep neural-network quantum state to a larger, broader, or deeper target neural-network quantum state is more effective and efficient than a cold start training of the target network.

Now that we have set the building blocks for the transfer learning to larger, broader, and deeper networks, we are exploring their combination for very large systems in different phases and for the identification of quantum critical points. Additional work is also necessary for the evaluation and tuning of the transfer protocols to other models than Ising.

Acknowledgment. This research is supported by the project "Complex quantum systems with neural networks: relaxation and quantum computing" (No. MOE-T2EP50120-0019) funded by the Singapore Ministry of Education. The computational work for this article was partially performed on resources of the National Supercomputing Centre (NSCC), Singapore (https://www.nscc.sg).

References

1. Ba, L.J., Caruana, R.: Do deep nets really need to be deep? In: Proceedings of the 27th International Conference on Neural Information Processing Systems, vol. 2, pp. 2654–2662 (2014)
2. Cai, Z., Liu, J.: Approximating quantum many-body wave functions using artificial neural networks. Phys. Rev. B **97**(3), 035116 (2018)
3. Carleo, G., Troyer, M.: Solving the quantum many-body problem with artificial neural networks. Science **355**(6325), 602–606 (2017)
4. Chakrabarti, B.K., Dutta, A., Sen, P.: Quantum Ising Phases and Transitions in Transverse Ising Models, vol. 41. Springer, Heidelberg (2008). https://doi.org/10.1007/978-3-540-49865-0
5. Chen, S.Y.C., Yoo, S.: Federated quantum machine learning. Entropy **23**(4), 460 (2021)
6. Choo, K., Carleo, G., Regnault, N., Neupert, T.: Symmetries and many-body excitations with neural-network quantum states. Phys. Rev. Lett. **121**(16), 167204 (2018)
7. Choo, K., Neupert, T., Carleo, G.: Two-dimensional frustrated J1–J2 model studied with neural network quantum states. Phys. Rev. B **100**(12), 125124 (2019)
8. Deng, D.L., Li, X., Sarma, S.D.: Machine learning topological states. Phys. Rev. B **96**(19), 195145 (2017)
9. Deng, D.L., Li, X., Sarma, S.D.: Quantum entanglement in neural network states. Phys. Rev. X **7**(2), 021021 (2017)
10. Du, S., Lee, J., Li, H., Wang, L., Zhai, X.: Gradient descent finds global minima of deep neural networks. In: International Conference on Machine Learning, pp. 1675–1685. PMLR (2019)
11. Efthymiou, S., Beach, M.J., Melko, R.G.: Super-resolving the ising model with convolutional neural networks. Phys. Rev. B **99**(7), 075113 (2019)
12. Gao, X., Duan, L.M.: Efficient representation of quantum many-body states with deep neural networks. Nat. Commun. **8**(1), 1–6 (2017)
13. Goodfellow, I., Bengio, Y., Courville, A., Bengio, Y.: Deep Learning, vol. 1. MIT Press, Cambridge (2016)
14. Gubernatis, J., Kawashima, N., Werner, P.: Quantum Monte Carlo Methods. Cambridge University Press, Cambridge (2016)
15. He, K., Zhang, X., Ren, S., Sun, J.: Deep residual learning for image recognition. In: Proceedings of the IEEE Conference on Computer Vision and Pattern Recognition, pp. 770–778 (2016)

16. Hibat-Allah, M., Ganahl, M., Hayward, L.E., Melko, R.G., Carrasquilla, J.: Recurrent neural network wave functions. Phys. Rev. Res. **2**(2), 023358 (2020)
17. Hinton, G., Srivastava, N., Swersky, K.: Neural networks for machine learning lecture 6a overview of mini-batch gradient descent (2012)
18. Hinton, G.E.: A practical guide to training restricted Boltzmann machines. In: Montavon, G., Orr, G.B., Müller, K.-R. (eds.) Neural Networks: Tricks of the Trade. LNCS, vol. 7700, pp. 599–619. Springer, Heidelberg (2012). https://doi.org/10.1007/978-3-642-35289-8_32
19. Hinton, G.E., Salakhutdinov, R.R.: Reducing the dimensionality of data with neural networks. Science **313**(5786), 504–507 (2006)
20. Hornik, K.: Approximation capabilities of multilayer feedforward networks. Neural Netw. **4**(2), 251–257 (1991)
21. Huang, Y., Moore, J.E.: Neural network representation of tensor network and chiral states. arXiv:1701.06246 (2017)
22. Jónsson, B., Bauer, B., Carleo, G.: Neural-network states for the classical simulation of quantum computing. arXiv:1808.05232 (2018)
23. Le Roux, N., Bengio, Y.: Representational power of restricted Boltzmann machines and deep belief networks. Neural Comput. **20**(6), 1631–1649 (2008)
24. Lecun, Y., Chopra, S., Hadsell, R., Ranzato, M.A., Huang, F.J.: A tutorial on energy-based learning. In: Predicting Structured Data. MIT Press (2006)
25. Liang, X., Liu, W.Y., Lin, P.Z., Guo, G.C., Zhang, Y.S., He, L.: Solving frustrated quantum many-particle models with convolutional neural networks. Phys. Rev. B **98**(10), 104426 (2018)
26. Liu, N., Zaidi, N.A.: Artificial neural network: deep or broad? An empirical study. In: Kang, B.H., Bai, Q. (eds.) AI 2016. LNCS (LNAI), vol. 9992, pp. 535–541. Springer, Cham (2016). https://doi.org/10.1007/978-3-319-50127-7_46
27. Lu, S., Gao, X., Duan, L.M.: Efficient representation of topologically ordered states with restricted Boltzmann machines. Phys. Rev. B **99**(15), 155136 (2019)
28. Mari, A., Bromley, T.R., Izaac, J., Schuld, M., Killoran, N.: Transfer learning in hybrid classical-quantum neural networks. Quantum **4**, 340 (2020)
29. Martin, G.: The effects of old learning on new in Hopfield and backpropagation nets. Microelectronics and Computer Technology Corporation (1988)
30. Montufar, G.F., Pascanu, R., Cho, K., Bengio, Y.: On the number of linear regions of deep neural networks. Adv. Neural Inf. Process. Syst. **27**, 2924–2932 (2014)
31. Nielsen, M.A.: Neural Networks and Deep Learning, vol. 25. Determination Press, San Francisco (2015)
32. Orús, R.: A practical introduction to tensor networks: matrix product states and projected entangled pair states. Ann. Phys. **349**, 117–158 (2014)
33. Pan, S.J., Yang, Q., et al.: A survey on transfer learning. IEEE Trans. Knowl. Data Eng. **22**(10), 1345–1359 (2010)
34. Poggio, T., Mhaskar, H., Rosasco, L., Miranda, B., Liao, Q.: Why and when can deep-but not shallow-networks avoid the curse of dimensionality: a review. Int. J. Autom. Comput. **14**(5), 503–519 (2017)
35. Pratt, L.Y.: Discriminability-based transfer between neural networks. In: Advances in Neural Information Processing Systems, pp. 204–211 (1993)
36. Raghu, M., Poole, B., Kleinberg, J., Ganguli, S., Sohl-Dickstein, J.: On the expressive power of deep neural networks. In: international Conference on Machine Learning, pp. 2847–2854. PMLR (2017)
37. Romero, A., Ballas, N., Kahou, S.E., Chassang, A., Gatta, C., Bengio, Y.: FitNets: hints for thin deep nets. arXiv preprint arXiv:1412.6550 (2014)

38. Saito, H., Kato, M.: Machine learning technique to find quantum many-body ground states of bosons on a lattice. J. Phys. Soc. Jpn. **87**(1), 014001 (2018)
39. Salakhutdinov, R., Hinton, G.: Deep Boltzmann machines. In: Artificial Intelligence and Statistics, pp. 448–455. PMLR (2009)
40. Sun, S., Chen, W., Wang, L., Liu, X., Liu, T.Y.: On the depth of deep neural networks: a theoretical view. In: Proceedings of the AAAI Conference on Artificial Intelligence, vol. 30 (2016)
41. Tan, M., Le, Q.: EfficientNet: rethinking model scaling for convolutional neural networks. In: International Conference on Machine Learning, pp. 6105–6114. PMLR (2019)
42. Thouless, D.J.: The quantum mechanics of many-body systems. Courier Corporation (2014)
43. Torlai, G., Mazzola, G., Carrasquilla, J., Troyer, M., Melko, R., Carleo, G.: Neural-network quantum state tomography. Nat. Phys. **14**(5), 447–450 (2018)
44. Weiss, K., Khoshgoftaar, T.M., Wang, D.D.: A survey of transfer learning. J. Big Data **3**(1), 1–40 (2016). https://doi.org/10.1186/s40537-016-0043-6
45. Yosinski, J., Clune, J., Bengio, Y., Lipson, H.: How transferable are features in deep neural networks? arXiv preprint arXiv:1411.1792 (2014)
46. Zagoruyko, S., Komodakis, N.: Wide residual networks. arXiv preprint arXiv:1605.07146 (2016)
47. Zen, R., et al.: Finding quantum critical points with neural-network quantum states. In: ECAI 2020–24th European Conference on Artificial Intelligence. Frontiers in Artificial Intelligence and Applications, vol. 325, pp. 1962–1969. IOS Press (2020)
48. Zen, R., et al.: Transfer learning for scalability of neural-network quantum states. Phys. Rev. E **101**(5), 053301 (2020)
49. Zhang, Y.H., Jia, Z.A., Wu, Y.C., Guo, G.C.: An efficient algorithmic way to construct Boltzmann machine representations for arbitrary stabilizer code. arXiv:1809.08631 (2018)

LGTM: A Fast and Accurate kNN Search Algorithm in High-Dimensional Spaces

Yusuke Arai[1]([⊠]), Daichi Amagata[1,2], Sumio Fujita[3], and Takahiro Hara[1]

[1] Osaka University, Osaka, Japan
{arai.yusuke,amagata.daichi,hara}@ist.osaka-u.ac.jp
[2] JST PRESTO, Tokyo, Japan
[3] Yahoo Japan Corporation, Tokyo, Japan
sufujita@yahoo-corp.jp

Abstract. Approximate k nearest neighbor (AkNN) search is a primitive operator for many applications, such as computer vision and machine learning. As these applications deal with a large set of high-dimensional points, a fast and accurate solution is required. It is known that graph-based AkNN search algorithms are faster and more accurate than other approaches, including hash- and quantization-based ones. However, existing graph-based AkNN search algorithms rely purely on heuristics, i.e., their performances are not theoretically supported. This paper proposes LGTM, a new algorithm for AkNN search, that exploits both locality-sensitive hashing and a proximity graph. The performance of LGTM is theoretically supported. Our experiments on real datasets show that LGTM outperforms state-of-the-art AkNN search algorithms.

Keywords: AkNN search · High-dimensional data

1 Introduction

The problem of k nearest neighbor (kNN) search on the Euclidean spaces is a fundamental problem in many applications, such as computer vision [7], machine learning [11], and databases [9]. This problem is formally defined as follows:

Definition 1 (k NEAREST NEIGHBOR SEARCH PROBLEM). *Given a set X of d-dimensional points, a query point q, and a result size k, the problem of k nearest neighbor search retrieves a set S of k points in X such that $\forall x \in S$, $\forall x' \in X - S$, $dist(x, q) \leq dist(x', q)$, where $dist(\cdot, \cdot)$ evaluates the Euclidean distance between two points (ties are broken arbitrarily).*

This problem has been extensively studied, and tree-based solutions, such as kd-tree [3], are known to be exact and efficient on low-dimensional spaces. However, recent applications usually deal with high-dimensional points (d is large), e.g., image feature vectors [10] and deep descriptions [1], and tree-based techniques are no more efficient on such data, due to the curse of dimensionality. It is

© Springer Nature Switzerland AG 2021
C. Strauss et al. (Eds.): DEXA 2021, LNCS 12924, pp. 220–231, 2021.
https://doi.org/10.1007/978-3-030-86475-0_22

actually hard to make exact kNN search on a high-dimensional space efficient [12], and applications allow for a small-error result if it can be fast obtained. Approximate kNN (AkNN) search, which aims at finding a high-recall kNN result quickly, is therefore receiving attention, and this paper focuses on the AkNN search problem.

Related Work. There are three main techniques for solving the AkNN search problem efficiently: hashing [4], quantization [8], and proximity graph [7]. Hashing is based on locality-sensitive hashing, and it accesses only points having the same hash values with a given query. Quantization is conceptually similar to hashing, because it reduces the dimensionality of the original points and searches for AkNNs of a given query from related arrays of quantized points in an inverted index. In a proximity graph, nodes correspond to points and there are edges between the nodes if their distances are small. Given a query and a start node of the proximity graph, graph-based algorithms traverse, in a greedy manner, those nodes that approach the query. Among the three techniques, graph-based algorithms show superior performance w.r.t. both computational efficiency and accuracy [6,9,11].

Our Contribution. However, state-of-the-art graph-based algorithms rely purely on heuristics, that is, their performances are not analyzable by theory. This is not desirable, because applications or users cannot know which parameter can *certainly* control the trade-off between efficiency and accuracy. (State-of-the-art algorithms [6,11] do not have such a parameter.) Furthermore, although analysts indeed need simple and efficient solutions, state-of-the-art algorithms require complex graph structures (e.g., hierarchical graph [11] and monotonic paths [6,7]) to obtain a high-recall AkNN result quickly. Nevertheless, these structures do not ensure access to points close to a given query with a small number of distance calculations.

To address the above issues, we make the following contributions:

- We devise a data structure that integrates hash tables and a proximity graph. Although this data structure is simple, it provides an *exact* kNN search algorithm with the same theoretical time as that of an existing *approximate* kNN search algorithm.
- The above data structure incurs high pre-processing and memory costs. We hence develop a technique that makes this theoretical algorithm practical by proposing LGTM. LGTM provides a high-recall AkNN result with a fast computation time, using only $O(n)$ space, where n is the cardinality of a given dataset. The trade-off between the efficiency and accuracy of LGTM is controllable by a single parameter. It is ensured that, if the parameter increases, the probability that LGTM accesses the exact kNN becomes high (i.e., the accuracy increases). Furthermore, different from the existing graph-based algorithms, LGTM can exploit multi-threading. Thanks to this advantage, even if the parameter increases, LGTM keeps computational efficiency.
- We conduct extensive experiments on real datasets SIFT, GIST, and Deep. The experimental results confirm that LGTM outperforms state-of-the-art AkNN search algorithms w.r.t. both efficiency and recall.

Organization. The rest of this paper is organized as follows. Section 2 introduces the theoretical motivation of LGTM, and Sect. 3 presents our algorithm LGTM. Section 4 describes our experimental result, and then we conclude this paper in Sect. 5.

2   Theoretical Motivation

Before we present our key idea, we introduce how to find AkNNs of a given query with an arbitrary graph, in Sect. 2.1. Then Sect. 2.2 establishes a new theorem that utilizes our key idea. (In Sect. 3, we propose LGTM which presents how to make the theory practical.)

2.1   Preliminary

We here introduce a greedy AkNN search algorithm, which is commonly used for graph-based algorithms [6]. We use $E(x)$ to denote the set of edges held by a point (i.e., node) $x \in X$. This algorithm maintains a temporal result set C, and the size of C is at most $\epsilon \cdot k$, where $\epsilon \geq 1$ is a heuristic parameter (if it is large, the accuracy tends to be larger but the efficiency degrades).

Given a start point s, the greedy algorithm inserts $\langle s, dist(s,q) \rangle$ into a priority queue Q. This algorithm checks the first element of Q, and let it be $\langle x, \cdot \rangle$. After the element is popped from Q, we check the edges $(x, x') \in E(x)$, where x' has not been visited. If $dist(x', q)$ improves the temporal result C, it is added to C and Q. (That is, this algorithm traverses points that can be closer to q.) This is repeated until we have $Q = \varnothing$. Algorithm 1 describes its detail.

Time Complexity. Let N be the number of points that are inserted into Q. Furthermore, let deg be the average degree of the given graph. The time complexity of Algorithm 1 is $O(N(deg + \log N))$. It is important to note that $N \ll n$.

2.2   Theoretical Foundation

To make Algorithm 1 fast, we observe that the start point has to be close to q (otherwise, many points are traversed and the number of distance calculations increases), as the above time complexity demonstrates. Figure 1, which assumes $k = 1$ and $\epsilon = 1$, gives this intuition. In the best case, where the start point is the exact nearest neighbor of a given query q, the number of distance calculations (edge or node traversal) is minimized, as shown in Fig. 1(a). On the other hand, when the start point is far from q, as in Fig. 1(b), the efficiency of Algorithm 1 degrades and the answer also tends to be wrong (due to local optima). Now we have to address the challenge: how to find such a start point that is sufficiently close to q. We overcome this challenge by utilizing LSH. Integrating LSH for the Euclidean space and a proximity graph is our key idea, and this yields a new finding (Theorem 2).

To present our new finding, we introduce the LSH scheme.

Algorithm 1: GREEDY-A*kNN-SEARCH

Input: X, a proximity graph, q, k, ϵ, and a start point s

1   $X_{visit} \leftarrow s$

2   $C \leftarrow \langle s, dist(s, q) \rangle$ (C is sorted by $dist(\cdot, q)$)

3   $Q \leftarrow C, \tau \leftarrow \infty$

4   **while** $Q \neq \varnothing$ **do**

5      $\langle x, \cdot \rangle \leftarrow$ the top of Q

6   Pop $\langle x, \cdot \rangle$ from Q

7   **for** *each* $(x, x') \in E(x)$ *s.t.* $x' \notin X_{visit}$ **do**

8         $X_{visit} \leftarrow X_{visit} \cup \{x'\}$

9         $\tau \leftarrow$ the $\epsilon \cdot k$-th distance in C

10   **if** $dist(x', q) < \tau$ **then**

11   Add $\langle x', dist(x', q) \rangle$ into C and remove the last one from C

12          $Q \leftarrow Q \cup \langle x', dist(x', q) \rangle$

13   Sort Q by $dist(\cdot, q)$

14   $S' \leftarrow$ the first k points in C

15   **return** S'

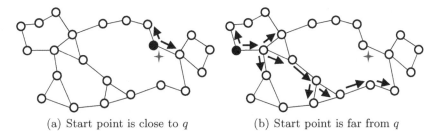

(a) Start point is close to q (b) Start point is far from q

Fig. 1. An example of Algorithm 1, where $k = 1$ and $\epsilon = 1$. Points represent nodes of a graph, and black points are start points. The arrows show edge traversal, and the red crosses are queries.

Definition 2 ((r, cr, p_1, p_2)-SENSITIVE HASHING [4]). *Given a radius r, an approximate ratio $c \geq 1$, and probability values p_1 and p_2, where $p_1 > p_2$, a family $\mathcal{H} = \{h : \mathbb{R}^d \to U\}$ is called (r, cr, p_1, p_2)-sensitive, if, for any $x, x' \in \mathbb{R}^d$, it satisfies the following conditions:*

- *If $dist(x, x') \leq r$, $\Pr[h(x) = h(x')] \geq p_1$.*
- *If $dist(x, x') \geq cr$, $\Pr[h(x) = h(x')] \leq p_2$.*

A well-known (r, cr, p_1, p_2)-sensitive hash function for the Euclidean space is

$$h(x) = \lfloor \frac{a \cdot x + b}{w} \rfloor,$$

where a is a d-dimensional point such that each dimension is drawn independently from a standard normal distribution $\mathcal{N}(0, 1)$, b is a random real value chosen from $[0, w)$, and w is a user-specified constant. Let $\theta = dist(x, x')$, and we have:

$$\Pr[h(x) = h(x')] = \int_0^w \frac{1}{\theta} \cdot f\left(\frac{t}{\theta}\right) \cdot \left(1 - \frac{t}{w}\right) dt, \tag{1}$$

where $f(z) = \frac{2}{\sqrt{2\pi}} e^{-\frac{z^2}{2}}$, i.e., $f(\cdot)$ is the normal probability distribution function. Equation (1) means that, as θ decreases, the collision probability increases. Note that this (r, cr, p_1, p_2)-sensitive hash family is efficient for processing a (r, c)-ball cover query, which is defined below:

Definition 3 $((r, c)$-BALL COVER QUERY$)$. *Given X, a query point q, a distance threshold r, and an approximate ratio c, $B(q, r)$ is defined as a ball centered at q with radius r. An (r, c)-ball cover query returns the following result:*

- *If $B(q, r)$ covers at least one point in X, it returns an arbitrary point in $B(q, cr)$.*
- *If $B(q, r)$ covers no points in X, it returns nothing.*

Let $G(x)$ be $(h_1(x), ..., h_m(x))$, i.e., $G(\cdot)$ is a compound hash function that concatenates m independent hash functions. Given $x \in X$, x is maintained in a bucket with $G(x)$, and a hash table is a set of the buckets. E2LSH [4] creates L hash tables and proves the following:

Theorem 1. *An (r, c)-ball cover query can be answered correctly in $O(n^{\frac{\ln 1/p_1}{\ln 1/p_2}})$ time with at least a constant probability by setting $L = 1/p_1^m$ and $m = \log_{1/p_2} n$.*

Let x^* be the nearest neighbor point of q. From the above theorem, via LSH, we can obtain a start point, which is close to q, with at least a constant probability in a time sub-linear to n, when we have $dist(x^*, q) \leq r$. However, this does not mean that we can obtain a high-recall AkNN result. This is because LSH gives a start point s such that $dist(s, q) \leq cr$, so the start point may not have any edges that can approach the kNN points of q. We overcome this limitation by introducing $(c + 1)r$-similarity graph.

Definition 4 $((c + 1)r$-SIMILARITY GRAPH$)$. *Given X, a radius r, and an approximation ratio $c \geq 1$, in a $(c + 1)r$-similarity graph, a node is a point $x \in X$ and there is an undirected edge between x and x' if and only if $dist(x, x') \leq (c + 1)r$.*

Now we present our new finding:

Theorem 2. *By using LSH and a $(c + 1)r$-similarity graph, we can obtain the exact nearest neighbor x^* of a given query q in $O(\max\{n^{\frac{\ln 1/p_1}{\ln 1/p_2}}, deg\})$ time with at least a constant probability, if $dist(x^*, q) \leq r$.*

PROOF. Assume that we have $dist(x^*, q) \leq r$. Through LSH, we have a point x such that $dist(x, q) \leq cr$ (see Definition 3). From triangle inequality, $dist(x, x^*) \leq r + cr = (1 + c)r$. Definition 4 guarantees that there is an edge between x and x^*. By using Algorithm 1, we can access x^* from x. Then, from Theorem 1 and the time complexity of Algorithm 1, this theorem is true. $\square$

Let $x^{*,k}$ be the k-th nearest neighbor of q. We furthermore have:

Corollary 1. *By using LSH and a $(c+1)r$-similarity graph, we can obtain the exact kNNs of a given query q in $O(\max\{n^{\frac{\ln 1/p_1}{\ln 1/p_2}}, deg^{O(1)}\})$ time with at least a constant probability, if $dist(x^*, q) \leq r$ and $dist(x^*, x^{*,k}) \leq (1+c)r$.*

PROOF. If $dist(x^*, q) \leq r$ and $dist(x^*, x^{*,k}) \leq (1+c)r$, Theorem 2 guarantees that Algorithm 1 reaches the i-th nearest neighbor of q for every $i \in [1, k]$ within a constant, i.e., $O(1)$, hop from a start point x provided by LSH, with at least a constant probability.   □

3   LGTM: From Theory to Practice

Section 2.2 proves that we can obtain the exact kNN in $O(\max\{n^{\frac{\ln 1/p_1}{\ln 1/p_2}}, deg^{O(1)}\})$ time, with at least a constant probability if $dist(x^*, q) \leq r$ and $dist(x^*, x^{*,k}) \leq (1+c)r$. Note that $O(\max\{n^{\frac{\ln 1/p_1}{\ln 1/p_2}}, deg^{O(1)}\}) = O(n^{\frac{\ln 1/p_1}{\ln 1/p_2}})$ in practice, because we usually have $deg \ll n$. That is, the exact kNN answer is obtained in the same time as required for processing a (r, c)-ball cover query. This is attractive, but a $(c+1)r$-similarity graph has two practical limitations. First, building it incurs $O(n^2)$ time. Although this is a pre-processing cost and the $(c+1)r$-similarity graph is general to any queries, this quadratic time is not tolerable for a large n. Second, the worst space complexity of a $(c+1)r$-similarity graph is also $O(n^2)$. Then, the challenge of this section is to design a practically-efficient AkNN search algorithm with a property similar to Theorem 2 without having the above limitations. It is important to note that the main property of LSH (suggested by Theorem 1) is that the success probability increases as we use more hash tables.

To address this challenge, we propose LGTM (Local Graph Traversal Method). LGTM still employs the key idea in Sect. 2.2, but it utilizes that idea in a more practical manner.

3.1   Pre-processing

To start with, we present the pre-processing of LGTM. This pre-processing is done only once. Details of this procedure are described in Algorithm 2.

LSH and Sampling. The first idea of LGTM is to remove $O(n^{\frac{\ln 1/p_1}{\ln 1/p_2}})$ time for determining a start point by sampling. In the pre-processing phase, LGTM constructs L hash tables. Different from E2LSH, each bucket maintains only a constant number of sampled points[1]. This approach determines a start point in $O(1)$ time, when a query is given, as shown later. In addition, the time and space complexities of constructing hash tables are both $O(n)$ by setting $L = O(1)$.

Using an Undirected AKNN Graph. The second idea of LGTM is to employ an approximate KNN graph (we do not use a KNN graph, because building it incurs $O(n^2)$ time). Note that K does not relate to k, as K is the degree of

---

[1] In our implementation, we maintain at most 50 points in each bucket.

Algorithm 2: PRE-PROCESSING OF LGTM

Input: X, L, and m
1 /* Hash table construction */
2 **for** *each $x \in X$* **do**
3   **for** *each $i \in [1, L]$* **do**
4   Maintain x in the bucket with $G_i(x) = (h_{i,1}(x), ..., h_{i,m}(x))$

5 **for** *each $i \in [1, L]$* **do**
6   **for** *each bucket in the i-th hash table* **do**
7   Sample $O(1)$ points in this bucket
8   Remove not sampled points from this bucket

9 /* Undirected AKNN graph construction */
10 Run NNDESCENT [5] on X
11 Convert each directed edge into undirected one

Algorithm 3: QUERY-PROCESSING OF LGTM (a single hash table case)

Input: X, q, k, ϵ, a hash table, and an AKNN graph
1 $s \leftarrow$ the nearest neighbor of q in the bucket with $G(q)$
2 Run Algorithm 1

---

this graph, and $deg = K = O(1)$. Therefore, the space complexity of LGTM (maintaining hash tables and an AKNN graph) is $O(n)$.

Similar to the $(c+1)r$-similarity graph, in an AKNN graph, a node is a point $x \in X$, and there is an edge between x and x', where $x' \in X$ is included in an AKNN of x among X. By utilizing a state-of-the-art AKNN graph construction algorithm, an AKNN graph can be built in nearly $O(n)$ time [5]. We convert each edge in the AKNN graph into undirected.

3.2 Online (Query) Processing

The query processing of LGTM is similar to Theorem 2 (see its proof). For ease of presentation, we first present LGTM in a single hash table case. After that, we introduce the general case.

Single Hash Table Case. Consider a case where $L = 1$. Given a query point q, LGTM first computes $G(q)$. LGTM next computes the nearest neighbor point s in the bucket with $G(q)$. (If this hash table does not have the bucket with $G(q)$, s is a random point in X.) LGTM uses s as a start point of Algorithm 1, then runs Algorithm 1 on the AKNN graph.

Discussion. Different from Corollary 1, LGTM does not guarantee that s can reach the kNNs of q, because of the sampling in Algorithm 2 and the AKNN graph. It is therefore necessary to increase the probability that a start point reaches the kNNs of q (i.e., a start point is close to q). The probability that a bucket maintains both x and x' in a hash table is

---

Algorithm 4: QUERY-PROCESSING OF LGTM

Input: X, q, k, ϵ, t hash tables (t is the number of available threads), and an AKNN graph

1 Prepare t copies of q, q_1, ..., q_t

2 Run Algorithm 3 for q_1, ..., q_t in parallel

3 Merge the AkNN results of q_1, ..., q_t

$$\Pr[G(x) = G(x')] = \prod_{i=1}^{m} \Pr[h_i(x) = h_i(x')].$$

Since each hash table is independently constructed, from the above equation, the probability that a bucket maintains both x and x' in an *arbitrary* hash table is

$$\Pr[\exists j \in [1, L], G_j(x) = G_j(x')] = 1 - (1 - \Pr[G(x) = G(x')])^L, \qquad (2)$$

where $G_i(\cdot)$ is the i-th compound hash function. Notice that, if we use L hash tables, we have L start points. Equation (2) demonstrates that the probability that at least one of the start points is close to q increases as L increases. Therefore, as we employ an AKNN graph, the probability that we can reach points closer to q increases. That is, *as L increases, the recall increases probabilistically*. This is the theoretical motivation for using multiple hash tables for LGTM. State-of-the-art graph-based algorithms [6,11] cannot have this advantage, because they fix the start point.

A straightforward approach that uses the above idea is to run Algorithm 3 multiple (e.g., L) times. One may consider that this approach is not efficient. This is true, but *we can remove this inefficiency* by using a multi-threading approach. Given a query q, consider multiple copies of q with different start points. These query copies can run *in parallel* with multi-threading, and we merge their AkNN results after their respective results are fixed. That is, as we have more available threads, the recall of this query tends to be higher while not losing computational efficiency.

General Case. Now we are ready to present the query processing of LGTM (Algorithm 4). Let t be the number of available threads (in the pre-processing, we set $L \geq t$). First, LGTM prepares t copies of a given query q, q_1, ..., q_t. LGTM runs Algorithm 3 for q_1, ..., q_t in parallel. Note that, for a query copy q_i, LGTM uses $G_i(\cdot)$ (i.e., each query copy uses a different hash table, to have a different start point from those of the others). Then, LGTM merges the AkNN results of the t query copies.

Time Complexity. Because each bucket in the hash tables has only a constant number of points, the start point is obtained in $O(1)$ time. Let N_i be the number of points inserted into Q in Algorithm 1 during processing of a query copy q_i. Since each query copy runs in parallel, the time complexity of Algorithm 4 is $O(\max_{i \in [1,t]} N_i(deg + \log N_i))$. Recall $deg = K = O(1)$, so $O(\max_{i \in [1,t]} N_i(deg + \log N_i)) = O(\max_{i \in [1,t]} N_i \log N_i)$.

4   Experiment

All experiments were conducted on a machine with Ubuntu 16.04 LTS OS, 18 core 3.0 GHz Core i9-9980XE, and 128 GB RAM.

Datasets. We used the following datasets.

- SIFT[2]: 1 million 128-dimensional points that represent image sift features.
- GIST: 1 million 960-dimensional points that represent image gist features.
- Deep [2]: 10 million 96-dimensional points representing deep image features.

We used the query points in their original sets (SIFT and Deep have 10,000 queries, while GIST has 1,000 queries).

Evaluated Algorithms. Our experiments evaluated the following algorithms:

- LGTM: our algorithm proposed in this paper. For the LSH in LGTM, m (w) is 4 (200), 5 (1), and 10 (1) on SIFT, GIST, and Deep, respectively. Besides, we generated 18 hash tables in the pre-processing of LGTM.
- HNSW [11]: a state-of-the-art graph-based algorithm that uses a hierarchical graph with an approximate small-world network model.
- NSG [6]: a state-of-the-art graph-based algorithm that adds monotonic paths to an AKNN graph.
- AKNNG: this algorithm runs Algorithm 1 on an undirected AKNN graph with random start points. This algorithm is used for demonstrating the effectiveness of our start point selection in LGTM.

(We do not consider the other algorithms, because [6,11] show that they are outperformed by HNSW and NSG, and [9] confirms that graph-based algorithms outperform the other approaches.) These algorithms are implemented in C++ and compiled by g++ 5.5 with -O3 flag. We used OpenMP for multi-threading, and SIMD instructions were also used. For HNSW and NSG, we used the fastest implementations[3][4].

Evaluation Criteria. We measured average (1) running time (i.e., time per a query) and (2) recall to evaluate the AkNN algorithms.

4.1   Comparison with AKNNG

To demonstrate the effectiveness of our hashing with proximity graph approach, we compare LGTM with AKNNG. Note that AKNNG also employs the multi-threading approach of LGTM. Table 1 depicts the average running time and recall ($k = 10$ and $t = 8$). We see that LGTM provides higher recall with shorter time than AKNNG. That is, our approach yields both efficiency and accuracy, whereas using simple random start points cannot maximize efficiency and accuracy.

---

[2] http://corpus-texmex.irisa.fr/.

[3] https://github.com/nmslib/hnswlib.

[4] https://github.com/ZJULearning/nsg.

Table 1. Average running time [μs] and recall comparison on SIFT, GIST, and Deep ($k = 10$ and $t = 8$)

Data	Algorithm	Time	Recall
SIFT	AKNNG	139	0.888
	LGTM	**115**	**0.944**
GIST	AKNNG	717	0.765
	LGTM	**699**	**0.845**
Deep	AKNNG	623	0.832
	LGTM	**224**	**0.906**

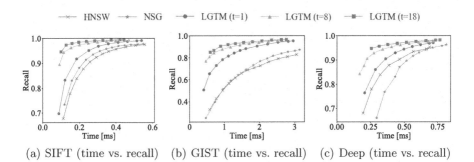

(a) SIFT (time vs. recall)   (b) GIST (time vs. recall)   (c) Deep (time vs. recall)

Fig. 2. Comparison with state-of-the-art. This figure compares LGTM (with different t) with HNSW and NSG, by plotting time–recall curves. Plots in upper-left space mean higher performances.

4.2 Comparison with State-of-the-art Algorithms

We next compare LGTM with two state-of-the-art algorithms, HSNW and NSG. Figure 2 describes the trade-off relationships between the efficiency and accuracy of LGTM ($t = 1$), LGTM ($t = 8$), LGTM ($t = 18$), HNSW, and NSG. We set $k = 10$, and Figs. 2(a)–(c) show time–recall curves.

Overall Observation. From Fig. 2, it is obvious that LGTM outperforms HNSW and NSG, *even when LGTM uses only a single hash table (thread)*. With the same computation time, LGTM returns a more accurate kNN result than those of HNSW and NSG. This result demonstrates that our integration of LSH and an AKNN graph functions better than complex graph structures.

Besides, it can be seen that each algorithm needs longer time on GIST, compared with the other datasets. GIST is a challenging dataset, because, in Algorithm 1, GIST has many points that update the temporal AkNN result. Nevertheless, LGTM ($t = 18$) obtains a high-recall AkNN result with much shorter time than HNSW and NSG.

Comparison with NSG. Now let us compare NSG with the other algorithms. Figure 2(c) shows that NSG is outperformed by the other algorithms by a large

Table 2. Average running time [μs] and recall comparison on SIFT, GIST, and Deep ($t = 8$ for LGTM)

Data	Algorithm	$k = 1$		$k = 10$		$k = 20$		$k = 30$		$k = 40$	
		Time	Recall	Time	Recall	Time	Recall	Time	Recall	Time	Recall
SIFT	NSG	69	0.832	119	0.734	165	0.685	218	0.707	255	0.694
	HNSW	68	0.766	119	0.617	175	0.640	248	0.657	312	0.677
	LGTM	**67**	**0.884**	**115**	**0.944**	**155**	**0.946**	**200**	**0.950**	**251**	**0.955**
GIST	NSG	415	0.604	832	0.459	1052	0.392	1378	0.375	1597	0.354
	HNSW	390	0.553	871	0.436	1162	0.377	1446	0.361	1732	0.345
	LGTM	**345**	**0.694**	**751**	**0.845**	**1008**	**0.850**	**1263**	**0.859**	**1539**	**0.867**
Deep	NSG	203	0.708	267	0.577	330	0.549	384	0.547	439	0.550
	HNSW	190	0.923	223	0.704	320	0.676	339	0.616	405	0.610
	LGTM	**184**	**0.931**	**214**	**0.906**	**262**	**0.913**	**338**	**0.913**	**403**	**0.917**

margin. NSG fixes its start node, and it is often far from a given query q, so NSG needs a large number of distance calculations to approach q. Furthermore, in cases where the start point is far from q, Algorithm 1 tends to fall into local optimum, rendering less accuracy. The performance of NSG is hence not good. This observation empirically confirms that a start point should be near q, to avoid both a large number of distance calculations and local optimum.

Comparison with HNSW. We here focus on LGTM and HNSW. We see that, *even with a single thread (hash table), LGTM outperforms HNSW*. This result implies that the way of start point selection in LGTM is better than that in HNSW. When LGTM uses 18 threads, its performance becomes clearly much than that of HNSW. For example, it reaches 90% recall much faster than HNSW in all the datasets. This multi-threading approach is one of the main advantages of LGTM. Complex graph structures like NSG and HNSW make the start point fixed. Such graphs cannot receive the benefits of our multi-threading approach[5]. Our simple yet carefully-designed approach functions better in practice.

Performance with Different k. An efficient AkNN algorithm should keep high performance even when k increase. Therefore, we investigated the robustness of LGTM, HNSW, and NSG. Table 2 depicts their performances with different k. LGTM keeps outperforming the other algorithms, i.e., provides higher recall with shorter running time on all the datasets. This result also suggests that LGTM is robust to k, and LGTM returns high recall when k is comparatively large.

5   Conclusion

This paper addressed the problem of approximate k nearest neighbor search in the high-dimensional Euclidean space. By using our idea that integrates

---

[5] NSG and HNSW can process different queries in parallel, but they *cannot* process a query in parallel.

Euclidean LSH and a proximity graph, we first presented a theorem that, with at least a constant probability, we can obtain the exact kNN result in a time sub-linear to n, under a reasonable assumption. This algorithm is theoretically sound, but has practical limitations. We therefore proposed LGTM by approximating the theoretical algorithm. We conducted extensive experiments using SIFT, GIST, and Deep, and the results demonstrate that LGTM outperforms the state-of-the-art algorithms.

Acknowledgments. This research is partially supported by JST PRESTO Grant Number JPMJPR1931, JSPS Grant-in-Aid for Scientific Research (A) Grant Number 18H04095, and JST CREST Grant Number JPMJCR21F2.

References

1. Babenko, A., Lempitsky, V.: Aggregating local deep features for image retrieval. In: ICCV, pp. 1269–1277 (2015)
2. Babenko, A., Lempitsky, V.: Efficient indexing of billion-scale datasets of deep descriptors. In: CVPR, pp. 2055–2063 (2016)
3. Bentley, J.L.: Multidimensional binary search trees used for associative searching. Commun. ACM **18**(9), 509–517 (1975)
4. Datar, M., Immorlica, N., Indyk, P., Mirrokni, V.S.: Locality-sensitive hashing scheme based on p-stable distributions. In: SoCG, pp. 253–262 (2004)
5. Dong, W., Moses, C., Li, K.: Efficient k-nearest neighbor graph construction for generic similarity measures. In: WWW, pp. 577–586 (2011)
6. Fu, C., Xiang, C., Wang, C., Cai, D.: Fast approximate nearest neighbor search with the navigating spreading-out graph. PVLDB **12**(5), 461–474 (2019)
7. Harwood, B., Drummond, T.: FANNG: fast approximate nearest neighbour graphs. In: CVPR, pp. 5713–5722 (2016)
8. Jegou, H., Douze, M., Schmid, C.: Product quantization for nearest neighbor search. IEEE Trans. Pattern Anal. Mach. Intell. **33**(1), 117–128 (2010)
9. Li, W., et al.: Approximate nearest neighbor search on high dimensional data-experiments, analyses, and improvement. IEEE Trans. Knowl. Data Eng. **32**(8), 1475–1488 (2020)
10. Lowe, D.G.: Distinctive image features from scale-invariant keypoints. Int. J. Comput. Vision **60**(2), 91–110 (2004)
11. Malkov, Y.A., Yashunin, D.: Efficient and robust approximate nearest neighbor search using hierarchical navigable small world graphs. IEEE Trans. Pattern Anal. Mach. Intell. **42**(4), 824–836 (2020)
12. Weber, R., Schek, H.J., Blott, S.: A quantitative analysis and performance study for similarity-search methods in high-dimensional spaces. In: VLDB, pp. 194–205 (1998)

TSX-Means: An Optimal K Search
Approach for Time Series Clustering

Jannai Tokotoko[1]([✉]), Nazha Selmaoui-Folcher[1][ID], Rodrigue Govan[1],
and Hugues Lemonnier[2]

[1] ISEA, University of New Caledonia, Nouméa, New Caledonia
{jannai.tokotoko,nazha.selmaoui}@unc.nc
[2] UMR ENTROPIE, IFREMER, Nouméa, New Caledonia
hugues.lemonnier@ifremer.fr

Abstract. Proliferation of temporal data in many domains has generated considerable interest in the analysis and use of time series. In that
context, clustering is one of the most popular data mining methods.
Whilst time series clustering algorithms generally succeed in capturing
differences in shapes, they most often fail to perform clustering based on
both shape and amplitude dissimilarities. In this paper, we propose a new
time series clustering method that automatically determines an optimal
number of clusters. Cluster refinement is based on a new dispersion criterion applied to distances between time series and their representative
within a cluster. That dispersion measure allows for considering both
shape and amplitude of time series. We test our method on datasets and
compare results with those from K-means time series (TSK-means) and
K-shape methods.

1 Introduction

Time series analysis is applied in many areas of business engineering, finance, economics, health care, etc. It serves various purposes such as subsequence matching, anomaly detection, pattern discovery, clustering, classification, etc. Our
study focuses on time series clustering There are two main approaches for time
series clustering. The first approach is based on feature construction. Series are
described by a vector of feature attributes [5], and instances are grouped using a
classical clustering method (*K-means, DBscan, ...*). The second one uses similarity
measures adapted to time series comparison, combined with basic approaches (e.g.
K-means) to cluster set of raw time series. Several similarity measures have been
suggested for time series clustering, such that DTW [8], SBD [7], $LCSS$ [10], and
ERP [1] measures. All those distance measures compare series considering only
effects of the temporal phase shift, and do not include amplitude drifts. However,
in some application domains, time series clustering should be done by considering
invariance and interval of measurements on the y axis as well as the shift of series
on the x axis. Indeed, the range of values on the y axis can strongly discriminate

This work is supported by PIL (Province of Loyalty Islands) in New Caledonia.

C. Strauss et al. (Eds.): DEXA 2021, LNCS 12924, pp. 232–238, 2021.
https://doi.org/10.1007/978-3-030-86475-0_23

between classes. For example, in agriculture or aquaculture domains, the range of y values in time series related to environmental data, such as changes in temperature, can significantly influences the growth and survival of living species. In this paper, we propose an approach based on shape analysis and that also takes into account the variance along the y axis. In addition, we develop a strategy that allows to automatically define an optimal number of clusters k using a new dispersion criterion applied on distances between instances and their representative within each cluster. Unlike most methods that normalize data, our approach can be applied to both normalized and raw time series. This new method is robust to the shifting of series on the x axis because we use metrics that take into account the distortion of series over time, in particular DTW, which is the most used for time series clustering [3, 11]. For the y shift, we consider a maximum interval over which those metrics vary. Section 2 presents notations and basic definitions. In Sect. 3, we present our contribution, in which a new dispersion measure of distance distribution is presented as well as the principle of our method. Section 4 gives results of experiments on several datasets and compared to those of TSK-means [4] and K-shape [9].

2   Notations and Definitions

Let s be a time series of length n where $s(i)$ corresponds to the value of the signal at time i. Let $T = \{s_1, s_2, \ldots, s_n\}$ a set of time series.

Clusters and Their Representatives: We call k-clustering C of T, the set $C = \{C_1, C_2, \ldots, C_k\}$ containing k homogeneous subsets of T (in relation to a measure of distance $Dist$), each having a representative noted R_{C_i} with $\forall i \in \{1, \ldots, k\}$, $C_i = \{s_{i_1}, s_{i_2}, \ldots, s_{i_{m_i}}\}$ and verifying the following criteria: **(1)** $T = \cup_{i=1}^k C_i$ and $C_i \cap C_j = \emptyset \ \forall i \neq j$ and **(2)** $Dist(R_{c_i}, s) < Dist(R_{c_j}, s) \ \forall s \in C_i$ and $j \neq i$. The representative of a cluster (called prototype) can be a centroid, medoid, etc.

Standard Deviation and Entropy of a Cluster: Let C_i be a cluster of C on T according to a measure of distance $Dist$. Let $Dist(C_i) = \{d_{i_1}, \ldots, d_{i_{m_i}}\}$ the set of values of the $Dist$ between an instance of C_i and its representative R_{C_i}. Let $\sigma(C_i)$ the standard deviation calculated on the distribution of values taken by $Dist(C_i)$, and $E(C_i)$ its entropy measure. $\sigma(C_i) = \sqrt{\frac{1}{m_i} \sum_{k=1}^{m_i} (d_{i_k} - \overline{d_i})^2}$ where $\overline{d_i}$ is the average of $Dist(C_i)$ and $E(C_i) = -\sum_{k=1}^{m_i} P(d_{i_k}) \times \log(P(d_{i_k}))$. In this paper, we used the distance measure DTW optimized by Kehog [6].

3   TSX-Means: A New Method for Time Series Clustering

Our approach mainly focuses on a new strategy for robust cluster refinement and automatic determination of the optimal number of clusters k. Any distance (or similarity) measure adapted to time series can be used in this approach. We tested it with different distance measures, such as measures derived from DTW. The method, based on a minimum number of clusters initially set to nb_min_clust

and a set of defined criteria, implements the principle of refining each cluster by revisiting all its instances. Instances that do not verify the criteria, in relation to the class they belong to, are put in a reject class. We then iterate the principle on that reject class (considered as a new set of series to be clustered) until the stopping conditions are verified. The criteria used in our approach are linked to the following thresholds: (**1**) *nb_min_inst*: the minimum number of instances allowed per cluster and (**2**) *seuil_disp*: the intra-cluster variability, defined from a new dispersion measure that depends on both the variability and the entropy measures of distances between each instance and its representative in cluster belongs to. In this contribution, we propose a new dispersion measure of distances between instances and their representative in a cluster. This dispersion measure, noted *disp*, is determined by the ratio between the standard deviation and the entropy of the distance values.

Definition 1 (measure of dispersion *disp***).** *Let C_i a cluster of the set T. We define its measure of dispersion by: $disp(C_i) = \frac{\sigma(C_i)}{E(C_i)}$.*

If the dispersion is minimal then the homogeneity is maximal. $disp(C_i)$ reflects the inner cluster variability. The smaller *disp* is, the smaller the variability around the representative is. That allows to select the nearest instances to a representative according to a fixed threshold, denoted s_d in the following.

Criteria for Selecting Cluster Instances: Let s_d a fixed threshold and C_i a cluster. A new associated cluster $C_i' \subset C_i$ is built, verifying the $disp(C_i') \leq s_d$. Computation of the dispersion measure requires at least two values. A minimum number of instances initially in the new C_i' cluster is thus provided by *nb_min_inst* in the algorithm. In order to determine those instances, $Dist(C_i)$ are ordered and saved in $Sort(Dist(C_i)) = \{v_1, v_2, \ldots, v_m\}$ with $\forall\, i < j,\, v_i \leq v_j$ (procedure *ApplyCriteria*). We integrate in C_i' the first *nb_min_inst* instances in the sorted list $Sort(Dist(C_i))$. If $disp(C_i') \leq s_d$ then other instances are added one by one in C_i', as long as the criterion remains true, otherwise instances that do not verify the criterion are put in the reject cluster. The value $disp(C_i')$ is updated each time an instance is added.

3.1   Principle of the Method

The algorithm takes as parameters thresholds *nb_min_clust*, *nb_min_inst*, and s_d and uses any *Dist*. As output, it provides a number of clusters determined automatically based on dispersion criteria, and a reject class noted CR. The principle of the algorithm is the following:

Step 1: Definition of Initial Clusters. Instances of T (set of time series) are partitioned into a minimum number of *nb_min_clust* clusters. To create those clusters, we apply the classic algorithm *TSK-Means* (or *K-shape*) with $k = nb\_min\_clust$ and a distance measure *Dist* (f.ex *DTW*, etc.). The procedure $[C, Dist(C)] = CreateInitialsClusters(T, nb\_min\_clust)$ of Algorithm 1 returns initial clusters.

Step 2: Refining Clusters by Applying the Dispersion Criterion. The procedure $[C', CR] = ApplyCriteria(C, Dist, s_d, nb\_min\_inst)$ consists in applying the homogeneity criterion to each cluster C_i to only keep instances verifying that criterion. The remaining instances are assigned to the reject class CR. If the number of instances of an initial cluster C_i is less than nb_min_inst, then this cluster is deleted and its instances are assigned to the reject class.

Step 3: Applying the Stopping Criterion. If the number of instances in the reject class is greater than nb_min_inst, then the initial step is repeated taking as new set T the rejected class. Otherwise, the algorithm stops.

3.2 *TSX-Means* Algorithm

At first call of our recursive method (Algorithm 1), the number of clusters to be determined $nbClust$, is initialized to 0, and the set of final clusters C_f to the empty set. At each call of the recursive algorithm, a new set of at most nb_min_clust clusters and the reject cluster are created from initial clusters obtained by the $CreateInitialsClusters$ method. The algorithm is therefore repeated as long as the reject cluster is not empty and the number of instances is greater than nb_min_inst. The method could assign to the reject cluster CR the same instances indefinitely if no admitted new cluster C_f was generated. The $recursiveCpt$ iteration counter allows to stop the algorithm when it reaches a maximum number of iterations provided by the user. Thus, it is possible to get a number of clusters lower than nb_min_clust or even no cluster at all. This occurs when the $ApplyCriteria$ method does not find any instance verifying the dispersion criterion in each of the initial clusters. This case is linked to a low value of the dispersion threshold. Nevertheless, increasing the threshold will integrate instances that are far from the representative and will lead to creating a cluster with high variability.

4   Experimental Results

The method has been tested on data of the UEA & UCR [2] archives. We tested our algorithm on 20 datasets. Chosen datasets have series of various lengths and a different number of classes. Most of them have a low number of classes (≤ 7), we say non complex data. In order to test our new method TSX-Means on more complex data, the last 7 datasets have a higher number of classes (≥ 24). For each dataset and for each distance used, we tested our method by varying the parameters s_d and nb_min_clust. Nb_min_inst was set to the number of instances of the smallest class of the dataset. Once the number k is found by our algorithm, we run *TSK-means* and *k-shape* with the same value of k to compare the performances between the 3 methods. We used different metrics (Accuracy, ARI and V-Measure (VM)) for performance comparison averaged for the tested parameters. Accuracy is calculated when the number of clusters is the actual number of classes. Otherwise, ARI and V-M are used. The parameter nb_min_clust has a greater impact on performance measures, and particularly

Algorithm 1. TSX-Means(T,*nb_min_clust*,s_d,*nbMaxIter*,*nbClust*,*recursifCpt*, *nb_min_inst*)

Output: - C_f set of clusters
1: **if** $nb\_min\_clust < p$ **then**
2:     $TmpCpt = 0$
3:     $\{C, Dist(C)\} = CreateInitialsClusters(T, nb\_min\_inst)$
4:     **for** i=1 to nb_min_clust **do**
5:         $\{C', CR\} = $ ApplyCriteria(C_i, $Dist$, $seuil\_disp$, nb_min_inst)
6:         **if** $C' \neq \emptyset$ **then**
7:             $C_f[nbClust] = C'$
8:             $T = T - C'$
9:             $nbClust = nbClust + 1$
10:         **else**
11:             $TmpCpt = TmpCpt + 1$
12:         **end if**
13:     **end for**
14:     **if** $TmpCpt == nb\_min\_clust$ **then**
15:         $recursifCPT = recursifCpt + 1$
16:     **end if**
17:     **if** $recursifCpt < nbMaxIter$ **then**
18:         $TSX\text{-}Means(T, nb\_min\_clust, s_d, nbClust, nbMaxIter, recursifCpt)$
19:     **end if**
20: **end if**
21: **return** C_f

V-Measure, than threshold s_d. The difference of ARI and V-M are, in average, 10% higher for complex data for *TSX-Means* than for *K-Shape* method. The new dispersion measure is a good indicator of cluster homogeneity. In general, *TSX-Means* method is more efficient than other methods, especially when the number of classes is very high. Table 1 shows results for accuracy scores. We noticed that dispersion measure improves clustering performance. Indeed, *TSX-Means* outperforms *TSK-Means* and *K-Shape* methods for the majority of data.

Table 1. Accuracy of *TSX-Means* with initial clusters from *TSK-Means*.

Dataset	Distances	k	TSX-Means	TSK-Means	Kshape	Reject TSX-means	Kmin
Car	sakoechiba	4	**0.446**	0.433	0.433	4	4
Fish	fast	7	**0.457**	0.440	0.391	0	5
Herring	itakura	2	**0.609**	0.594	0.509	0	2
LargeKitchen	sakoechiba	3	0.517	0.453	**0.521**	0	3
Meat	classic	3	**0.782**	0.653	0.750	1	3
Refrigeration	fast	3	**0.363**	0.361	0.360	0	3
SmallKitchen	itakura	3	0.417	**0.460**	0.407	0	3
WormsTwoClass	fast	2	**0.575**	0.511	0.602	0	2

We noticed that the dispersion measure improves clustering performance with the set of measures derived from *DTW*. Indeed, *TSX-Means* outperforms *TSK-Means* and *k-shape* methods for the majority of data. Table 2 shows accuracy scores of *TSX-Means* using *K-Shape* as initial clusters generator. Our method outperforms *k-Shape* for 5/7 data.

Table 2. Accuracy of *TSX-Means* and the *K-Shape* with initial clusters from *K-Shape*

Dataset	k	X-Shape	K-Shape	nb_min_clust
Computers	2	**0.616**	0.548	2
Meat	3	**0.700**	0.683	3
OSULeaf	6	0.420	0.435	6
OliveOil	4	**0.800**	0.433	4
RefrigerationDevices	3	**0.416**	0.368	3
ScreenType	3	0.357	0.360	3
Yoga	2	**0.487**	0.480	2

5   Conclusion and Perspectives

We proposed a new dispersion measure in a cluster, and designed a new method *TSX-Means* for time series clustering, allowing to automatically determine an optimal number of clusters. This measure allows to refine clusters initially generated by existing clustering methods. Performance of *TSX-Means* was compared to *TSK-Means* and *K-Shape* methods on a set data. Quality measures of clustering performance showed that *TSX-Means* method outperforms *TSK-Means* and *K-Shape*, especially for data with a very large number of clusters.

References

1. Chen, L., Ng, R.T.: On the marriage of LP-norms and edit distance. In: VLDB, pp. 792–803. Morgan Kaufmann (2004)
2. Dau, H.A., et al.: The UCR time series archive. IEEE/CAA J. Automatica Sinica **6**(6), 1293–1305 (2019)
3. Dilmi, M.D., Barthès, L., Mallet, C., Chazottes, A.: Iterative multiscale dynamic time warping (IMs-DTW): a tool for rainfall time series comparison. Int. J. Data Sci. Anal. **10**(1), 65–79 (2019). https://doi.org/10.1007/s41060-019-00193-1
4. Huang, X., Ye, Y., Xiong, L., Lau, R.Y., Jiang, N., Wang, S.: Time series k-means: a new k-means type smooth subspace clustering for time series data. Inf. Sci. **367–368**, 1–13 (2016)
5. Kalpakis, K., Gada, D., Puttagunta, V.: Distance measures for effective clustering of ARIMA time-series. In: ICDM, pp. 273–280 (2001)
6. Keogh, E., Pazzani, M.: Derivative dynamic time warping. First SIAM-ICDM **1**, 1–11 (2002)
7. Meesrikamolkul, W., Niennattrakul, V., Ratanamahatana, C.A.: Shape-based clustering for time series data. In: PaKDD, pp. 530–541 (2012)

8. Müller, M.: Dynamic time warping. In: Information Retrieval for Music and Motion,, pp. 69–84. Springer, Heidelberg (2007). https://doi.org/10.1007/978-3-540-74048-3_4

9. Paparrizos, J., Gravano, L.: k-shape: efficient and accurate clustering of time series. In: ACM SIGMOD ICMD, pp. 1855–1870 (2015)

10. Vlachos, M., Kollios, G., Gunopulos, D.: Discovering similar multidimensional trajectories. In: ICDE 2002, pp. 673–684 (2002)

11. Zhang, Z., Tavenard, R., Bailly, A., Tang, X., Tang, P., Corpetti, T.: Dynamic time warping under limited warping path length. Inf. Sci. **393**, 91–107 (2017)

A Globally Optimal Label Selection Method via Genetic Algorithm for Multi-label Classification

Tianqi Ji, Jun Li, and Jianhua Xu$^{(\boxtimes)}$

School of Computer and Electronic Information, School of Artificial Intelligence,
Nanjing Normal University, Nanjing, Jiangsu 210023, China
182202009@stu.njnu.edu.cn, {lijuncst,xujianhua}@njnu.edu.cn

Abstract. Multi-label classification addresses a special pattern classification problem where an instance could belong to multiple class labels simultaneously. Nowadays, a pivotal challenging issue is to deal with high-dimensional label space case, which generally deteriorates classification performance and increases computation burdens for traditional multi-label classifiers. To this end, dimensionality reduction originally for feature space is also applied to label space, deriving two kinds of techniques: label embedding and label selection. Boolean matrix decomposition (BMD) approaches to the original binary matrix via two low-rank binary matrix Boolean multiplication. Without any constraint on low-rank matrices, this BMD could result in a label embedding method. When the left matrix consists of a column subset of original matrix, interpolative decomposition is implemented, which corresponds to a label selection technique. In this paper, we propose a globally optimal label selection approach, which consists of two stages: to remove a few uninformative labels via an exact BMD without any approximated loss, and to select most informative labels of fixed number via globally optimal genetic algorithm. Our experiments on four benchmark data sets with more than 100 labels show that our proposed method is superior to three existing approaches, according to two performance evaluation metrics (precision and discounted gain@n) for high-dimensional label space.

Keywords: Multi-label classification · Data mining · Label selection · Boolean matrix decomposition · Genetic algorithm

1 Introduction

Multi-label classification deals with a particular supervised classification issue in which any instance could be associated with multiple class labels at the same time and the classes are not mutually exclusive [6]. Its data mining applications include text categorization, computer vision, bioinformatics, and so on. Given a

Supported by Natural Science Foundation of China (NSFC) under Grants 62076134 and 61703096.

multi-label training set, the multi-label classification task is to learn a classifier from input feature space to output label space, which then is used to predict a set of relevant labels for an unseen instance.

As various applications continuously occur in data mining, the dimensionality of label space also increases correspondingly, which greatly deteriorates classification performance and increases computational complexity for many traditional multi-label classifiers. To cope with this situation, dimensionality reduction originally for feature space is also accepted for high-dimensional label space to improve multi-label classification efficiency and effectiveness [11]. In this case, besides a label reduction operator in the training stage, it is needed to provide a recovery operator, which is used in the testing phase to recover a low-dimensional label vector back to the original high-dimensional label space. Therefore, how to design label reduction and recovery operators becomes a principal issue.

Now, the existing methods are divided into two kinds: label embedding (LE) and label selection (LS). The LE is to convert original high-dimensional label space into a low-dimensional real or binary one linearly or nonlinearly, and the LS is to select some most informative labels to hold label physical meanings.

The first LE algorithm is ML-CS [7] based on compressed sensing, which uses a random matrix as its projection matrix. However, it is needed to solve a complicated L_1 minimization problem as its reconstruction step, resulting in an extremely high complexity. After that, besides efficient projection procedure, much work pursues more efficient recovery step. In PLST [13], principal component analysis (PCA) is executed on label matrix to create an orthonormal projection matrix and an efficient recovery matrix. To fusion feature information, canonical correlation analysis (CCA) is modified into an orthonormal form, which then is combined with PCA to construct CPLST [5]. Hilbert-Schmidt independence criterion (HSIC) is applied to achieve projection and recovery matrices in [4]. Besides PCA and HSIC, via adding a local covariance matrix for each instance, ML-mLV [14] proposes a trace ratio optimization problem to find out projection and recovery matrices. To execute LS task, in MOPLMS [1], an L_1 and $L_{1,2}$ regularized least square regression problem is solved, but an independent recovery stage based on Bayesian rule is added. In ML-CSSP [3], the LS task is regarded as column subset selection problem (CSSP), which is efficiently implemented via a randomized sampling way with a natural recovery matrix.

Boolean matrix decomposition (BMD) [2] decomposes a binary matrix into two low-rank binary matrix Boolean multiplication. Without any constraints on two decomposed matrices, BMD could induce some LE methods, e.g., MLC-BMaD [15]. If the left matrix comprises a subset of original label matrix, such a BMD is named as interpolative form, to induce some LS methods exactly. So far, an exact BMD [12] is realized to remove a few uninformative labels without any approximated loss, which is directly used to implement a LS methods (MLC-EBMD) [10]. Generally, there are many informative labels from MLC-EBMD in a crude rank, so selecting fewer informative labels becomes just sub-optimal.

In this study, we construct an additional selection phase with genetic algorithm (GA) to select most informative labels among those remained labels from

EBMD, which can theoretically induce a globally optimal selection according to theoretical basis and analysis [9]. Therefore a novel globally optimal label selection algorithm is proposed in this paper, which consists of two stages: to remove uninformative labels via EBMD and to select most informative labels using GA. The experiments illustrate that our proposed method is more effective, compared with one LE method (MLC-BMaD [15]) and two LS techniques (MLC-EBMD [10] and ML-CSSP [1]) on four data sets with more than 100 labels.

2   Preliminaries

The traditional multi-label classification is to learn a classifier: $\mathbf{y} = f(\mathbf{x}) : R^D \rightarrow \{0, 1\}^C$ according to a multi-label training data set of size N as follows:

$$\{(\mathbf{x}_1, \mathbf{y}_1), \ldots, (\mathbf{x}_i, \mathbf{y}_i), \ldots (\mathbf{x}_N, \mathbf{y}_N)\} \tag{1}$$

where, for the i-th instance, $\mathbf{x}_i \in R^D$ is its D-dimensional real feature column vector, and $\mathbf{y}_i \in \{0, 1\}^C$ its C-dimensional binary label column vector whose 1 and 0 components imply the relevant and irrelevant labels, respectively.

Further, we define a label data matrix $\mathbf{Y} \in \{0, 1\}^{N \times C}$:

$$\mathbf{Y} = [\mathbf{Y}_{ij} | i = 1, ..., N; j = 1, ..., C] = [\mathbf{y}_i | i = 1, ..., N]^T = [\mathbf{y}^j | j = 1, ..., C] \tag{2}$$

where $\mathbf{y}^j$ corresponds to the j-th label in $\mathbf{Y}$.

For $\mathbf{Y}$, given an integer c, its Boolean matrix decomposition (BMD) is represented as Boolean multiplication of two low-rank binary matrices (i.e., label matrix $\mathbf{Z}$ and label correlation matrix $\mathbf{B}$), i.e.,

$$\mathbf{Y} \approx \mathbf{Z} \circ \mathbf{B} = \sum_{i=1}^{c} (\mathbf{z}^i \circ \mathbf{b}_i) \tag{3}$$

with

$$\begin{aligned} \mathbf{Z} &= [\mathbf{z}^1, ..., \mathbf{z}^i, ..., \mathbf{z}^c] = [\mathbf{z}_1, ..., \mathbf{z}_i, ..., \mathbf{z}_N]^T \in \{0, 1\}^{N \times c} \\ \mathbf{B} &= [\mathbf{b}^1, ..., \mathbf{b}^i, ..., \mathbf{b}^C] = [\mathbf{b}_1, ..., \mathbf{b}_i, ..., \mathbf{b}_c]^T \in \{0, 1\}^{c \times C} \end{aligned} \tag{4}$$

where "$\circ$" indicates Boolean multiplication, and $\mathbf{z}_i$ is the low-dimensional label vector of $\mathbf{y}_i$. This BMD is realized via minimizing the following error,

$$E(\mathbf{Y}, \mathbf{Z} \circ \mathbf{B}) = \|\mathbf{Y} - \mathbf{Z} \circ \mathbf{B}\|_1 = \sum_{i=1}^{N} \sum_{j=1}^{C} |\mathbf{Y}_{ij} - (\mathbf{Z} \circ \mathbf{B})_{ij}| \tag{5}$$

where $\|\cdot\|_1$ is 1-norm of matrix to count the number of "$+1/-1$" elements.

When the matrix $\mathbf{Z}$ consists of a subset of $\mathbf{Y}$, this decomposition becomes interpolative matrix decomposition (IMD). While the relation $\|\mathbf{Y} - \mathbf{Z} \circ \mathbf{B}\|_1 = 0$ holds true, such a BMD is called as exact BMD (EBMD) [12], where the smallest $c << C$ is referred to as the Boolean rank of $\mathbf{Y}$.

For multi-label classification, the low-rank matrix $\mathbf{Z}$ from BMD and IMD respectively correspond to LE and LS, which convert the original C-label multi-label problem into a c-label one ($c << C$). The matrix $\mathbf{B}$ is regarded as a recovery matrix to recover the low c-dimensional vector back to the high C-dimensional one. With this label dimensionality reduction, a general framework for multi-label classification can be summarized as two more complicated procedures: **Training Stage**: a) to conduct matrix decomposition $\mathbf{Y} \approx \mathbf{Z} \circ \mathbf{B}$ to obtain $\mathbf{Z}$ and $\mathbf{B}$; b) to learn a multi-label classifier $\mathbf{z} = f(\mathbf{x})$. **Testing Stage**: a) to calculate the low-dimensional binary label vector $\mathbf{z}$ for a testing instance $\mathbf{x}$ using $f(\mathbf{x})$; b) to reconstruct the high-dimensional label vector $\mathbf{y} = \mathbf{B}^T \circ \mathbf{z}$.

3 The Proposed Method

In this section, we propose a novel label selection method combining with exact Boolean matrix decomposition (EBMD) and genetic algorithm (GA).

3.1 Uninformative Label Reduction via EBMD

The EBMD [12] is to find a smallest Boolean rank c to satisfy $\mathbf{Y} = \mathbf{Z} \circ \mathbf{B}$ exactly, where $\mathbf{Z}$ is a column subset of $\mathbf{Y}$. In this sub-section, this EBMD is used to remove some uninformative labels in $\mathbf{Y}$ to execute a primary LS step.

We summarize a heuristic EBMD algorithm [12] as follows: (i) to derive an entire recovery matrix $\mathbf{B}^T = \overline{\mathbf{Y}}^T \circ \mathbf{Y}$ of size $C \times C$, where $\overline{\mathbf{Y}}$ is a "not" logic operation; (ii) to estimate $\sigma_i = \left\| \mathbf{z}^i \right\|_1 \left\| \mathbf{b}_i \right\|_1$ to indicate the number of "1" components in $\mathbf{z}^i \circ \mathbf{b}_i$, which is to evaluate the contribution of $\mathbf{z}^i \circ \mathbf{b}_i (i = 1, ..., C)$ to $\mathbf{Y}$; (iii) to rearrange the columns of $\mathbf{Y}$ and rows of $\mathbf{B}$ in the ascending order of $\sigma_i (i = 1, ..., C)$; (iv) to initialize the label selection matrix $\mathbf{Z} = \mathbf{Y}$, the product matrix $\mathbf{T} = \mathbf{ZB}$, and the number of selected columns $k = C$; (v) for $i = 1, ..., C$, we execute: (a) to calculate $\mathbf{T}_i = \mathbf{z}^i \circ \mathbf{b}_i$; (b) if $\|\mathbf{T}\|_1 = \|\mathbf{T} - \mathbf{T}_i\|_1$, we remove the i-th column in $\mathbf{Z}$ and the i-th row in $\mathbf{B}$, and then let $\mathbf{T} = \mathbf{T} - \mathbf{T}_i$ and $k = k - 1$; and (vi) we obtain a selected label matrix $\mathbf{Z}_{N \times k}$ and its recovery matrix $\mathbf{B}_{k \times C}$.

This procedure has been used to construct a multi-label LS algorithm MLC-EBMD in [10], which only remove those uninformative labels in $\mathbf{Y}$.

3.2 Most Informative Label Selection via GA

Through the above EBMD, we could remove the $(C - k)$ uninformative labels and remain k informative labels. Generally, k is much larger than c (the number of selected labels), i.e., $k >> c$. It is worth noting that the aforementioned algorithm could provide a crude rank for the remained labels according to σ_i. However, a more precise rank is needed to improve classification performance further. In this sub-section, we apply GA to obtain the c most informative labels, which theoretically could achieve a globally optimal solution [9].

The GA is a randomized search algorithm that simulates the natural genetic mechanisms of biological evolutional process. Through three genetic operations: selection, crossover and mutation, a population of individuals are evolved one by one generation and their fitness function values are optimized gradually. After several generations are executed or a proper stopping criterion is reached, an individual with the optimal fitness is chosen an optimal solution [9]. When this GA is applied to execute our LS task, it is needed to fix following three aspects:

Encoding Chromosomes. Binary coding technique is used to encode the M chromosomes of length k: $\mathbf{q}_i = [q_{i1}, ..., q_{ij}, ..., q_{ik}]^T \in \{0, 1\}^k$ $(i = 1, ..., M)$, where $q_{ij} = 1$ indicates that the j-th label is selected. Since the c most informative labels will be selected, each individual is initialized to have the c "1" components only.

Adding an Auxiliary Mutation Operation. Generally, after three genetic operations are conducted, the number of "1" components in each chromosome is changed, i.e., $\|\mathbf{q}_m\|_1 \neq c$. To this end, we add an auxiliary mutation operation to fix the number of selected labels to be c exactly.

Defining Fitness Function. We construct a relative reconstruction accuracy (RRA) as our fitness function to measure the approximation ability from the reconstructed matrix $(\mathbf{Z} \circ \mathbf{B})$ to original matrix $(\mathbf{Y})$:

$$RRA(\mathbf{Y}, \mathbf{Z} \circ \mathbf{B}) = \frac{1}{\|\mathbf{Y}\|_1} \sum_{i=1}^{N} \sum_{j=1}^{C} |\{(i, j) | \mathbf{Y}_{ij} = (\mathbf{Z} \circ \mathbf{B})_{ij}\}| \qquad (6)$$

where the numerator indicates how many "1"s in $\mathbf{Y}$ are recovered by $\mathbf{Z} \circ \mathbf{B}$.

3.3 Label Selection Algorithm Combining EBMD and GA

Based on the previous work, we build an LS method for multi-label classification, which consists of two stages: uninformative label reduction with EBMD in Subsect. 3.1, and most information label selection via GA in Subsect. 3.2. This procedure is referred to as a global optimal LS method for multi-label classification, via combining EBMD with GA, simply, LS-BMDGA.

4 Experiments

In this section, we will compare our LS-BMDGA with three existing methods (MLC-EBMD [10], ML-CSSP [3] and MLC-BMaD [15]) on four benchmark data sets.

4.1 Basic Experimental Settings

In order to evaluate the performance of four aforementioned label reduction techniques, we choose four benchmark multi-label data sets with more than 100 class labels from Mulan library[1], including Mediamill, Bibtex, Corel16k-s6, and Bookmarks. Their detailed statistics are listed in Table 1.

In order to fit the high-dimensional label applications, two evaluation metrics are utilized in our experiments, i.e., precision@n, and (DisCounted Gain) DCG@n $(n = 1, 2, 3, ...)$. Please refer to [8] to see their detailed definitions. It is desired to achieve high values for both of metrics for a well-performed method.

---

[1] http://mulan.sourceforge.net/datasets-mlc.html.

Table 1. Statistics of selected multi-label data sets.

Dataset	#Train	#Test	#Features	#Labels	#Cardinality
Mediamill	30993	12914	120	101	4.376
Bibtex	4880	2515	1836	159	2.402
Corel16k-s6	5192	1737	500	162	3.136
Bookmarks	57985	29871	2150	208	2.028

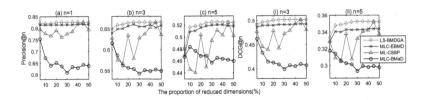

Fig. 1. Precision@n(n = 1, 3, 5) and DCG@n(n = 3, 5) from four methods on Mediamill.

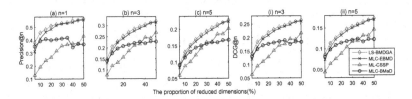

Fig. 2. Precision@n(n = 1, 3, 5) and DCG@n(n = 3, 5) from four methods on Bibtex.

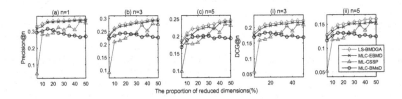

Fig. 3. Precision@n(n = 1, 3, 5) and DCG@n(n = 3, 5) from four methods on Corel16k-s6.

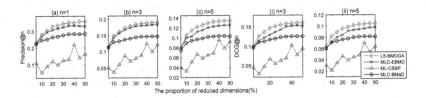

Fig. 4. Precision@n(n = 1, 3, 5) and DCG@n(n = 3, 5) from four methods on Bookmarks.

Table 2. The number of wins for each method and metric across four data sets.

Metric	MLC-BMaD	ML-CSSP	MLC-EBMD	LS-BMDGA
Precision@1	0	0	1	**39**
Precision@3	0	1	1	**38**
Precision@5	0	0	1	**39**
DCG@3	0	0	1	**39**
DCG@5	0	0	1	**39**
Total wins	0	1	5	**194**

In our experiments, we compare four methods via a train-test way, in which random forest with 100 trees is considered as our baseline classifier. For MLC-BMaD, its threshold for calculating label association matrix is set to 0.7. Further, on our LS-BMDGA, its key parameters are assigned as follows: the population size $p = 200$, evolutional generations $g = 400$, selection probability $p_s = 0.8$, crossover probability $p_c = 0.8$, and mutation probability $p_m = 0.01$.

Since precision@1 is equal to DCG@1, we set $n = 1, 3, 5$ for precision@n, and $n = 3, 5$ for DCG@n. In order to verify and compare classification performance extensively for the above four approaches, the number of reduced labels (i.e., c) is varied from 5% to 50% of the number original labels (i.e., C) with a step 5%.

4.2 Experimental Results and Analysis

At first, we regard two evaluation metrics (i.e., precision@n and DCG@n) as two functions of the different proportions of reduced labels (i.e., c/C), respectively. The experimental results on four data sets in Table 1 are shown in Figs. 1, 2, 3 and 4.

From these four figures, it is observed that: (i) For MLC-BMaD and ML-CSSP, their performance is unstable, e.g., ML-CSSP and MLC-BMaD achieve the worst performance on Bookmarks and Mediamill, respectively. Additionally, overall, MLC-EBMD has a better performance than MLC-BMaD and ML-CSSP. (ii) At most of label proportions, our LS-BMDGA works best, compared with three existing methods, whose main reason is that we find out an optimal label subset form label selection with Boolean interpolative matrix decomposition.

To compare four approaches more accurately, we adopt the "win" index in [4] to represent the number of each technique reaches the best metric values across four data sets and ten label proportions (200 wins, totally), as shown in Table 2. Our LS-BMDGA reaches the best value of 194 times for two metrics over four data sets, which greatly exceeded the number of wins from summation (6 wins) of the other three methods.

On the basis of the above experimental analysis, it is concluded that our proposed LS method (LS-BMDGA) performs the best, compared with three existing approaches (MLC-BMaD, ML-CSSP, and MLC-EBMD).

5   Conclusions

In this paper, we propose a new label selection technique to deal with some multi-label applications with many labels, which could preserve selected label physical meanings effectively. Our proposed method consists of two effective stages. One is to apply an exact Boolean matrix decomposition to remove some uninformative labels and the other is to apply genetic algorithm to select most informative labels of the fixed number. Theoretically, such an approach could implement a globally optimal label selection strategy. The detailed experiments show that our proposed method is superior to three existing techniques in terms of precision and discounted gain. In future work, we will conduct some experiments on more benchmark data sets, and compare our method with more existing approaches, to validate multi-label classification performance of our proposed method further.

References

1. Balasubramanian, K., Lebanon, G.: The landmark selection method for multiple output prediction. In: 29th International Conference on Machine Learning (ICML 2012), pp. 283–290. OmniPress, Madison (2012)
2. Belohlavek, R., Outrata, J., Trnecka, M.: Toward quality assessment of Boolean matrix factorizations. Inf. Sci. **459**, 71–85 (2018)
3. Bi, W., Kwok, J.: Efficient multi-label classification with many labels. In: 30th International Conference on Machine Learning (ICML 2013), pp. 405–413. Omni-Press, Madison (2013)
4. Cao, L., Xu, J.: A label compression coding approach through maximizing dependance between features and labels for multi-label classification. In: 27th International Joint Conference on Neural Networks (IJCNN 2015), pp. 1–8. IEEE Press, New York (2015)
5. Chen, Y.N., Lin, H.T.: Feature-aware label space dimension reduction for multi-label classification. In: 26rd Annual Conference on Neural Information Processing System (NIPS 2012), pp. 1538–1546. MIT Press, Cambridge (2012)
6. Herrera, F., Charte, F., Rivera, A.J., del Jesus, M.J.: Multilabel Classification: Problem Analysis, Metrics and Techniques. Springer, Cham (2016)
7. Hsu, D.J., Kakade, S.M., Langford, J., Zhang, T.: Multi-label prediction via compressed sensing. In: 23rd Annual Conference on Neural Information Processing System (NIPS 2009), pp. 772–780. MIT Press, Cambridge (2009)
8. Jain, H., Prabhu, Y., Varma, M.: Extreme multi-label loss functions for recommendation, tagging, ranking & other missing label applications. In: 22nd ACM SIGKDD International Conference on Knowledge Discovery Data Mining (KDD 2016), pp. 935–944. ACM Press, New York (2016)
9. Katoch, S., Chauhan, S.S., Kumar, V.: A review on genetic algorithm: past, present, and future. Multimedia Tools Appl. **80**(5), 8091–8126 (2020). https://doi.org/10.1007/s11042-020-10139-6
10. Liu, L., Tang, L.: Boolean matrix decomposition for label space dimension reduction: Method, framework and applications. J. Phys. Conf. Ser. **1345**(5), 052061–052066 (2019)
11. Siblini, W., Kuntz, P., Meyer, F.: A review on dimensionality reduction for multi-label classification. IEEE Trans. Knowl. Data Eng. **33**(3), 839–857 (2021)

12. Sun, Y., Ye, S., Sun, Y., Kameda, T.: Exact and approximate Boolean matrix decomposition with column-use condition. Int. J. Data Sci. Anal. 1(3-4), 199–214 (2016). https://doi.org/10.1007/s41060-016-0012-3
13. Tai, F., Lin, H.T.: Multilabel classification with principal label space transformation. Neural Comput. **24**(9), 2508–2542 (2012)
14. Wang, X., Li, J., Xu, J.: A label embedding method for multi-label classification via exploiting local label correlations. In: Gedeon, T., Wong, K.W., Lee, M. (eds.) ICONIP 2019. CCIS, vol. 1143, pp. 168–180. Springer, Cham (2019). https://doi.org/10.1007/978-3-030-36802-9_19
15. Wicker, J., Pfahringer, B., Kramer, S.: Multi-label classification using Boolean matrix decomposition. In: 27th Annual ACM Symposium on Application Computing (SAC 2012), pp. 179–186. ACM Press, New York (2012)

Semantic Web and Ontologies

Discovering HOI Semantics from Massive Image Data

Mingguang Zheng[1,2], Shouhong Wan[1,2], and Peiquan Jin[1,2(✉)]

[1] University of Science and Technology of China, Hefei, China
zhmg@mail.ustc.edu.cn, {wansh,jpq}@ustc.edu.cn
[2] Key Laboratory of Electromagnetic Space Information, CAS, Hefei, China

Abstract. Human-Object Interaction (HOI) plays an important role in human-centric scene understanding. However, the commonly used two-stage methods have large computational costs and a slow inferring speed. The existing one-stage methods detect HOIs by detecting the central points or the union boxes of human and objects, which need to process a large scale of regions and many unnecessary features. In this paper, we propose a novel one-stage method for discovering HOI semantics from massive image data. In particular, we present two new designs in our method, namely *action classification* and *displacement prediction*. Further, we design a special HOI score calculation strategy, which can decay the HOI score of the results that have bad matching result. We evaluate our method on the popular HICO-DET benchmark and compare our proposal with a number of existing approaches. The results show that our method outperforms existing methods in discovering HOI semantics. *abstract* environment.

Keywords: Human-Object Interaction · One-stage detection · Action classification · Displacement prediction

1 Introduction

In recent years, human-object interaction (HOI) has attracted much attention. Given an image, HOI detection aims to localize the human and objects, and detect the relationship between them. As human beings always play a central role in real-world scenes, HOI detection has been an important way to extract high-level semantics from image data and become an useful tool for many image understanding tasks, such as image classification [12], object detection [17], and image retrieval [14,18].

Previous work on HOI detection mainly used two-stage methods, which contained two stages. In the first stage, a pre-trained object detector is used to detect all human and objects. In the second stage, all the possible triplets <human, verb, object> are analyzed by sending all the relevant features into an action classification network. However, this kind of methods completely separate HOI detection into two parts, yielding poor performance in real applications.

© Springer Nature Switzerland AG 2021
C. Strauss et al. (Eds.): DEXA 2021, LNCS 12924, pp. 251–263, 2021.
https://doi.org/10.1007/978-3-030-86475-0_25

(a) UnionDet   (b) PPDM

(c) Ours

Fig. 1. Comparisons of various interaction detection methods. (a) UnionDet [6] detect the union box of human and object in an interaction. (b) PPDM [8] detect the central point of the human and its corresponding object. (c) Our method classifies the actions for each human and object, and combine them to produce the HOI.

Recently, some researchers developed one-stage methods for HOI detection. UnionDet [6] detected the union box of human and box as shown in Fig. 1(a). Wang et al. [16] treated the central point of the human and its corresponding object as interaction points, making HOI detection become an interaction detection problem. They used an individual object detector as well as an interaction point detector, and detected human, objects, and interaction in parallel. PPDM [8] used a common feature extractor for all the detector. However, detecting the central point or the union box of human and objects needs to handle the whole union region of human and objects, which needs a complex network to handle so much information.

In this paper, we propose a novel one-stage method for discovering HOI semantics from massive image data. As shown in Fig. 1(c), our architecture do not directly localize the interaction like Wang et al. [16] and PPDM [8]. Instead, we directly classify the actions of the human at his bounding-box's center point and find what he interacts with by predicting the displacements from him to his corresponding objects. Specially, we predict a displacement for each action category for the reason that a person can interact with many objects. In addition, one displacement for one category can make each prediction network focus on its category-relevant feature. However, determining the categories of actions that only consider the human feature may lead to some wrong judgement, especial when the human pose has multiple corresponding verbs. We observe that the number of possible interaction is limited when the object categories are known. For example, a chair can only be sited in most real-world scene. And some actions such as *catch* and *throw* can be inferred by observing whether there is a hand around the object. Therefore, we perform the action classification at the central location of an object region and determine the action categories considering both the action classification result of human and objects.

When producing the HOI triplets, we use the distance between the predicted location and the ground truth location of the interacted object to match the human and object. The previous PPDM method does not consider such a distance when calculating the HOI scores, which will cause some wrong human-object matching. On the other hand, When a human interacts with multiple objects or the human and corresponding object is far away in image, the predicted displacement from human to object would be inaccurate. To solve this problem, we design an novel HOI score calculating strategy, which introduces a decay parameter to decay the score of the HOI triplet of wrong matching results.

Briefly, we make the following contributions in this paper:

(1) we propose a novel one-stage framework for discovering HOI semantics from massive image data. Two designs, including action classification and displacement prediction, are presented in our method. Compared to existing methods, our method needs not to perform an interaction localization step and is more effective in discovering HOIs.
(2) To avoid the bias of HOI score calculation, we design a new HOI score calculating strategy, which can solve the inaccuracy problem in displacement prediction.
(3) We conduct extensive experiments on a popular benchmark HICO-DET and compare our proposal with various existing methods, and the results suggest the effectiveness of our method.

The rest of the paper is structured as follows. Section 2 summarizes related work. Section 3 details the framework of discovering HOI semantics from image data. Section 4 reports the experimental results, and finally, in Sect. 5, we conclude the whole paper.

2   Relate Work

2.1   Two-Stage Methods for HOI Detection

Previous HOI detection methods mainly employed a two-stage strategy. In the first stage, a pre-trained object detector (most likely to be Faster-RCNN [11]) is used to get the bounding-box of all the possible human and objects in the input image. In the second stage, all the possible human-object pairs would be fed into a multi-stream architecture to predict the interaction categories. The multi-stream architecture consist of at least two stream: the human stream and the object stream. Both of them take the visual feature of the human or object bounding-box as input [5].

Most of the previous works concentrate on improving the second stage, and the most popular method is to introduce addition features. Bansal et al. [1] and PDNet [20] introduced word embedding to get the language prior knowledge. Wan et al. [13] introduced human pose information to help the interaction detection. Besides, some researcher focused on catching the relation between the instances or interactions in image. iCAN [5] utilize the spatial relation of the instance through adding a two-channel binary feature. DRG [4], CHG [15], RPNN [21] used the graph networks to model the relation between the interactions.

Although the two-stage methods have high performance, they always have complicate architecture, and expensive calculating cost as well as a slow inferring speed, due to their sequential and separated two-stage framework.

2.2   One-Stage Methods for HOI Detection

Recently, some researcher begin to explore the one-stage methods for HOI detection. Wang et al. [16] detect the interaction through detecting the center points of interacting human and objects and finding the corresponding by predicting a displacement. But they still completely separated the object and interaction detector. PPDM [8] use the similar way, but they used a common feature extractor and trained the network end-to-end. But they didn't consider the human-object matching result when calculating the final HOI score. UnionDet [6] detect the union box of human and object to detect the interaction, which need a large receptive field to cover the large boxes. DIRV [3] design a rule of where the interaction regions are and detect interaction base on it.

Unlike all the above methods, our methods does not detect the interaction directly. We take good use of the object detector and classify the action of the detected human and object. And we also modify the HOI score calculation formula to produce better HOI scores.

3   Framework for HOI Detection

3.1   Architecture

As shown in Fig. 2, our architecture composes of three parts: an common feature extract networks, object detection branch and interaction matching branch. The feature extract networks take image $I \in R^{H \times W}$ as input and produce the common featuremap $F \in R^{H/d \times W/d}$. Then the featuremap F is fed into the following two branch.

In the object detection branch, we detect the central points and the size of human and object bounding-boxes. In the interaction matching branch, we classifies the actions at each location and predict the displacement to the corresponding human or object. Then we use a human-based matching strategy to match the human to its corresponding object to generate the HOI triplets and use our novel decay calculation strategy to produce the final confident score for each HOI.

3.2   Object Detection Branch

The object detection branch follows the setting of CenterNet [22], which treats object detection as a standard keypoint estimation problem. In CenterNet,

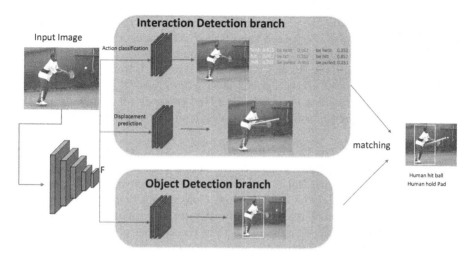

Fig. 2. Architecture of our HOI detection. It consists of three parts: (a) Common feature extract networks, which take the image as input and produces the common feature map. (b) The object detection branch uses convolution modules to produce the bounding box of human and objects. (c) The interaction matching branch consist of two parts: the action classification part classifies the action for each human and object, and the displacement prediction part predict the displacement to the relevant human or object for all the human and objects. After getting the output of the two branch, we use a human-based matching strategy to match the detected human and object to produce the HOIs.

a bounding-box is represented as (x, y, w, h) ,where (x, y) is the center point coordinate of the box and (w, h) denote the weight and height of the box. The object detector branch takes $F \in R^{H/d \times W/d}$ as input and outputs a center point heatmap $V \in R^{H/d \times W/d}$ for each object category and the height and weight of the predicted box at each location of the input feature map F.

During the inference, a 3×3 max-pooling will be used to extract the points whose value is greater to its 8-connected neighbors, which plays a similar role as non-maxima suppression (NMS) in convenient object detection methods. The extracted points will be set as the center points of bounding-box and their values on the heatmap will be the confidence score of the boxes after playing a sigmoidal transformation.

3.3   Interaction Detection Branch

The interaction detection branch is composed of two part: action classification and displacement prediction. Both of them take feature map F as input.

Action Classification. We classify the verb categories for all the locations of the input feature map F. Supposed there are N verb categories in dataset, we treat action classification as N binary classification problems. Taking the feature map F as input, we employ a 3 × 3 convolution layer with ReLU, followed by a 1 × 1 convolutional layer and a Sigmoid layer to produce 2N heatmap, each of which is the binary classification result for a HOI category for human or object. We train this branch with focal loss [9], treating it similar to a keypoint detection problem.

Displacement Prediction. Similar to the action classification, we employ a 3×3 convolution layer with ReLU, followed by a 1 × 1 convolutional layer and a Sigmoid layer to predict the displacement. Considering that a human always interact with multi objects but an object always interact with only one person, we predict $N+1$ displacement $d_1, d_2 \cdots d_{N+1}$ (N is the number of action categories) for each location of the feature map F, where $d_i = (x_i, y_i) \quad i \in (1 \cdots N)$ denotes the displacement from human to the corresponding object of the ith category action and $d_{N+1} = (x_{N+1}, y_{N+1})$ denotes the displacement from object to the corresponding human. During training, we only compute the L1 loss at each location of the ground truth human points or object points.

3.4   Loss and Inference

The final loss function is shown in Eq. 1.

$$L = L_p + L_c + L_d \tag{1}$$

L_p is the loss for the object detection branch, which all follow the setting of CenterNet [22], the Lc is the loss for action classification and the L_d is the L1 loss for displacement prediction. The whole network is trained end-to-end.

Through the object detection branch, we get the bounding-box of all the possible human and object in images and their confident score S_h, S_o. And the action classification branch give each human or object N action classification scoreS_h^a ,S_o^a and a displacement (x_{ho}, y_{ho}) or (x_{oh}, y_{oh}) to the corresponding human or object. We define the confidence score of a human-verb pair <human, verb> as $S_{h,a} = S_h \times S_h^a$. During the interference, we first calculate the confidence score of all the human-verb pair and get the K pair of the highest human-verb confidence scores.

Then, for each human-verb pair, we match them with r objects to produce HOI triplets. The matching strategy is as follows: Supposed there are M detected objects in the image, for each human-verb pair, we calculate its matching score with each detected object. The calculation formula is given in Eq. 2.

$$(x_o', y_o') = (x_h + x_{ho}, y_h + y_{ho})$$
$$(x_h', y_h') = (x_{o_i} + x_{oh}, y_{o_i} + y_{oh})$$
$$S_{dis} = \frac{1}{Min\left\{D(x_o', y_o', x_{o_i}, y_{o_i}), D(x_h', y_h', x_h, y_h)\right\} + 1} \tag{2}$$
$$S_{h,a}^{o_i} = S_o \cdot S_{o,a} \cdot S_{dis}$$

Here, D is the distance measurement function, which we finally choose the Euclidean distance, and (x_h, y_h), (x_{o_i}, y_{o_i}) is the coordinates of the center points of human and the ith object. We use both the displacement predicted from the human and object to calculate the score and take the higher one. Because when the human have same action with more than one object, the predicted displacement will be inaccurate and some interaction is more easy to be inferred from the object. Thus, this method can produce more reliable matching scores. Then, we choose r objects with the highest matching score as the human-verb pair's matching objects to generate the HOI triplet.

When calculating the HOI score, we should take the displacement prediction into account and the most simply way is to use the S_{dis}. However, the size of S_{dis} is quite different in different case. If we directly use the S_{dis}, it would make some HOI have inappropriate low confidence scores. Therefore, we use a special decay calculation strategy, which decays the HOI score with the decay of its matching score. Specially, for a HOI, we define its matching decay parameter by Eq. 3.

$$\lambda = \frac{S_{h,a}^{o_i}}{\alpha \cdot Max(S_{h,a}^{o_{j_k}})} \quad k = 1, 2 \cdots r \tag{3}$$

where $O_{j_k}(k = 1, 2 \cdots r)$ represents the objects a human-verb pair match with in the previous stage, and α is a hyper-parameter which denotes the base decay ratioλ. We finally set α to 2 (to 1 for the objects of highest matching scores) in our experiments to decay the HOI scores that do not get the highest matching score. This strategy assumes that the HOI with the highest matching score has a similar level in distance. If objects have similar S_o and $S_{o,a}$, the decay ratio λ will mainly depend on S_{dis} to suppress the object that is far away to the predicted location. In addition, when objects have higher S_o and $S_{o,a}$ but lower S_{dis}, the displacement prediction branch may have inaccurate output. In this case, the decay ratio would be smaller because of the higher S_o and $S_{o,a}$.

The final HOI score is calculated by Eq. 4.

$$S_{HOI} = \lambda S_h \cdot S_o \cdot Min(S_{h,a}, S_{o,a}) \tag{4}$$

AS shown in Eq. 4, we consider both the result of object detection and interaction detection, and use the minimum value of $S_{h,a}$ and $S_{o,a}$ as the confidence score of the action. In our experiment, we find that the network is more likely to produce inaccurate positive classification results. Therefore, using minimum value of $S_{h,a}$ and $S_{o,a}$ can help us get a more accurate confidence score for the action.

4 Performance Evaluation

4.1 Experimental Setting

Datasets. We conduct experiments on HICO-DET [2] benchmark to evaluate the proposed methods. The HICO-DET dataset consist of 38,118 images for training and 9658 images for testing. In this dataset, there are 600 classes of different interactions, corresponding to 80 object categories and 117 action verbs. Note that those 117 action verbs include the 'no interaction' class.In the dataset, HOIs which appear less than 10 times are considered as the rare set, and the rest 462 kinds of HOIs form the non-rare set.

Metrics. We follow the standard evaluation protocols, as in [2], use the mean average precision(mAP) as evaluation metric for our experiment. A predicted HOI triplet is considered as a true positive sample when it meet the following two requirements: (1) the predicted interaction category is the same type as the ground truth. (2) both the human and object bounding-boxes of the HOI triplet have IoUs (intersection-over-union) greater than 0.5 with the respective ground-truth.

4.2 Implementation Details

Our network is modified from the CenterNet [22] and use the DLA-34 [19]and the Hourglass104 [10] as the feature extractor, both of which are pre-trained on COCO dataset. The object detection branches are also initialized with the

Table 1. Comparison on HICO-DET

Methods	Backbone	Detector	Pose	Language	Full	Rare	NonRare
Two-stage methods							
Shen	VGG-19	COCO			6.46	4.24	7.12
InteractNet	ResNet-50-FPN	COCO			9.94	7.16	10.77
GPNN	ResNet-101	COCO			13.11	9.34	14.23
iCAN	ResNet-50	COCO			14.84	10.45	16.15
PMFNet-Base	ResNet-50-FPN	COCO			14.92	11.42	15.96
TIN	ResNet-50	COCO	✓		17.22	13.51	18.32
PMFNet	ResNet-50-FPN	COCO		✓	17.46	15.65	18.00
CHG	ResNet-50	COCO			17.57	16.85	17.78
Peyre et al.	ResNet-50-FPN	COCO		✓	19.40	14.63	20.87
VSGNet	ResNet152	COCO			19.80	16.05	20.91
FCMNet	ResNet-50	COCO	✓	✓	20.41	17.34	21.56
ACP	ResNet-152	COCO	✓	✓	20.59	15.92	21.98
PD-Net	ResNet-152	COCO		✓	20.81	15.90	22.28
One-stage methods							
IPNet	Hourglass	COCO			19.56	12.79	21.58
PPDM	DLA-34	HICO-DET			20.29	13.06	22.45
PPDM	Hourglass	HICO-DET			21.73	13.78	24.10
Ours	DLA-34	HICO-DET			21.16	12.72	23.68
Ours	Hourglass	HICO-DET			**22.35**	12.82	**25.92**

COCO pre-trained weight. We apply random flip and random crop data augmentation approaches to our model and apply the Adam optimizer [7] to optimize the loss function. During training and inference, the input resolution is 512 × 512 and the output is 128 × 128. We train the DLA-based model 120 epochs with a batch size of 84 and the Hourglass104-based model 130 model with a batch size of 30. All experiments are carried out on 8 GPUs.

4.3   Results and Comparison

We compare our methods with other HOI detection methods on the HICO-DET dataset. The compared methods most used a two-stage HOI detector, which first use a pre-train Faster-RCNN to detect all the possible human and object in the images and then apply the action classification network on all the human-object pairs. Besides, many previous works utilize other additional knowledge such as human poses and language priors to improve the performance of the networks. Differently, our method doesn't use any addition knowledge. Table 1 shows the result and comparison on the HICO-DET dataset. As shown in Table 1, our method outperforms the other methods.

4.4   Ablation Study

In this section, we do some ablation study to explore the influence of different components on the final performance. All the experiment here are apply on the HICO-DET dataset.

HOI Score Decay Strategy. In this part, we examine the superiority of our HOI scores decay strategy. We apply experiments using different decay parameter $/lambda$ in Eq. (4) and different hyper-parameter r in matching stage on the HICO-DET dataset. r is the number of selected matching objects for an human-verb pair and higher r produce more possible HOI triplets, which will increase the importance of HOI score calculation strategy and the infer time of the network. We set the hyper-parameter α in Eq. (3) to 2 in all the experiments. As shown in Table 2, directly using the distance score would lead to the worst performance. Calculating HOI score without decaying performs well when r is 1, but its performance quickly go bad with the increasing of r. Fixed decay ratio only works well when r is low, too. And decaying the HOI score considering the whole matching result can take the best used of the additional HOI prediction to increase the performance.

Combination of Various Components. We evaluate the combination of the four components in our methods: the action classification at human location and objects location and the displacement prediction from human and objects. The results in Table 3 shows that only using the classification result of the object is hard to predict correct HOI, but combined with the classification result of human, it can help to get better performance. Each of the displacements prediction perform well, and considering the number of parameters, the displacement prediction from objects is better in practice use because it needs to predict only a displacement for each location. When pursuing the best performance, we can apply all of the components.

Table 2. Ablation study of decay strategy

λ	r=					
	1	2	3	4	5	6
S_{dis}	18.02	17.90	17.81	17.79	17.77	17.78
1	20.45	19.90	19.60	19.41	19.29	19.28
0.5	20.45	20.62	20.59	20.56	20.52	20.55
decay base on S_{dis}	20.45	20.76	20.80	20.83	20.85	20.88
decay base on $S_o \cdot S_{dis}$	20.45	20.78	20.90	20.95	20.99	21.03
decay base on $S_{o,a} \cdot S_{dis}$	20.45	20.77	20.86	20.90	20.92	20.96
decay base on $S_o \cdot S_{o,a} \cdot S_{dis}$	20.45	20.91	21.04	21.07	21.11	**21.16**

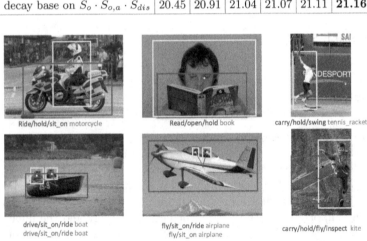

Fig. 3. Visualization results on HICO-DET dataset. The yellow boxes presents the detected human and the blue boxes presents the objects. If a subject has interaction with an object, they will be linked by a line and its action categories will be written below using the same color. (Color figure online)

Table 3. Ablation study of the components

Method						
cls.H	cls.O	Dis.H	Dis.O	Full	Rare	NonRare
✓		✓		20.73	12.29	23.34
✓			✓	20.76	12.15	23.42
	✓	✓		19.62	11.53	22.11
	✓		✓	19.53	11.24	22.05
✓	✓	✓		20.93	12.54	23.47
✓	✓		✓	20.95	12.44	23.52
✓	✓	✓	✓	**21.16**	12.72	23.68

Visualization Results. Figure 3 shows some examples of our HOI detection on HICO-DET dataset. Our method can find out most of the interaction in images. But our methods have difficulty in dealing with the interactions that human and object is far away such as the third image in Fig. 3, which we miss the interaction with the ball. Besides our method doesn't consider the influence between the interaction, such as in the fourth image in Fig. 3, only one person can be driving the boat but our method is likely to predict the action "drive" for both person. These problem will be study in our future works.

5   Conclusion

In this paper, we present a novel one-stage method for discovering HOI semantics from massive image data. Our method produces HOI triplets through action classification and displacement prediction. More specifically, we first classify the human's and object's action categories and then predict a displacement. Further, to get more accurate HOI scores, we propose a new decay calculation strategy to suppress the HOI triplets that have low matching scores. We conducted experiments on a popular benchmark and compared our proposal with various competitors. The results suggested the effectiveness of our method.

Acknowledgments. This paper is supported by the National Science Foundation of China (grant no. 62072419).

References

1. Bansal, A., Rambhatla, S.S., Shrivastava, A., Chellappa, R.: Detecting human-object interactions via functional generalization. In: Proceedings of AAAI, vol. 34, pp. 10460–10469 (2020)
2. Chao, Y.W., Wang, Z., He, Y., Wang, J., Deng, J.: Hico: a benchmark for recognizing human-object interactions in images. In: Proceedings of ICCV, pp. 1017–1025 (2015)

3. Fang, H.S., Xie, Y., Shao, D., Lu, C.: DIRV: dense interaction region voting for end-to-end human-object interaction detection. arXiv preprint arXiv:2010.01005 (2020)

4. Gao, C., Xu, J., Zou, Y., Huang, J.-B.: DRG: dual relation graph for human-object interaction detection. In: Vedaldi, A., Bischof, H., Brox, T., Frahm, J.-M. (eds.) ECCV 2020. LNCS, vol. 12357, pp. 696–712. Springer, Cham (2020). https://doi.org/10.1007/978-3-030-58610-2_41

5. Gao, C., Zou, Y., Huang, J.B.: ican: instance-centric attention network for human-object interaction detection. arXiv preprint arXiv:1808.10437 (2018)

6. Kim, B., Choi, T., Kang, J., Kim, H.J.: UnionDet: union-level detector towards real-time human-object interaction detection. In: Vedaldi, A., Bischof, H., Brox, T., Frahm, J.-M. (eds.) ECCV 2020. LNCS, vol. 12360, pp. 498–514. Springer, Cham (2020). https://doi.org/10.1007/978-3-030-58555-6_30

7. Kingma, D.P., Ba, J.: Adam: A method for stochastic optimization. arXiv preprint arXiv:1412.6980 (2014)

8. Liao, Y., Liu, S., Wang, F., Chen, Y., Qian, C., Feng, J.: Ppdm: parallel point detection and matching for real-time human-object interaction detection. In: Proceedings of CVPR, pp. 482–490 (2020)

9. Lin, T.Y., Goyal, P., Girshick, R., He, K., Dollár, P.: Focal loss for dense object detection. In: Proceedings of ICCV, pp. 2980–2988 (2017)

10. Newell, A., Yang, K., Deng, J.: Stacked hourglass networks for human pose estimation. In: Leibe, B., Matas, J., Sebe, N., Welling, M. (eds.) ECCV 2016. LNCS, vol. 9912, pp. 483–499. Springer, Cham (2016). https://doi.org/10.1007/978-3-319-46484-8_29

11. Ren, S., He, K., Girshick, R.B., Sun, J.: Faster R-CNN: towards real-time object detection with region proposal networks. In: Proceedings of NIPS (2015)

12. Tian, Q., Wan, S., Jin, P., Xu, J., Zou, C., Li, X.: A novel feature fusion with self-adaptive weight method based on deep learning for image classification. In: Hong, R., Cheng, W.-H., Yamasaki, T., Wang, M., Ngo, C.-W. (eds.) PCM 2018. LNCS, vol. 11164, pp. 426–436. Springer, Cham (2018). https://doi.org/10.1007/978-3-030-00776-8_39

13. Wan, B., Zhou, D., Liu, Y., Li, R., He, X.: Pose-aware multi-level feature network for human object interaction detection. In: Proceedings of ICCV, pp. 9469–9478 (2019)

14. Wan, S., Jin, P., Yue, L.: An approach for image retrieval based on visual saliency. In: Proceedings of IASP, pp. 172–175 (2009)

15. Wang, H., Zheng, W., Yingbiao, L.: Contextual heterogeneous graph network for human-object interaction detection. In: Vedaldi, A., Bischof, H., Brox, T., Frahm, J.-M. (eds.) ECCV 2020. LNCS, vol. 12362, pp. 248–264. Springer, Cham (2020). https://doi.org/10.1007/978-3-030-58520-4_15

16. Wang, T., Yang, T., Danelljan, M., Khan, F.S., Zhang, X., Sun, J.: Learning human-object interaction detection using interaction points. In: Proceedings of CVPR, pp. 4116–4125 (2020)

17. Yang, X., Wan, S., Jin, P.: Domain-invariant region proposal network for cross-domain detection. In: Proceedings of ICME, pp. 1–6 (2020)

18. Yang, X., Wan, S., Jin, P., Zou, C., Li, X.: MHEF-TripNet: mixed triplet loss with hard example feedback network for image retrieval. In: Zhao, Y., Barnes, N., Chen, B., Westermann, R., Kong, X., Lin, C. (eds.) ICIG 2019. LNCS, vol. 11903, pp. 35–46. Springer, Cham (2019). https://doi.org/10.1007/978-3-030-34113-8_4

19. Yu, F., Wang, D., Shelhamer, E., Darrell, T.: Deep layer aggregation. In: Proceedings of CVPR, pp. 2403–2412 (2018)

20. Zhong, X., Ding, C., Qu, X., Tao, D.: Polysemy deciphering network for human-object interaction detection. In: Vedaldi, A., Bischof, H., Brox, T., Frahm, J.-M. (eds.) ECCV 2020. LNCS, vol. 12365, pp. 69–85. Springer, Cham (2020). https://doi.org/10.1007/978-3-030-58565-5_5
21. Zhou, P., Chi, M.: Relation parsing neural network for human-object interaction detection. In: Proceedings of ICCV, pp. 843–851 (2019)
22. Zhou, X., Wang, D., Krähenbühl, P.: Objects as points. arXiv preprint arXiv:1904.07850 (2019)

Fuzzy Ontology-Based Possibilistic Approach for Document Indexing Using Semantic Concept Relations

Kabil Boukhari[✉] and Mohamed Nazih Omri[✉]

MARS Research Laboratory, LR17ES05, University of Sousse, Sousse, Tunisia
Mohamednazih.omri@eniso.u-sousse.tn

Abstract. To overcome the weaknesses of current information retrieval system and to utilize the strengths knowledge extraction a novel approach based on fuzzy ontology and possibility theory is proposed for indexing documents. Possibility theory allows to model and quantify the relevance of a document given a controlled vocabulary through two measures: necessity and possibility. Fuzzy ontology is used to improve the indexing process on information retrieval by means of external resource. Besides, the fuzzy approach has been proposed in order to make the availability of terms representations in a document more flexible. It allows a formal representation of a knowledge domain in the form of a hierarchical terminology provided with semantic relationships. As a result, the proposed approach has been made on different corpora, had better performance than other indexing approaches and it prove important results.

Keywords: Knowledge extraction · Possibility theory · Data analysis · Fuzzy ontology · Information retrieval · Document indexing

1 Introduction

The amount of data and textual information on the web is constantly increasing and the manual processing of information is too expensive and time consuming, especially when it comes to a specific domain such as medical domain. Thus, any automatic indexing process is at the heart of documentary research. Classic information search models are based only on words found in the document. However, these which are not systematically relevant share a set of words with the query, in the form of a set of words, which can thus be returned to browsers. The success of a documentary retrieval system depends mainly on three tasks: (i) document representation (indexing), (ii) query representation and (iii) query/document correspondence. Document indexing tends to select and extract words that match the content of the document, making it easier to find information. Much research has been proposed in this context to improve information retrieval models.

Several reasons that motivated us to opt for the integration of the theory of possibilities and fuzzy ontology. Possibility Theory is an efficient and robust

C. Strauss et al. (Eds.): DEXA 2021, LNCS 12924, pp. 264–269, 2021.
https://doi.org/10.1007/978-3-030-86475-0_26

method for uncertain processing tasks such as matching documents and external resources in a reliable and efficient manner. This theory is known as a robust method for dealing with uncertain tasks. As for the use of fuzzy ontology, this makes it possible to increase the performance of information search since they offer semantic links between the different concepts handled.

The rest of this paper is organized as follows: Sect. 2 presents the related work. Section 3, details the description of the proposed approach. In the next Sect. 4 we describe the experimental evaluations and we analyze the obtained results. Section 5 summarizes different steps of the proposed approach and gives the main prospects for this work.

2   Related Work

Several approaches have been proposed for knowledge extraction and documents indexing [4,5]. We can classify these approaches into two families: (i) Approaches based on free terms extraction and (ii) Approaches based on the controlled vocabulary.

Approaches based on free terms extraction use only terms existing in the document [12]. In [8] an approach has been proposed to extract complex terms from texts, they exploited statistical and linguistic methods. In [15], authors presented an approach for extracting keywords based on CRF (Conditional Random Fields). The work in [6] used Natural Language Processing (NLP) to extract keywords from a document. In [14] the authors proposed an approach to extract automatically keywords from scientific documents. It generates the candidate expressions based on a word expansion algorithm and introduces document frequency feature. The work in [10] has proposed a keyword extraction approach which only uses the document to index without using the whole collection.

Approaches based on controlled vocabulary use external resources and exploits specific terminology for indexing document. The author of [13] presented the approach "QuickUMLS" which uses a dictionary to extract medical concept based on approximate matching. In [9], authors used more than 200 ontologies to facilitate the space of medical discoveries by providing to scientists unified view of this diverse information. In [7], the authors proposed an indexing approach for biomedical documents by using the MeSH thesaurus, the basic idea is to use the VSM method for extracting concepts, and combines a static and a semantic method to estimate the concept's relevance for a given document.

3   Fuzzy Ontology-Based Possibilistic Proposed Approach

The proposed approach consists of 3 stages. (i) Pre-processing, (ii) concept extraction and (iii) filtering task.

Pre-processing. The pre-processing step consists of 5 tasks: (i) divide the document into sentences, (ii) remove punctuation, (iii) remove stop words, (iv) de-suffix words and (v) divide each sentence into words. The four tasks (ii), (iii),

(iv) and (v) are also applied to each term of the MeSH thesaurus. We used the RAID algorithm [3] an improved version of SAID [1], for stemming.

Concepts Extraction. In this part, we present the concepts extraction phase which is based on the two methods: Possibility theory and Fuzzy ontology.

Possibilistic Concepts Extraction. In this part, the concepts extraction is done first to extract the terms that compose a concept. For this, we use possibility theory that allows us to calculate a score for each term. The extracted candidate terms are those having a non-zero score. Besides, the corresponding concepts are assigned to the terms and the concept score corresponds to the score of its term. If a concept matches more than one term among the candidate terms, we associate it the maximum score. The term having the same score as its concept is denoted representative term.

The conditional possibility allows to calculate two measures: The document possibility giving a term (see Eq. 1) and The document necessity giving a term (see Eq. 2).

Equations 1 and 2 are calculated between the document to be indexed D_i and each term T_k. Terms with non-zero values are ranked according to the score Sc (Eq. 1). The score Sc in term k and in a document i is the sum of the possibility and necessity values. The addition of the necessity and possibility measures has already been adapted in other approaches for the relevance calculation. Indeed, we have exploited this method to disambiguate words and the measures combination gave an interesting result.

$$Sc(T_k, D_i) = \prod(D_i|T_k) + N(D_i|T_k)$$

$$\prod(D_i|T_k) = \frac{\prod(D_i \wedge T_k)}{\prod(T_k)} \tag{1}$$

$$N(D_i|T_k) = 1 - \prod(\overline{d_i}|T_k) \tag{2}$$

We represent word in the document in two cases: (i) More a word is frequent in the document, more it is likely to be a representative for this document. (ii) More a word is frequent in the document and it is less frequent in other documents of the collection, more it is necessarily to be a representative for this document.

In this step, we calculate the concept score from the terms that it compose. This is the higher value of its terms scores, the term having the maximum score is considered as representative term of document.

$$Score(C) = \max_{T_k \in (C)} (Sc(T_k)) \tag{3}$$

T(C): Set of Concept Terms

A representative term take the same values of the concept possibility and necessity.

Fuzzy Concepts Extraction. The creation of the fuzzy ontology goes through four stages: (i) Corpus/External vocabulary pre-treatment, (ii) Concept identification, (iii) Fuzzy membership assignment and fuzzy ontology creation and (iv) Concept extraction based on fuzzy ontology.

This phase of "Corpus/External vocabulary pre-treatment" is described in the section "Pre-processing".

To identify the relationships among the concepts we have used the MeSH thesaurus. In this work we have fuzzified the semantic relationships (Synonyms, described by, preferred term, non-preferred term, preferred concept, non-preferred).

The relationship between two concepts is assigned to the given relationship in the range of [0–1]. In fuzzy ontology, membership scores are assigned to the relationships to show its robustness. Here we defined the weights to the type of semantic relationship between the concepts.

All relationships and scores are assigned based on the MeSH thesaurus. For each concept C_1 having relationship R with concept C_2 and Fuzzy Membership, we create two nodes C_1 and C_2 by adding the semantic relation on the link between the nodes and the weight. Finally, now our fuzzy ontology has been created.

In the last part we use the fuzzy ontology to find the most associated concept, those having the highest membership. We give the document word to find the most related concept for the proposed word using the fuzzy ontology. In this work, we select the top semantically related concepts from the ontology, given a document words, those having the highest membership in between [0,1].

Filtering and Final Ranking. To refine this approach, we divided this bag of the candidate concepts into two others: (i) Principal concepts bag: contains all concepts in common and those which all their words are present in the document. (ii) Secondary concepts bag: here we find concepts that some words of their term are present in the document. To build the final list of concepts, we start by filtering the two lists by adding concepts from the secondary list to the main list by exploiting the Unified Medical Language System (UMLS) medical terminology and the MeSH thesaurus architecture.

4   Analysis of Experimental Results and Discussion

To evaluate the proposed approach, we used a subset of 150,000 documents of the OHSUMED 88 collection relating to scientific articles from Pubmed [1]. Moreover, we used three evaluation measures (i) the Precision that represents the ratio between the Number of Correct Concepts and the total Number of Extracted Concepts, (ii) the Recall defined as the ratio between the Number of Correct Concepts and the Number of concepts that correspond to Manual Indexing and (iii) the F-score that combines precision and recall with an equal weight concepts.

---

[1] http://trec.nist.gov/data/t9_filtering.html.

For the evaluation, the approaches are classified according to their families: those based on controlled language using controlled vocabulary [2], those based on partial correspondence (MaxMatcher [16]/QuickUMLS [13]) and others based on semantic algorithm with exact matching (BioAnnotator [11]). By analyzing

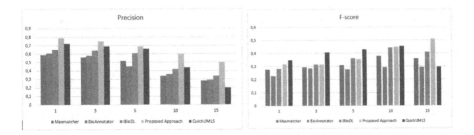

Fig. 1. Precision and F-score value for different approaches

Fig. 1, we notice that the proposed approach gives high performance and good results comparing to other approaches. The precision value shows the robustness of our approach in different ranks. These results highlight the interest of using the fuzzy logic and the possibility theory. A fuzzy membership is assigned for all type of semantic relationships to deal with the vocabulary mismatch problem among the concepts. In addition, the possibility theory gives more relevance in the concept extraction phase. Indeed, it is modeled by the necessity degree which contributes to improve the both extraction and classification of the relevant concepts. The recall value of our approach is higher than the recall of MaxMatcher approach, although the base keeps all the concepts partially matched to the document. This result is due to the filtering step applied to the proposed approach which is not the case for MaxMatcher approach.

The filtering step represents a robust solution to minimize incorrect concepts generated by partial matching. Also, the stemming process is applied to the proposed approach only on words with the stem length greater than 4.

5   Conclusion and Prospects

In this paper, we proposed a new document indexing approach by exploiting fuzzy ontology and possibility theory. The obtained results and the evaluation test show clearly the importance of our approach, which give more representative concepts compared to other approaches especially with the use of the fuzzy logic and the possibilistic theory.

The future works can be articulated around two new directions. The first is to conduct a more in-depth comparative study. In a second direction, we plan to exploit new source of controlled vocabulary as external terminologies. Moreover, we are working to evaluate the proposed approach on Big Data corpora.

References

1. Boukhari, K., Omri, M.N.: SAID: a new stemmer algorithm to indexing unstructured document. In: The International Conference on Intelligent Systems Design and Applications, pp. 59–63 (2015)
2. Boukhari, K., Omri, M.N.: Information retrieval based on description logic: application to biomedical documents. In: Conference: International Conference on High Performance Computing and Simulation (HPCS 2017), vol. 15, pp. 1–8 (2017)
3. Boukhari, K., Omri, M.N.: RAID: robust algorithm for stemming text document. Int. J. Comput. Inf. Syst. Ind. Manage. Appl. **8**, 235–246 (2016)
4. Boukhari, K., Omri, M.N.: Approximate matching-based unsupervised document indexing approach: application to biomedical domain. Scientometrics **124**(2), 903–924 (2020). https://doi.org/10.1007/s11192-020-03474-w
5. Boukhari, K., Omri, M.N.: DL-VSM based document indexing approach for information retrieval. J. Ambient Intell. Human. Comput. 1–25 (2020)
6. Bracewell, D., Ren, F., Kuroiwa, S.: Multilingual single document keyword extraction for information retrieval. In: Proceedings of Natural Language Processing and Knowledge Engineering (NLP-KE), pp. 517–522 (2005)
7. Chebil, W., Soualmia, L.F., Omri, M.N., Darmoni, S.J.: Biomedical concepts extraction based on possibilistic network and vector space model, pp. 227–231 (2015)
8. Fkih, F., Omri, M.N.: Complex terminology extraction model from unstructured web text based linguistic and statistical knowledge. Int. J. Inf. Retrieval Res. **2**(3), 1–18 (2012)
9. Jonquet, C., et al.: NCBO resource index: ontology-based search and mining of biomedical resources. Web Semant. **9**(3), 316–324 (2011)
10. Matsuo, Y., Ishizuka, M.: Keyword extraction from a single document using word co-occurrence statistical information. Int. J. Artif. Intell. Tools **13**, 1–13 (2004)
11. Mukherjea, S., et al.: Enhancing a biomedical information extraction system with dictionary mining and context disambiguation. IBM J. Res. Dev. **48**(5–6), 693–702 (2004)
12. Omri, M.N., Chenaina, T.: Uncertain and approximative knowledge representation to reasoning on classification with a fuzzy networks based system. In: IEEE International Fuzzy Systems Conference, pp. 1632–1637 (1999)
13. Soldaini, L., Goharian, N.: QuickUMLS: a fast, unsupervised approach for medical concept extraction. In: MedIR Workshop, SIGIR, pp. 1–4 (2016)
14. You, W., Fontaine, D., Barthès, J.P.: An automatic keyphrase extraction system for scientific documents. Knowl. Inf. Syst. **34**(3), 691–724 (2013)
15. Zhang, C., Wang, H., Liu, Y., Wu, D., Liao, Y., Wang, B.: Automatic keyword extraction from documents using conditional random fields. J. Comput. Inf. Syst. **4**(3), 1169–1180 (2008)
16. Zhou, X., Zhang, X., Hu, X.: MaxMatcher: biological concept extraction using approximate dictionary lookup*. In: Pacific Rim International Conference on Artificial Intelligence, pp. 1145–1149 (2006)

Multi-Objective Recommendations and Promotions at TOTAL

Idir Benouaret[1(✉)], Mohamed Bouadi[2], and Sihem Amer-Yahia[1]

[1] CNRS, Univ. Grenoble Alpes, Grenoble, France
{idir.benouaret,sihem.amer-yahia}@univ-grenoble-alpes.fr
[2] SAP Labs, Mougins, France
mohamed.bouadi@sap.com

Abstract. In this paper, we revisit the semantics of recommendations and promotional offers using multi-objective optimization principles. We investigate two formulations of product recommendation that go beyond traditional settings by optimizing simultaneously two conflicting objectives: *Budget-Reco* optimizes two customer-centric goals, namely utility and budget, and *Business-Reco* optimizes utility, a customer-centric goal, and profit margin, a business-oriented goal. To capture those objectives, we formulate knapsack problems and propose adaptations of exact and approximate algorithms. We also propose *Group-Promo*, the problem of generating product promotions that we model as a group discovery problem with multiple objectives and develop a Pareto-based solution. Our experiments on our TOTAL datasets demonstrate the importance of multi-objective optimization in the retail context, as well as the usefulness of our solutions when compared to their exact baselines. The results are valuable to TOTAL's marketing department that has been improving hand-crafted strategies by launching several promotional campaigns using our algorithms.

Keywords: Product recommendations · Promotional offers · Case study

1 Introduction

Traditionally, recommendation systems were designed to learn a user's interest to suggest items that maximize the user's utility, items the user is most likely to appreciate. Collaborative filtering emerged as the most common method to estimate item utilities. In practice, several other dimensions are considered when designing recommendations especially in the context of retail and e-commerce platforms. On one hand, users only purchase items that are within their budget, on the other hand, the business impact of recommendations in terms of revenue

Supported by TOTAL Data Science Services. We are very grateful to Christiane Kamdem-Kegne and Jalil Chagraoui from the Marketing Department of TOTAL, France.

loss or gain, plays a major role in determining which products to recommend to which customers. Similarly, when designing promotional campaigns to reward their customers, retailers consider the business value of each customer. Our study focuses on product recommendations and promotions that combine customers' interests with business-oriented goals such as profit margin and revenue.

We formalize optimization problems coming from two real-world application scenarios. We revisit the semantics of recommendations and promotional offers using multi-objective optimization principles. We investigate two formulations of product recommendation that go beyond traditional settings by optimizing simultaneously two conflicting objectives: *Budget-Reco* optimizes both utility and budget, two customer-centric goals, and *Business-Reco* optimizes utility to serve customers' interests, and profit margin, a business-oriented goal. We formalize knapsack problems and adapt exact and approximate algorithms [6,10]. We also propose *Group-Promo*, the problem of providing customers with product promotions that we formalize as a group discovery problem with multiple objectives and develop a solution based on Pareto plans ([20]). In summary, we make the following contributions:

1. We formalize three new problems: *Budget-Reco*, *Business-Reco* and *Group-Promo*;
2. We develop exact and approximation algorithms that efficiently solve our optimization problems;
3. We conduct a set of experiments with a real dataset on TOTAL customers in collaboration with our data analysts at the marketing department.

The paper is organized as follows. Section 2 provides the data model and preliminaries. In Sect. 3, we formalize our problems and discuss the proposed algorithms. In Sect. 4, we subject our algorithms to a set of experiments. In Sect. 5, we discuss related works. Finally, Sect. 6 concludes the paper.

2   Data Model and Preliminaries

Let $\mathcal{U} = \{u_1, u_2, .., u_m\}$ be the set of customers and $\mathcal{I} = \{i_1, i_2, .., i_n\}$ be the set of products. For a given customer u, $\mathcal{H}_u \subset \mathcal{I}$ denotes the purchase history of u. We note $\mathcal{X}$ the boolean purchase matrix of m customers in $\mathcal{U}$ over the n products in $\mathcal{I}$. We now provide the definitions that we use in designing our multi-objective recommendation and promotion framework.

1. *Utility:* The function $Utility(u, i)$, measures the interestingness of a product i for a customer u. We use a classical Item-based Collaborative Filtering (IBCF) approach [23] which calculates a similarity $sim(i, j)$ between each pair of products i and j using cosine similarity. Our model accommodates other similarity functions and recommendation strategies such as association rules [3]. We choose IBCF for its better precision on our dataset [1].
2. *Budget:* Different customers have different spending habits. The budget is a customer specific upper-bound on the cumulated cost of multiple products.

For a given customer u, we denote β_u her budget. In our experiments in Sect. 4, we estimate the budget using the customer's average spending on past purchases.

3. *Profit Margin:* For a given product i, its profit margin $Margin(i)$ is calculated as follows: $Margin(i) = \alpha \times sp(i)$, where $sp(i)$ is the selling price of product i, and α is a value between 0 and 1 that depends on the product category and the location at which it is purchased (in our dataset, different gas stations).

$$Margin(u, i) = (F_u^{COCO} \times \alpha_{COCO} + F_u^{CODO} \times \alpha_{CODO} + F_u^{DODO} \times \alpha_{DODO}) \times sp(i) \tag{1}$$

where $F_u^{COCO}, F_u^{CODO}, F_u^{DODO}$, correspond to the frequency the customer u visits station types COCO, CODO and DODO respectively. These station types are determined by the Marketing department of TOTAL and depend on the size and location of the station as well as the traffic and number of customers in that station.

4. *Cost Price:* For a given product i, $CostPrice(i)$ is defined as the price of product i from which is subtracted its profit margin.

5. *User Generosity:* The generosity of a customer u, denoted Gen_u, is estimated using a business rule that is used by TOTAL's marketing department to run promotional campaigns. The user generosity consists of a small percentage of the total revenue (usually ranging from 1% to 5%) that is generated by the customer in a given period.

3   Multi-Objective Recommendations and Promotions

3.1   Budget Recommendations

Generally, customers have finite budgets, a.k.a. spending power, and they will spend more on products they prefer buying. It is therefore natural to account for budget in generating recommendations. To do that, we propose, *Budget-Reco*, a new formulation that accounts for both utility and budget. Given a target customer u with purchase history $\mathcal{H}_u$ and a budget β_u, the set of all available products $\mathcal{I}$, select a set of k items $\mathcal{S} \subset \mathcal{I}$ such that:

$$\sum_{i \in \mathcal{S}} Utility(u, i) \text{ is maximized}$$

$$subject\ to$$

$$\sum_{i \in \mathcal{S}} sp(i) \leq \beta_u \tag{2}$$

$$|\mathcal{S}| = k$$

$$\mathcal{S} \cap \mathcal{H}_u = \emptyset$$

Budget-Reco is a variant of the well-known 0–1 knapsack problem [8] where the values of items correspond to their utilities and the capacity of the knapsack corresponds to the budget β_u. A naive approach to solving this problem is to generate the set of all possible combinations of k items and then choose the k items achieving the highest utility whose cumulative costs is less than β_u. However, this approach is prohibitively expensive. To address that, we use two exact algorithms (branch-and-bound and dynamic programming) and a greedy heuristic:

- **Branch-and-Bound.** The general algorithmic concept of branch-and-bound is based on an intelligent enumeration of the solution space since in many cases only a small subset of the feasible solutions are enumerated explicitly. It is however guaranteed that the parts of the solution space which are not considered explicitly cannot contain the optimal solution [11,24].
- **Dynamic Programming.** Another popular method to compute an optimal solution for our problem is dynamic programming. Instead of enumerating the solution space of the problem, this algorithm starts by solving a small subproblem and then extends the solution iteratively until an overall optimal solution is found [11,15].
- **Greedy Algorithm.** We implemented a greedy item-based collaborative filtering which selects the items that yield the next highest utility as long as the budget constraint β_u is not achieved.

3.2   Business Recommendations

Given a target customer u with purchase history $\mathcal{H}_u$, the set of all available products $\mathcal{I}$, an integer constant k, *Business-Reco* selects all sets $\mathcal{S} \subset \mathcal{I}$ satisfying:

$$\sum_{i \in \mathcal{S}} Utility(u, i) \text{ is maximized}$$

$$\sum_{i \in \mathcal{S}} Margin(u, i) \text{ is maximized} \tag{3}$$

$$\text{subject to}$$

$$|\mathcal{S}| = k$$

$$\mathcal{S} \cap \mathcal{H}_u = \emptyset$$

The output is k-sets of product recommendations, where each set $\mathcal{S}$ satisfies the conditions above.

The main challenge in designing an algorithm for *Business-Reco* is the conflicting nature of the two objectives (utility and margin). In our recent work, we proposed a bi-objective approach based on dynamic programming to generate the set of all non-dominated k-sets [2]. However, computing all k-sets in polynomial time is not feasible. In real world applications, it is not necessary to find an optimal solution but a "good" solution which can be computed in reasonable time. As a consequence, we formulated *Business-Reco* as a multi-objective knapsack problem and implemented a Fully Polynomial Time Approximation Scheme (FPTAS) [7] algorithm with a performance guarantee that returns a suboptimal solution that is close within a factor of $(1 - \epsilon)$ to the optimal solution but with a much faster response time.

3.3   Promotional Offers

Business experts at the marketing department of TOTAL frequently run promotional offers to reward the fidelity of their customers. The set of products that are targets of a promotional campaign are determined by experts. The retailer's goal could be to sell out the stock of some products or to promote new products that are not popular yet. To address that, we propose to formalize *Group-Promo* whose goal is to find which group of customers to match with which subset of products so that: (1) the number of targeted customers is maximized, (2) overlap between group members is minimized, (3) the average utility of products across group members is maximized, and (4) the cost price of the promoted products is closest to the generosity of customers.

Problem Definition. Given k sets of items in $\mathcal{I}$ and the set of customers $\mathcal{U}$, *Group-Promo* forms k pairs (S, G), where $G \subseteq \mathcal{U}$ is a group, for each set $S \subseteq \mathcal{I}$ such that:

$$\frac{|\bigcup G|}{|\mathcal{U}|} \text{ is maximized;}$$

$$|\bigcap G| \text{ is minimized;}$$

$$\max_{i \in S} avg_{u \in G} \, Utility(u, i) \text{ is maximized;} \tag{4}$$

$$\max_{i \in S} \max_{u \in G}(|Gen(u) - CostPrice(i)|) \text{ is minimized;}$$

Pareto Formulation. When optimizing more than one objective, there may be many incomparable group-sets. For instance, for a target set of items, we can form two groups, each group has its own advantage: the first one has higher utility and the second optimizes better for overlap between group members. In this section, we borrow the terminology of Multi-Objective Optimization introduced in [20] and define these concepts.

Definition 1. *Local Plan.* *A local plan p, associated with a group G for a target set of items S, is a tuple $< Utility(G, S), Convergence(G, S) >$ where*
$Utility(G, S) = \max_{i \in S} avg_{u \in G} \, Utility(u, i)$
$Convergence(G, S) = \max_{i \in S} \max_{u \in G}(|Gen(u) - CostPrice(i)|)$

A local plan consists of objectives that have to be satisfied within each group.

Definition 2. *Global Plan.* *A global plan p, associated with a set of k groups is a tuple $< Coverage, Intersection >$ where*
$Coverage = \frac{|\bigcup G|}{|\mathcal{U}|}$
$Intersection = |\bigcap G|$

A global plan consists of objectives that have to be satisfied across all groups.

Algorithm 1: Promotional Offer Algorithm to solve *Group-Promo*

    Input: k sets of items in $\mathcal{I}$, a set of customers $\mathcal{U}$
    Output: Pair (S, G) for each set $S \subseteq \mathcal{I}$
1 $\mathcal{P} \leftarrow \emptyset$
2   **foreach** *itemset S* **do**
3 $G_s \leftarrow$ Construct user groups with LCM
4 $P_G \leftarrow \emptyset$
5   **for** *all user groups g in G_s* **do**
6 $p \leftarrow$ Local-Plan(g,S)
7   **if** *p is not dominated by any other plan in P_G* **then**
8 $P_G.add(p)$
9   **end**
10   **end**
11 **end**
12 **for** *all k-groupset combinations from P_G* **do**
13   Keep non-diminated plans and add them to $\mathcal{P}$
14 **end**
15 return($\mathcal{P}$)

Definition 3. *Dominance.* *A plan p_1 dominates a plan p_2 if p_1 has better or equivalent values as p_2 for every objective. The term "better" is equivalent to "greater than" for maximization objectives, and "lower than" for minimization objectives. Furthermore, plan p_1 strictly dominates p_2 if p_1 has better values than p_2 for every objective.*

Definition 4. *Pareto Plan.* *A plan p is Pareto if no other plan strictly dominates p. The set of all Pareto plans is denoted as $\mathcal{P}$.*

Promotional Offer Algorithm. The algorithm described in 1 first constructs a set of possible groups for each set of items, based on the description of the frequent items purchased by customers. We use the LCM [28] mining algorithm for this task (Line 3). This step produces a large number of candidate groups that share common buying patterns. We compute, for each group, a local plan and compare the plans to discard all the dominated groups (Lines 5–11). After that, we generate different combinations of k groups where each group is associated with one of the set of items in the input. We keep the group-sets that are not dominated by any other (Lines 12–14).

4 Experiments

The purpose of our experiments is to study the balance between recommendation accuracy and response time for our three problem formulations. While our datasets are proprietary, our code is made available on GitHub[1]

---

[1] github.com/multiobjective-recos/Biobjective_RecSys.

4.1 Experimental Protocol

We split our dataset into training and test sets using the temporal global strategy [17]. We used purchase records from January 2017 to December 2018 for training and records from January 2019 to December 2019 for testing. We discarded cold start customers having fewer than 10 purchases.

4.2 Evaluation Measures

The goal is to evaluate the effectiveness of our recommendations with respect to accuracy and the generated profit margin in the case of product recommendations. In the case of promotions, we measure the induced costs and the average utility. We use the following evaluation measures and report results as averages over all test customers.

- *Precision*: measures the percentage of relevant recommendations among the top-k recommendations.

$$precision_u@k = \frac{card(S_u@k \cap test_u)}{k}$$

where $S_u@k$ is the top-k recommendations and $test_u$ is the target test set of customer u

- *Margin*: measures the average profit margins generated by the top-k recommendations.

$$margin_u@k = \frac{\sum_{i \in S_u@k} Margin(u,i)}{k}$$

where $S_u@k$ is the set of top-k recommendations.

- *Utility* measures the average utility generated by the products offered to the customer.

$$Utility_u = \frac{\sum_{i \in S} Utility(u,i)}{card(S)}$$

where, given a customer u, S is the set of product promotions for u.

- *Costs* measures the average costs generated by the products offered to a customer.

$$Costs_u = \frac{\sum_{i \in S} CostPrice(i)}{card(S)}$$

where, given a customer u, S is the set of product promotions for u.

Fig. 1. Precision@10 for all algorithms and all customer segments (*Budget-Reco*)

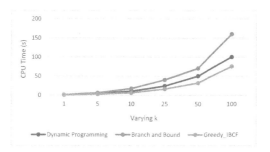

Fig. 2. Response time of all algorithms for solving *Budget-Reco* as a function of recommendation size k

4.3   Recommendation Experiments

Budget Recommendations. Since users generally have different spending habits, we customize each customer's budget to better reflect the value of the budget constraint β_u. We calculate the average spending of each customer in a single transaction based on past purchases. We segment them into 4 different subsets: $[0.20[$, $[20.50[$, $[50, 100[$ and ≥ 100. For example, the segment $[0.20[$ groups together customers who spend between 0 and 20 euros on average. Recommendations are then tested on each segment independently. Figure 1 shows the achieved *precision*@10 values of all algorithms and for each customer segment. Precision results for different values of k were similar and are omitted due to lack of space.

Results show that for all customer segments the greedy-IBCF performs very closely to the two exact algorithms (branch-and-bound and dynamic programming) with a marginally small decrease on precision. Greedy-IBCF is much faster than exact algorithms especially when the number of recommendations k increases.

Business Recommendations. We use the same experimental setting to evaluate business recommendations. We report the achieved results on precision and profit margin with different values of k. We vary the approximation parameter

ϵ in $\{0.1, 0.3, 0.5\}$ and compare the results against a baseline *Pareto* which is an exact algorithm based on dynamic programming [2].

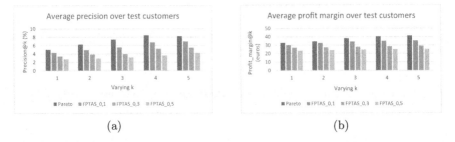

(a)   (b)

Fig. 3. Average precision (a) and profit margin (b) for different values of k (*Business-Reco*)

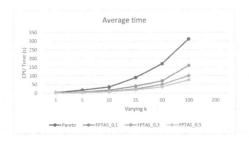

Fig. 4. Response time of all algorithms for solving *Business-Reco* as a function of recommendation size k

Results are reported in Fig. 3. Without much surprise the approximation algorithm performs worse than *Pareto* and the values of precision and profit margin decrease as ϵ increases. However, in terms of response time, the approximation method is always faster as is shown in Fig. 4. The gap between the response time needed by *Pareto* and the approximation algorithm increases with the number of recommendations. These results have pushed our collaborators at TOTAL to consider using the exact algorithm for testing their models offline and to use the approximation algorithm that yields an acceptable response time without hurting much precision and profit margin.

4.4   Promotion Experiments

To validate our solution for *Group-Promo*, we measure the average utility of the promoted products and the average costs across all customers. We compare our solution with one of the hand-crafted promotional campaigns that was deployed by analysts at TOTAL. In such campaigns, marketers select a set of items for each customer independently and then determine the subset of products that minimize the difference between the price of the products and the generosity of the customer. Our results in Table 1 for different values of k, show that our solution provides customers with much more interesting products (higher utility scores). While having an average cost per customer that is very close to the average cost induced by applying our *Group-Promo* algorithm.

Table 1. Promotions results

	Group-Promo		TOTAL's offers	
	Utility (%)	Costs (Euros)	Utility (%)	Costs (Euros)
k = 10	11.61	3.191	5.19	2.657
k = 20	19.191	5.779	8.72	6.155
k = 50	28.77	9.550	10.27	10.450
k = 100	32.91	11.789	11.85	11.053

5   Related Work

Our work is related to several others in its aim and optimization mechanism. We survey the closest literature and emphasize similarities and differences.

Multi-objective Optimization. There exist different approaches to solve a multi-objective problem [26,27]. Scalarization, where all objectives are combined into a single objective. Another popular method is ϵ-constraints where one objective is optimized and others are constrained [21]. This formulation could be used in our context in cases where a specific profit amount is desired and products must be chosen accordingly. Another approach is Multi-Level Optimization [18] which needs a meaningful hierarchy between objectives. In our case, all objectives are independent and conflicting, hence using this mechanism is not feasible. While we focus on recommendations, it overlaps with a well-known problem in combinatorial optimization which is the knapsack problem [9,11] where the goal is to optimize an objective function given some capacity constraints.

Recommendations and Promotions. Our work is obviously related to the rich area of recommendations [22]. Recommendation approaches can be broadly classified into collaborative filtering (CF) and content-based (CB) [22]. CF approaches started with user-based and then more attention was paid to item-based to address the sparsity and scalability challenges in user-based CF [23].

Promotions play a central role in shopping [12,13]. Instead of focusing solely on determining relevant recommendations, we aim to identify promotions that are also considered interesting [14,19]. Among other things, customers use promotions for budgeting and planning, whereas retailers use promotions to accelerate purchase cycles, stimulate sales of complementary products and attract new customers [5,16]. The motivation for focusing on existing promotions instead of product recommendations is that results from a user study that investigated customers' preferences regarding features in an intelligent mobile grocery assistant indicate that customers are more interested in information about relevant and actual special offers than suggestions for additional products [4].

6   Conclusion

In this paper, we revisited multi-objective optimization and applied it in the context of retail to generate product recommendations and promotional offers. We proposed new formulations and appropriate solutions that adapt existing algorithms. Our empirical validation on real datasets confronts exact and approximate solutions as well as a hand-crafted promotional campaign against our results, showing the importance of multi-objective recommendation approaches in the retail world.

Our work lays the ground for a number of short-term and medium-term directions. Our immediate line of work is to launch a large-scale experiments and analyses study on a variety of publicly-available retail datasets such as Amazon. Our second immediate course of action is to gather the results of the recommendations and promotional offers that were deployed recently at our partner's premises and confront the results to our empirical evaluation.

In the medium-term, we would like to characterize users' generosity more finely for different product categories. That would require to formulate new problems that combine different generosities for the same customer. The problem of promotional offers also needs to be revisited to combine products with different constraints in the same promotional package. We would also like to study the applicability of our framework in a changing context where for instance, customers' generosity evolves and users may migrate from one group to another. This will necessitate the design of adaptive algorithms to handle such changes.

References

1. Benouaret, I., Amer-Yahia, S.: A comparative evaluation of top-n recommendation algorithms: case study with total customers. In: 2020 IEEE International Conference on Big Data (Big Data), pp. 4499–4508. IEEE (2020)
2. Benouaret, I., Amer-Yahia, S., Kamdem-Kengne, C., Chagraoui, J.: A bi-objective approach for product recommendations. In: 2019 IEEE International Conference on Big Data (Big Data), pp. 2159–2168. IEEE (2019)
3. Benouaret, I., Amer-Yahia, S., Roy, S.B., Kamdem-Kengne, C., Chagraoui, J.: Enabling decision support through ranking and summarization of association rules for total customers. In: Transactions on Large-Scale Data-and Knowledge-Centered Systems XLIV, pp. 160–193. Springer, Berlin (2020)
4. Bhattacharya, S., et al..: Ma $$ ive-an intelligent mobile grocery assistant. In: 2012 Eighth International Conference on Intelligent Environments, pp. 165–172. IEEE (2012)

 5. Blattberg, R.C., Briesch, R., Fox, E.J.: How promotions work. Market. Sci. **14**(3_supplement), G122–G132 (1995)
 6. Chekuri, C., Khanna, S.: A polynomial time approximation scheme for the multiple knapsack problem. SIAM J. Comput. **38**(3), 1 (2006)
 7. Erlebach, T., Kellerer, H., Pferschy, U.: Approximating multiobjective knapsack problems. Manag. Sci. **48**(12), 1603–1612 (2002)
 8. Kellerer, H., Pferschy, U., Pisinger, D.: Introduction to NP-completeness of knapsack problems. In: Knapsack Problems. Springer, Berlin, pp. 483–493. Springer (2004). https://doi.org/10.1007/978-3-540-24777-7_16
 9. Pisinger, D., Toth, P.: Knapsack problems. In: Du, D.Z., Pardalos, P.M. (eds.) Handbook of Combinatorial Optimization. Springer, Boston (1998) . https://doi.org/10.1007/978-1-4613-0303-9_5
10. Kellerer, H., Pferschy, U., Pisinger, D.: Other knapsack problems. In: Knapsack Problems, pp. 389–424. Springer, Heidelberg (2004)
11. Kellerer, H., Pferschy, U., Pisinger, D.: Knapsack Problems with 33 Tables. Springer, Berlin (2010)
12. Lal, R., Matutes, C.: Retail pricing and advertising strategies. J. Busi. **67**, 345–370 (1994)
13. Lohse, G.L., Spiller, P.: Electronic shopping. Commun. ACM **41**(7), 81–87 (1998)
14. Makhijani, R., Chakrabarti, S., Struble, D., Liu, Y.: Lore: a large-scale offer recommendation engine with eligibility and capacity constraints. In: Proceedings of the 13th ACM Conference on Recommender Systems. pp. 160–168 (2019)
15. Martello, S., Pisinger, D., Toth, P.: Dynamic programming and strong bounds for the 0–1 knapsack problem. Manag. Sci. **45**(3), 414–424 (1999)
16. Mauri, C.: Card loyalty. a new emerging issue in grocery retailing. J. Retail. Consum. Serv. **10**(1), 13–25 (2003)
17. Meng, Z., McCreadie, R., Macdonald, C., Ounis, I.: Exploring data splitting strategies for the evaluation of recommendation models. In: Fourteenth ACM Conference on Recommender Systems, pp. 681–686 (2020)
18. Migdalas, A., Pardalos, P.M., Värbrand, P.: Multilevel optimization: algorithms and applications, vol. 20. Springer Science & Business Media (2013). https://doi.org/10.1007/978-1-4613-0307-7
19. Nurmi, P., Salovaara, A., Forsblom, A., Bohnert, F., Floréen, P.: Promotionrank: ranking and recommending grocery product promotions using personal shopping lists. ACM Trans. Inter. Intell. Syst. (TiiS) **4**(1), 1–23 (2014)
20. Omidvar-Tehrani, B., Amer-Yahia, S., Dutot, P.F., Trystram, D.: Multi-objective group discovery on the social web. In: Joint European Conference on Machine Learning and Knowledge Discovery in Databases. pp. 296–312. Springer (2016)
21. Papadimitriou, C.H., Yannakakis, M.: On the approximability of trade-offs and optimal access of web sources. In: Proceedings 41st Annual Symposium on Foundations of Computer Science, pp. 86–92. IEEE (2000)
22. Park, D.H., Kim, H.K., Choi, I.Y., Kim, J.K.: A literature review and classification of recommender systems research. Exp. Syst. Appl. **39**(11), 10059–10072 (2012)
23. Sarwar, B., Karypis, G., Konstan, J., Riedl, J.: Item-based collaborative filtering recommendation algorithms. In: Proceedings of the 10th international conference on World Wide Web, pp. 285–295 (2001)
24. Shih, W.: A branch and bound method for the multiconstraint zero-one knapsack problem. J. Oper. Res. Soc. **30**(4), 369–378 (1979)
25. Su, X., Khoshgoftaar, T.M.: A survey of collaborative filtering techniques. Adv. Artif. intell. **2009**, 421425 (2009)

26. Trummer, I., Koch, C.: Approximation schemes for many-objective query opti-
    mization. In: Proceedings of the 2014 ACM SIGMOD International Conference on
    Management of Data, pp. 1299–1310 (2014)
27. Tsaggouris, G., Zaroliagis, C.: Multiobjective optimization: Iiproved FPTAS for
    shortest paths and non-linear objectives with applications. Theor Comput. Syst.
    45(1), 162–186 (2009)
28. Uno, T., Asai, T., Uchida, Y., Arimura, H.: An efficient algorithm for enumerating
    closed patterns in transaction databases. In: Suzuki, E., Arikawa, S. (eds.) DS
    2004. LNCS (LNAI), vol. 3245, pp. 16–31. Springer, Heidelberg (2004). https://
    doi.org/10.1007/978-3-540-30214-8_2

An Effective Algorithm for Classification of Text with Weak Sequential Relationships

Qiqiang Xu[1], Ji Zhang[2(✉)], Ting Yu[3], Wenbin Zhang[4], Mingli Zhang[5], Yonglong Luo[6], Fulong Chen[6], and Zhen Liu[7]

[1] Nanjing University of Aeronautics and Astronautics, Nanjing, China
[2] University of Southern Queensland, Toowoomba, Australia
Ji.Zhang@usq.edu.au
[3] Zhejiang Lab, Hangzhou, China
yuting@zhejianglab.com
[4] Carnegie Mellon University, Pittsburgh, USA
wenbinzhang@cmu.edu
[5] Montreal Neurological Institute, Mcgill University, Montreal, Canada
mingli.zhang@mcgill.ca
[6] Anhui Normal University, Wuhu, China
ylluo@ustc.edu.cn, long005@mail.ahnu.edu.cn
[7] Guangdong Pharmaceutical University, Guangzhou, China
liu.zhen@gdpu.edu.cn

Abstract. Text classification is a fundamental task that is widely used in various sub-domains of natural language processing, such as information extraction, semantic understanding, etc. For the general text classification problems, various deep learning models, such as Bi-LSTM, Transformer, BERT, etc. have been used which achieved good performance. In this paper, however, we consider a new problem on how to deal with a special scenario in text classification which has a weak sequential relationship among different classification entities. A typical example is in the block classification of resumes where there are sequential relationships existing amongst different blocks. By fully utilizing this useful sequential feature, we in this paper propose an effective hybrid model which combines a fully connected neural network model and a block-level recurrent neural network model with feature fusion that makes full use of such a sequential feature. The experimental results show that the average F1-score value of our model on three 1,400 real resume datasets is 5.5–11% higher than the existing mainstream algorithms.

Keywords: Text classification · Recurrent neuron network · Feature fusion · Dynamic decision · Hybrid model

1 Introduction

Text classification (TC) is a fundamental task [14,15] widely involved in various sub-domains in NLP. Yet, different from the general classification task,

ⓒ Springer Nature Switzerland AG 2021
C. Strauss et al. (Eds.): DEXA 2021, LNCS 12924, pp. 283–294, 2021.
https://doi.org/10.1007/978-3-030-86475-0_28

we consider the text classification problem under a special scenario where the samples are orderly organized in a semi-structured document such as in resumes. Resume block classification (called RBC for short), which will be investigated as the instance of this problem in this paper, is by nature a text classification task with the goal being to classify every information block in the resume correctly according to the text content. Among different resume blocks, there are some sequential relationships. For example, the "basic information" block always appears in the first place, followed by "educational background", and "self-evaluation" is likely to be positioned in the last place, etc. Nevertheless, such sequential relationships are weak as they are not fixed nor stable due to the casual writing style and the diversity of templates be used for resumes. For such a special text classification task (called TC-WSR for short), the general classification algorithms such as [3,10] fail to sufficiently consider the contextual sequential relationship among different classification entities and their classification accuracy can thus be further improved.

For resume block classification, we firstly proposed a Block-Level Bidirectional RNN model (B-BRNN) in [21] to attempt the utilization of the sequential feature among different blocks, and successfully improves the F1-score by nearly 6%. However, B-BRNN is based on such an idealized assumption that the sequence relationship between resume blocks has a strong regularity, actually, the sequence relationship among resume blocks is only partially ordered, which is caused by the large number of fixed format resume templates, while not everyone strictly follows a uniform template, due to the casual writing style of different people, the sequence relationship among resume blocks shows instability. When predicting the resume sample which has a huge difference in the sequence feature distribution with that of training set (we call such kind of resume as abnormal resume), the performance of the model will be seriously affected. The weak and unstable sequential relationships will lead to the high variance for the model.

To further improve the performance of resume block classification in dealing with such a weak sequential feature, we in this paper focus on the research of a new hybrid model which is able to achieve better generalization for the abnormal resumes mentioned above, meanwhile, can maintain a good performance for the normal resumes as possible. To this end, we firstly use the feature fusion strategy [1] to improve the performance of B-BRNN. Secondly, we design an additional fully connected neural network model (FNN), which is leveraged to classify each resume block individually as the FNN model is not effected by the sequence characteristics among different resume blocks. Finally, we use the hybrid model strategy in ensemble learning to fuse these two models. Different from the existing hybrid model integration methods, the novelty of our hybrid model lies in that it can automatically adjust the weight of each sub-model to the current classification decision based on the input samples, whereby the classification advantages and disadvantages of each sub-model under different samples can be well complemented to further optimize the classification performance.

Specifically, the main contributions of this paper are summarized as follows: 1) we propose an improved B-BRNN with feature fusion, and design a dynamic decision hybrid model which combines the FNN and our improved B-BRNN to improve the model's adaptability in dealing with new resume samples, and enhance the generalization ability of the model; 2) We propose an innovative dynamic weighting strategy which is different from all the existing hybrid strategies in ensemble learning. Our proposed strategy can dynamically adjust the weighting of each sub-model according to the input samples, whereby further improving the classification accuracy of hybrid model.

The remainder of this paper is organized as follows. Section 2 discusses the related work. Section 3 introduces our proposed model for the problem of resume block classification. Section 4 introduces the experimental design for evaluating our model and the detailed analysis of the experimental results. The last section presents the conclusion and some future research directions.

2 Related Work

In the past research, the algorithms used for general TC task have experienced the traditional ML-based algorithms represented by support vector machines (SVM), random forest (RF), gradient boosting tree (GBT), and DL-based algorithms represented by CNN [11], RNN [5,7], Transformer [20], Bert [6], etc. [12]. Among them, the deep learning based models have achieved the state-of-the-art performance in general TC task. However, there hasn't been earlier work specifically on the TC-WSR task except us, and in the past, all the researchers regarded the RBC as a general TC task [3,9,10], we in [21] pointed that these models are all general purposed, which only take one entity as the input, and output the single entity classification result individually, ignoring the correlation among different entities, while in the RBC problem, the arrangement of blocks has a certain order, sufficiently utilizing this sequence feature can improve the accuracy of RBC, therefore, we proposed a B-BRNN model to obtain this sequence feature, experiment results showed that the F1-score is 6% higher than the existing models. However, we exaggerate the role of this sequence feature, actually, there is only a weak sequence feature among different blocks, when facing the abnormal resumes mentioned above, the classification performance of B-BRNN will decline sharply, different from the previous B-BRNN, our proposed hybrid model in this paper can not only maintain the high F1-score under normal resumes but a high F1-score under abnormal resumes.

In addition, we investigate the existing hybrid model strategies in ensemble learning. Previous strategies includes Bagging such as Random Forest [13], Boosting such as [2], Blending such as [4], and Stacking such as [18]. But in essence, the importance of each basic sub model in these hybrid models has been fixed in the training stage, and in the prediction stage, for any test sample, the hybrid model will assign the same static weight of each sub based model learned in the training stage to produce the final classification result, while in this paper, our novelty lies in dynamic decision, given a specific test sample, our hybrid model will judge the importance of each sub model for the current classification task, and dynamically assign a new weight for each sub model to produce the final classification result.

3   Our Model

In order to make full use of the extra sequence feature among the blocks of resume, improve the performance of RBC, and at the same time, improve the adaptability of the model to abnormal data, a dynamic decision hybrid model which combines the B-BRNN model improved by feature fusion with the FNN model is proposed in this paper. Figure 1 is the overview of our proposed model, the input of this hybrid model is a sequence representing all the blocks in a resume with their original order, and each block is embedded by word2vec [8], the output of this model is the labels of all blocks in a resume sample, in addition, the improved B-BRNN and FNN model can be pre-trained separately, which can save training time by multi-thread parallel training.

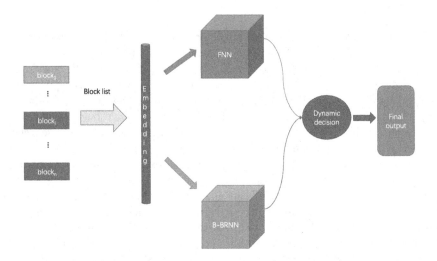

Fig. 1. Dynamic decision hybrid model overview

3.1   B-BRNN Improved by Feature Fusion

$Input = \{block_0, ..., block_i, ..., block_{n-1}\}$ is the block sequence, where n is the maximum number of blocks, and $block_i$ is composed of $\{word_0, ..., word_i, ..., word_{m-1}\}$, where m is the maximum size of words in one block. The words are embedded by word2vec (word2vec technology is more mature, more portable than BERT [6], less demanding on computer performance, and easy to experiment, although BERT can achieve better representation effect than word2vec after a large number of corpus pre training, but BERT also brings about the big increase of computation and memory cost), then after word embedding, a specific x_i can be represented as: $\frac{1}{m} \sum_{j=0}^{m-1} embedding(word_j)$, which represents the feature vector of $block_i$. The input of our model is $X = \{x_0, ..., x_i, ..., x_{n-1}\}$. After the Bi-GRU recurrent unit [5], the hidden output $H = \{h_0, ..., h_i, ..., h_{n-1}\}$

is obtained, then we use a feature fusion strategy to concatenate the input X and the hidden output of Bi-GRU, H, and the purpose of this is to let the model not only pay attention to the context order features of each resume block, but also give some weights to the text content features of resume block itself, so as to weaken the sensitivity of B-BRNN to abnormal resume data with different context sequence characteristics. More specifically, the details of feature fusion can be described as: when get the $X = \{x_0, \cdots, x_i, \cdots, x_{n-1}\}$, and $H = \{h_0, \cdots, h_i, \cdots, h_{n-1}\}$, we design two dense layers: $dense_1, dense_2$, which have a $relu$ activate function, after $dense_1$, $F_1 = \{f_1^0, \cdots, f_1^i, \cdots, f_1^{n-1}\}$ is obtained, and as the same, after $dense_2$, $F_2 = \{f_2^0, \cdots, f_2^i, \cdots, f_2^{n-1}\}$ is obtained, then, we concatenate F_1 and F_2 together:

$$F = \{f^0, \cdots, f^i, \cdots, f^{n-1}\} \tag{1}$$

$$= \{C(f_1^0, f_2^0), \cdots, C(f_1^i, f_2^i), \cdots, C(f_1^{n-1}, f_2^{n-1})\} \tag{2}$$

where $C(f_1^i, f_2^i)$ means that the two vectors are connected end to end to form a new vector. Then, given a specific f_i in F, we mark the output of dense layer as $o_i = dense(f_i) = (o_i^0, o_i^1, ..., o_i^{k-1})^T$, where k is the category counts. The final output y_i in $Y = \{y_0, ..., y_i, ..., y_{n-1}\}$ can be described as follows:

$$y_i = softmax(o_i) = (p_i^{(0)}, \cdots, p_i^{(j)}, \cdots, p_i^{(k-1)})^T \tag{3}$$

$$= (\frac{e^{o_i^0}}{\sum\limits_{j=0}^{k-1} e^{o_i^j}}, \frac{e^{o_i^1}}{\sum\limits_{j=0}^{k-1} e^{o_i^j}}, \cdots, \frac{e^{o_i^{k-1}}}{\sum\limits_{j=0}^{k-1} e^{o_i^j}})^T \tag{4}$$

Then, we choose cross entropy as the loss function, the total loss of one batch is described as Eq. (6):

$$Loss = \frac{1}{N} \sum_{s=1}^{N} \sum_{i=0}^{n-1} L_s(y_i, y_i') \tag{5}$$

$$= -\frac{1}{N} \sum_{s=1}^{N} \sum_{i=0}^{n-1} \sum_{j=0}^{k-1} p_{i(s)}'^{(j)} \log p_{i(s)}^{(j)} \tag{6}$$

where N is the batch size, $L(y_i, y_i')$ is the cross entropy of one time step in the improved B-BRNN, so the total loss is the sum of loss in each time step.

3.2 Dynamic Decision Hybrid Model

Inspired by the idea of hybrid model in ensemble learning, such as random forest (RF) [13], since the hybrid strategy can ensemble each sub model that has advantages and disadvantages at different aspects into a model with strong learning ability, why not use this idea to combine our improved B-BRNN model with a general classification model which is not be affected by the sequence feature among different resume blocks? Based on this idea, we further design a dynamic decision hybrid model which combines a Fully connected Neuron

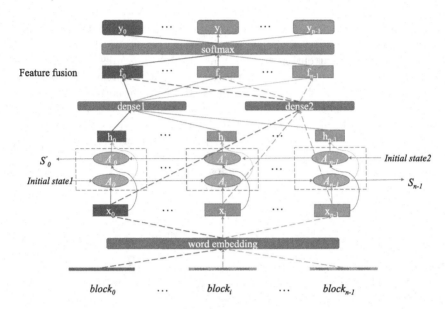

Fig. 2. B-BRNN structure improved by feature fusion

Network (FNN) with our improved B-BRNN model through a dynamic decision hybrid strategy. And the reason why we choose FNN to do model integration is that FNN has low complexity and computational cost, meanwhile, it can achieve a good classification result (Fig. 2).

Different from the previous hybrid strategy, the novelty of our hybrid model lies in "dynamic decision", which means that our hybrid model can analyze the characteristics of samples according to the input samples, automatically judge the importance of sub models for the current input, adjust the weight of each sub model, and make the best classification decision. Since the FNN model is not affected by the sequence feature of resume blocks, and a high classification accuracy can be achieved overall (even though lower than B-BRNN in normal situation that don't contain abnormal resumes), and the B-BRNN model can achieve a high F1-score when facing the normal resume data, but a low F1-score when facing abnormal resume data (i.e. FNN has a good stability, while B-BRNN has a large variance), we hope that the hybrid model can determine the confidence level of B-BRNN model prediction according to some evaluation criteria, and use it as the weight assignment of B-BRNN in final classification decision. In this way, when abnormal data appears, FNN will play a major role, when facing normal resume data, the weight of B-BRNN will be increased, so as to improve the generalization ability of the model and reduce the model variance.

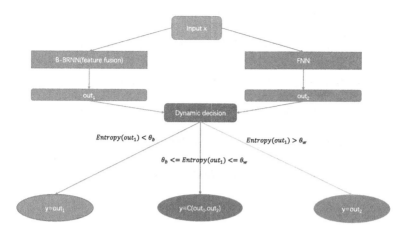

Fig. 3. The dynamic decision strategy

The specific way to implement "dynamic decision" is described as Fig. 3.

We use the entropy [16] to judge the confidence level of B-BRNN (with feature fusion), entropy is calculated by the Eq. (7):

$$Entropy = -\sum_{i=1}^{k} p_i \log p_i \tag{7}$$

where k is the number of categories and p_i is the probability of the occurrence of the ith category. For a multi classification problem, entropy describes the confusion level of the model for classification results, high entropy value means that the model is more uncertain for the current classification. Therefore, entropy is suitable to evaluate the classification confidence level of our improved B-BRNN model. Then we can set two entropy thresholds: θ_b, θ_w, representing the best and worst confidence levels of B-BRNN for the current input resume sample, respectively. When the entropy value of B-BRNN is below the threshold of the best confidence θ_b, we can fully trust B-BRNN, when between θ_b and θ_w, we take the method of concatenating this two model's classification result together, more specifically, $out_1 = (p_1^{(1)}, \cdots, p_1^{(i)}, \cdots, p_1^{(k)})$, and $out_2 = (p_2^{(1)}, \cdots, p_2^{(i)}, \cdots, p_2^{(k)})$, the concatenating operation is described as the follows:

$$C(out_1, out_2) \qquad = \frac{(out_1 + out_2)}{2} \tag{8}$$

$$= (\frac{p_1^{(1)} + p_2^{(1)}}{2}, \cdots, \frac{p_1^{(i)} + p_2^{(i)}}{2}, \cdots, \frac{p_1^{(k)} + p_2^{(k)}}{2}) \tag{9}$$

while higher than the worst confidence θ_w, we fully discard the result of B-BRNN, adopt the result of FNN as the final classification decision. As for how to determine this two thresholds, we use the proposed B-BRNN (improved by feature fusion) to predict the test set of 700 resume samples with normal order and the test set after random disorder respectively, and count the average entropy of all samples in this two test sets for 10 times, and take the average of these 10 times as the final best confidence and worst-case confidence threshold. According to the experiments statistic data, θ_b is close to 0.4, and θ_w is close to 0.9.

4   Experimental Evaluation

4.1   Experiment Design

We used three real Chinese resume datasets which are used in [21], in these datasets, we divide the resume blocks into 10 categories:base info, education background, skill, job experience, project experience, self comment, honour, school experience, sci experience, other. The size of each resume dataset is 1,400. These datasets have already segmented each resume into a block sequence arranged by the original order and provides the type label of each block, which are the input of the proposed hybrid model. These three datasets can be available at: https://github.com/xqqhelloword/resume-block-classification/tree/master/resume_project/resume_dataset.

In this paper, two experiments are designed, the first is to compare the classification performance of each model on the normal resume test set, and the second is to test the adaptability of each model to the resume sample with abnormal sequence features.

In the first experiment, the size of training set varies from 100 resumes (about 650 blocks) to 700 resumes (about 4,500 blocks), with a increased step size of 100. The test set is 400 resumes (about 3,500 blocks), validation set is 100 resumes. SVM [19], TextCNN [11], W-BRNN (Bi-GRU) [5], FNN [17], Bert [6], B-BRNN [21] are chosen as the algorithms for performance comparison with our proposed B-BRNN (feature fusion), dynamic decision hybrid model. We select F1-score as the evaluation metrics.

In the second experiment, we choose FNN, B-BRNN, B-BRNN improved by feature fusion (ours), static weight hybrid model (i.e., only taking a average strategy on the outputs of B-BRNN and FNN without considering the weight change of sub-models when facing different test resume sample, as described in Fig. 3: $C(out_1, out_2)$), dynamic decision hybrid model (ours) to train under the same training set of 700 normal resumes, and observe the change of classification performance of these models under the test set which composed of 700 resumes when the counts of abnormal resumes accounting for 0% to 100%.

4.2   Experimental Results

Table 1 shows that the F1-score of all our proposed models including B-BRNN improved by feature fusion and dynamic decision hybrid model are 5.5% to 11% higher respectively than the existing models including SVM, TextCNN and W-BRNN, FNN, BERT (Chinese-base version) and in the existing models, the FNN model performs better than other models (only lower than BERT, but Bert is a huge pre training model, which has complex structure and high resource cost), which is the reason we use the FNN to fuse our improved B-BRNN model.

Table 1. Classification performance comparison

Compare models	Resumes-A	Resumes-B	Resumes-C	Average
SVM	84.9 ± 0.1	86.0 ± 0.1	85.3 ± 0.1	85.4 ± 0.1
TextCNN	85.9 ± 0.3	88.0 ± 0.2	86.8 ± 0.3	86.9 ± 0.27
WBRNN (Bi-GRU)	89.1 ± 0.1	90.1 ± 0.2	87.8 ± 0.2	89.0 ± 0.17
FNN	89.6 ± 0.1	90.0 ± 0.1	91.0 ± 0.2	90.2 ± 0.13
BERT (base-Chinese)	90.6 ± 0.1	90.9 ± 0.15	91.3 ± 0.15	90.9 ± 0.13
B-BRNN	**96.5 ± 0.3**	**96.9 ± 0.2**	**96.7 ± 0.2**	**96.7 ± 0.23**
B-BRNN (feature fusion)	95.8 ± 0.3	96.8 ± 0.1	**96.7 ± 0.2**	96.4 ± 0.2
Hybrid model (dynamic decision)	95.3 ± 0.1	95.8 ± 0.1	95.5 ± 0.2	95.5 ± 0.13

While compared with the B-BRNN in [21], and the result of dynamic decision hybrid model on the normal dataset is slightly lower than BRNN (very closed to BRNN), which shows that our hybrid model didn't discard the sequence feature when facing normal dataset.

Another experiment's result is shown as the Fig. 4, from this figure, we can see the F1-score of FNN (wathet line) is close to 91%, and is not effected by the sequence feature between different resume blocks which means FNN is the most stable classification model even though it isn't the most accurate in normal resume data (the B-BRNN is the most accurate, about 96.7% when facing normal data), while the B-BRNN (navy blue line) has a big variance when the proportion of abnormal resume samples changes from 0% to 100%, at about 35%, the F1 score is lower than FNN, and when at 100%, the F1-score of B-BRNN has dropped to 81%, which is not a ideal result. Meanwhile, it is worth mentioning that the experimental results in Fig. 4 also prove that the measures proposed by us to improve the adaptability of B-BRNN for abnormal resume data are effective, B-BRNN improved by feature fusion (orange line) is slightly higher than B-BRNN in general, which proves that F1-score is improved under abnormal resume, while our dynamic decision hybrid model (yellow line) is generally higher than FNN, and it is close to FNN (90%) even when the proportion of abnormal resume reaches 100%, which fully proves that our hybrid model is very effective, and increases the adaptability of the model to abnormal resume. In addition, in order to prove the advantage of our proposed "dynamic decision" method, we also test the static weight hybrid model that only contat the outputs of FNN and improved B-BRNN together, result is shown as the gray line in Fig. 4, which suggests that, the performance of our dynamic decision hybrid model is much better than the static weight hybrid model.

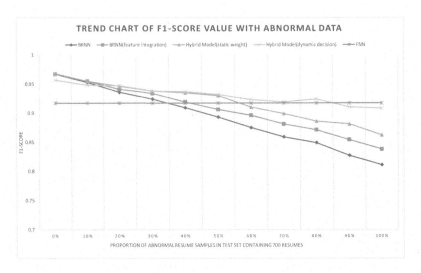

Fig. 4. F1-score under different proportion of abnormal resume samples in the 700 resume size test set (Color figure online)

5   Conclusions and Future Work

In this paper, we investigated the problem of resume block classification, a special problem of text classification which features a weak sequential relationship among different classification entities. To solve this problem, we propose an effective hybrid model which combines a fully connected neural network model and a block-level recurrent neural network model with feature fusion that makes full use of such a sequential feature. The experimental results demonstrate a better classification performance of our proposed method compared with other mainstream general classification models.

The following insights can be drawn from the experimental results.

(1) In specific classification scenarios like RBC which has a weak sequence feature among different classification entities, we can sufficiently utilize this useful feature to further improve the performance of classification models, however, this unstable and weak sequence feature didn't always work well due to the casual writing style of writers. Given a specific resume sample, we need to weigh the importance of the sequence features and the features of text content itself reasonably, make a comprehensive consideration on the influence of these two features for the final text classification decision.

(2) Through the feature fusion, and dynamic decision hybrid strategy, we successfully solve the problem of B-BRNN in [21] that it performs bad when facing the resume sample which has a huge difference with the sequence features B-BRNN has learned, improves the adaptability of B-BRNN and reduces the model variance. Actually, the essence of feature fusion is to weaken the context features from horizontal time step accumulation in B-BRNN, and

strengthen the content features of vertical input since the root cause of this problem is that B-BRNN relies too much on the empirical distribution of context sequence features among resume blocks and ignores the content of resume blocks itself. And why the hybrid strategy work is because that we fully utilize the advantages and disadvantages of each sub model since the FNN is not effected by the context order characteristics between different resume blocks, and B-BRNN (with feature fusion) can achieve a much better result when facing the normal resume data.

In the future, we will focus on the following several directions. 1) Instead of calculating two prior thresholds based on statistics, we plan to propose a mechanism to make the hybrid model learn the rule of generating dynamic weight of each sub model by itself. 2) Since the idea of pre-train and fine-tune can achieve a ideal classification result with a very small size training set, we will consider fine tuning BERT to extract features from the input resume block.

Acknowledgement. The authors would like to thank the support from Zhejiang Lab (111007-PI2001) and Zhejiang Provincial Natural Science Foundation (LZ21F030001).

References

1. Chaib, S., Liu, H., Gu, Y., Yao, H.: Deep feature fusion for VHR remote sensing scene classification. IEEE Trans. Geosci. Remote Sens. **55**(8), 4775–4784 (2017)
2. Chakrabarty, N., Kundu, T., Dandapat, S., Sarkar, A., Kole, D.K.: Flight arrival delay prediction using gradient boosting classifier. In: Abraham, A., Dutta, P., Mandal, J., Bhattacharya, A., Dutta, S. (eds.) Emerging Technologies in Data Mining and Information Security. AISC, vol. 813, pp. 651–659. Springer, Singapore (2019). https://doi.org/10.1007/978-981-13-1498-8_57
3. Chen, J., Zhang, C., Niu, Z.: A two-step resume information extraction algorithm. Math. Probl. Eng. **2018** (2018)
4. Chen, P.L., et al.: A linear ensemble of individual and blended models for music rating prediction. In: Proceedings of KDD Cup 2011, pp. 21–60 (2012)
5. Chung, J., Gulcehre, C., Cho, K., Bengio, Y.: Empirical evaluation of gated recurrent neural networks on sequence modeling. arXiv preprint arXiv:1412.3555 (2014)
6. Devlin, J., Chang, M.W., Lee, K., Toutanova, K.: BERT: pre-training of deep bidirectional transformers for language understanding. arXiv preprint arXiv:1810.04805 (2018)
7. Gers, F.A., Schmidhuber, J., Cummins, F.: Learning to forget: continual prediction with LSTM (1999)
8. Goldberg, Y., Levy, O.: word2vec explained: deriving Mikolov et al.'s negative-sampling word-embedding method. arXiv preprint arXiv:1402.3722 (2014)
9. Gu, N., Feng, J., Sun, X., Zhao, Y., Zhang, L.: Chinese resume information automatic extraction and recommendation algorithm. Comput. Eng. Appl. **53**, 141–148 (2017)
10. Jiang, Z., Zhang, C., Xiao, B., Lin, Z.: Research and implementation of intelligent Chinese resume parsing. In: 2009 WRI International Conference on Communications and Mobile Computing, vol. 3, pp. 588–593. IEEE (2009)
11. Kim, Y.: Convolutional neural networks for sentence classification (2014)

12. Li, Q., et al.: A survey on text classification: from shallow to deep learning. arXiv preprint arXiv:2008.00364 (2020)
13. Liaw, A., Wiener, M., et al.: Classification and regression by randomForest. R News **2**(3), 18–22 (2002)
14. Pham, T., Tao, X., Zhang, J., Yong, J.: Constructing a knowledge-based heterogeneous information graph for medical health status classification. Health Inf. Sci. Syst. **8** (2020). Article number: 10. https://doi.org/10.1007/s13755-020-0100-6
15. Pham, T., Tao, X., Zhang, J., Yong, J., Zhang, W., Cai, Y.: Mining heterogeneous information graph for health status classification. In: The 5th International Conference on Behavioral, Economic, and Socio-Cultural Computing (2018)
16. Rényi, A., et al.: On measures of entropy and information. In: Proceedings of the Fourth Berkeley Symposium on Mathematical Statistics and Probability, Volume 1: Contributions to the Theory of Statistics. The Regents of the University of California (1961)
17. Rumelhart, D.E., Hinton, G.E., Williams, R.J.: Learning representations by back-propagating errors. Nature **323**(6088), 533–536 (1986)
18. Sun, W., Trevor, B.: A stacking ensemble learning framework for annual river ice breakup dates. J. Hydrol. **561**, 636–650 (2018)
19. Suykens, J.A.: Support vector machines: a nonlinear modelling and control perspective. Eur. J. Control. **7**(2–3), 311–327 (2001)
20. Vaswani, A., et al.: Attention is all you need. In: Advances in Neural Information Processing Systems, vol. 30, pp. 5998–6008 (2017)
21. Xu, Q., Zhang, J., Zhu, Y., Li, B., Guan, D., Wang, X.: A block-level RNN model for resume block classification. In: 2020 IEEE International Conference on Big Data (Big Data), pp. 5855–5857. IEEE (2020)

PatRIS: Patent Ranking Inventive Solutions

Xin Ni[1]($\boxtimes$), Ahmed Samet[2], Hicham Chibane[1], and Denis Cavallucci[1]

[1] ICUBE/CSIP, INSA de Strasbourg, 24 Boulevard de la Victoire,
67084 Strasbourg, France
{xin.ni,hicham.chibane,denis.cavallucci}@insa-strasbourg.fr
[2] ICUBE/SDC, INSA de Strasbourg, 300 Bd Sebastien Brant, 67412 Illkirch, France
ahmed.samet@insa-strasbourg.fr

Abstract. Patent document is the most suitable element containing inventive knowledge. How to efficiently use this knowledge towards R&D activities has been becoming a significant challenge for engineers. In particular, in patent documents, corresponding solutions of different domains problems might be a kind of latent inventive solutions for target problems when these problems are similar enough. With the help of our previous works, latent inventive solutions from different domains patents are able to retrieve by IDM-SIM model based on LSTM neural networks and IDM-Matching model relying on XLNet neural networks. Nevertheless, we also notice that several solutions with the same similarity value are generated when a large size of input patents. It produces an obstacle for engineers to choose the most suitable inventive solutions. In this work, we propose thus an inventive solutions ranking model called PatRIS relying on the multiple-criteria decision analysis (MCDA) approach to rank inventiveness of latent solutions via corresponding patent indicators and the semantic similarity indicator. We postulate that latent solutions inventiveness is a positive correlation with corresponding patents inventiveness. The final case study over the real-world patent dataset illustrates the effectiveness of our method. This work aims to help engineers easily achieve the most possible inventive solutions from a large size of patents. To our best knowledge, no other such work proposes a way to rank inventive solutions from patent documents. Moreover, it illustrates a promising usage of automatically achieving latent inventive solutions from a large size of patent documents for engineers.

Keywords: Multiple Criteria Decision Analysis · Patent mining · Inventive solution ranking · TOPSIS · TRIZ

1 Introduction

Patent documents are significant intellectual resources of protecting interests of individuals, organizations, and companies [19]. They also provide valuable information to solve engineering problems and enhance inventiveness. The innovative knowledge in patents always tends to present the latest problem-solving

© Springer Nature Switzerland AG 2021
C. Strauss et al. (Eds.): DEXA 2021, LNCS 12924, pp. 295–309, 2021.
https://doi.org/10.1007/978-3-030-86475-0_29

solutions. The inventive knowledge contained in patents could be defined as *problems* that it aims to solve. Problem describes unsatisfactory features of existing methods or situations. For instance, for the touch pen use case[1], *non-conductive material like plastic could hampers users to operate the pen with wearing gloves, having very dry skin, or some situations in which the user does not make good conductive contact with the device to the touch screen.* This problem could show up especially when the environment is cold. The patent, therefore, proposes *partial solutions* which provide improvements or changes to the defined problems. A partial solution could be *replacing the inner molding built by non-conductive material of touch pen with a ideally metal material device so that the stylus tip operates even when held by an extremely good insulator.*

It has always been a challenge for engineers without a broad understanding of different domains' knowledge to make full use of invention knowledge in patents. However, exploring several patents by an expert turns to be an arduous task. Various machine learning tools and techniques have been developed to assist experts to analyze patents and to automatize the process of mining knowledge. Indeed, Natural Language Processing (NLP) techniques have witnessed a major leap forward. A few works provided frameworks to automatically mine inventive solutions in patent documents [19–21]. Unfortunately, these approaches suffer several drawbacks. The number of mined solutions is too high to be analysed by an expert. Even for the high value of similarity threshold between problem sentences, the number of retrieved solutions remains too excessive. In addition, ranking the mining patent solutions inventiveness remains an unsolved issue.

In this paper, we aim to address these limits and develop a framework for mining the most possible inventive solutions to engineers. We introduce PatRIS to rank inventive solutions according to their latent inventiveness. We assume that the inventiveness of a solution is associated with its corresponding patent inventiveness and similarity value of corresponding problems towards the target problem. PatRIS is based on multiple-criteria decision analysis (MCDA) method called TOPSIS with six inventiveness-related indicators. The final detailed case study illustrates the effectiveness of PatRIS and latent perspective on reality. For especially emphasizing, PatRIS is the first that combined with the MCDA method, which is with several indicators to rank latent inventive solutions containing in different patents for the target problem. This ranking approach is different from the direct ranking approach based only on sentence similarity value [19–21]. Additionally, PatRIS differs from patent classical ranking methods that are limited to a few indicators, or to specific patent fields [1, 2, 4, 12, 23, 26]. Furthermore, to our best knowledge, this work is the first to rank latent inventive solutions from different domains for the target problem from a large number of patents.

---

[1] Reader may refer to this link for the full patent https://patents.google.com/patent/US8847930B2/.

2   Related Work

In patent solution mining, Ni et al. have introduced SAM-IDM [21] and IDM-Matching [20] models that are used to find similar problems and link them to the corresponding solutions from patents of different domains via LSTM neural networks [11] and XLNet neural networks [29]. However, these works fail to address the issue of ranking several inventive solutions with the same similarity value when a large size of input patents. In addition, there is lack of approaches for evaluating inventiveness of solutions contained in patent documents, but several approaches have been proposed to measure the value and inventiveness of patents. Abrams et al. [1] have shown that the relationship between citations and patent value forms an inverted-U, with fewer citations at the high end of value than in the middle. Besides, patent citations are recognized to reflect the economic and technological importance of innovations [4]. There are also a few research works about patent inventiveness based on specific patent fields. For instance, Huenteler et al. [12] describe the pattern of innovation in the energy technologies by analyzing patent-citation networks in solar PV and wind power. Through an empiric data, Park et al. [23] propose a framework relying on patent indexes to support company merger and acquisition. The work of Squicciarini et al. [26] proposes a wide array of indicators, from patent scope, patent family size, grant lag, citations, to different indexes, capturing the technological and economic value of patented inventions in detail. It contributes to the definition and measurement of patent quality. They also mentioned that different indicators like forward citations and grant lags may generate different effect for different countries' patents. Moreover, several research works [7,18,27] further explore the effect of different patent features like grant decision, renewal, and opposition as well as indicators to the patent inventiveness.

We notice that, through these related works, several research works keep focusing on evaluating patent inventiveness in the same specific domain. Several works [1,4,7,18,23,26,27] manage to explore the effect of different patent indicators towards the patent value. However, to our best knowledge, no other works aim to evaluate the solution inventiveness in patents. Furthermore, there is no work managing to use different domains solutions as a type of inventive solutions to solve the target problem when these problems are similar enough. However, these works about patent indicators inspire us to build our solution inventiveness evaluation approach via patent indicators.

3   Typical Patent Inventiveness Indicators

Our work collects several patent inventiveness indicators in order to build our inventive solution ranking approach, as shown in Table 1.

The earliest priority date (first application of the patent worldwide) is recommended as one of reflections for the inventive performance of patents [5,25]. Patent citation is another significant indicator of the patent innovation. It makes possible to track the knowledge flow among different patents. There are basically two types of citations, cited-forward citations and cited-backward citations.

Cited-forward citations are citations subsequently received by patents. It can be used to assess the technological impact of inventions [17,24]. Cited-backward citations are patents referenced during the application process of patents. It is usable to track the inventive knowledge spillovers in technology [22]. Especially, a patent value as well as the number and quality of its forward citations have repeatedly been found to be correlated [13]. Meanwhile, patents that received more citations than the average are more likely to be renewed [9]. In addition, when a patent from the patent family (e.g. a divisional application or a continuation application) cites a patent instead of the patent itself, this is a family citation, because the family cites other patents. We separate thus citations as within family or not. Number of IPC classes is the number of technical classes the patent involved. The positive and sizeable correlation between the firm market value and the number of IPC classes of its patents is proposed as well [15]. Nevertheless, there is limited evidence about a correlation between the number of IPC classes and inventiveness of a patent. Lanjouw et al. [14] and Harhoff et al. [10] also keep opposite opinions. Number of inventors may present the cost of the research behind the invention. Besides, the number of inventors listed in the patent is associated with the economical and technological value of patents [3,8]. Moreover, the more resources involved, the more research-intensive and expensive the project. The size of patent families is proxied by the number of patent offices at which a given invention has been protected [26], due to the international patenting is much more costly than domestic applications. In addition, patent holders want to secure patent protection in various countries and regions. It implies they contain a higher expectation of return from the patent [18].

Table 1. Patent indicators and explanations

Patent indicator	Explanation
Number of Inventors (NI)	The number of inventors involved in the patent
Cited-Forward Citations with no Family (CFCNF)	Forward citations that are not family-to-family cites
Cited-Forward Citations with Family (CFCF)	Forward citations that are family-to-family cites
Cited-Backward Citations with no Family (CBCNF)	Backward citations that are not family-to-family cites
Cited-Backward Citations with Family (CBCF)	Backward citations that are family-to-family cites
Number of IPC Classes (NIPCC)	The number of technical classes
Priority Date (PD)	The earliest filing date in a family of patent applications
Family Size (FS)	The number of countries in which the same invention is patented

In this work, the number of IPC classes is not chosen as a patent inventiveness indicator due to several opposite research opinions. In addition, for family size, different patents definitions may matter for the 25% of patent families with complex structures and lead to different family compositions, which have an impact on family size as a proxy of patent value [16]. We also notice that priority date of the patent is not suitable to this work. We aim to rank inventiveness among different solutions. Corresponding patents' priority dates are not in the same comparing time level. Furthermore, the solution of the earlier priority date's patent does not present more inventiveness than others. Thus, these two indicators, priority date and family size, are also not chosen in this work. Overall, five patent indicators, cited-forward citations with no family, cited-forward citations with family, cited-backward citations with no family, cited-backward citations with family, and number of inventors are used to be indicators in this work.

4 Patent Ranking Inventive Solutions: PatRIS

As shown in Fig. 1, an inventive solutions ranking model called PatRIS is proposed. In detail, patent documents P_{at} from different domains, as the input, firstly flow to LSTM neural networks to achieve similar problems set $\mathcal{P}$ to the target problem P_{target} and corresponding similarity values (SV). After that, XLNet neural networks extract solutions $\mathcal{S}$ for given similar problems compared to the target problem according to the context information in related patent documents. These solutions from different domains patents are seen as latent inventive solutions for the target problem.

Nevertheless, when the input number of patents is high, several inventive solutions with the same similarity value will be also generated to the list. It contributes the obstacle to rank them only based on corresponding similarity values. As shown in Table 2, we can see that there are three latent inventive solutions with the same similarity value 0.86 in the real-world U.S. patent sample. The assumption that inventiveness of latent solutions is related to inventiveness of corresponding patents provides a solution to address the issue. PatRIS makes use of patent inventiveness indicators in Sect. 3 and the similarity value indicator SV to build the inventive solution ranking model based on multiple-criteria decision analysis method, in order to rank these inventive solutions according to patent features and semantic similarity eventually.

Moreover, multiple-criteria decision analysis (MCDA) is a sub-discipline of the operation research that explicitly evaluates multiple criteria in the decision making to aid in understanding the inherent trade-off [6]. PatRIS relying on a MCDA method called TOPSIS contains the first five patent indicators in Table 1 and semantic similarity value (SV) as sixth criterion to build the ranking system. It aims to rank mined solutions based on latent inventiveness by combining

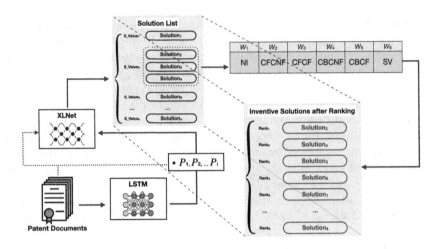

Fig. 1. The framework of PatRIS

Table 2. Inventive solutions from different domains

Target problem	Similar problems	Patent number	Similarity value	Domain	Inventive solutions
Patent Number: US9534284 Domain: C The second metal layer is not provided	The web page is not captured normally	US9535571	0.86	G	Associated with a broadcasting application according to an exemplary embodiment of the present invention
	If the wfe instruction is not intended for agent discovery purposes	US9535772	0.83	G	Are accessible to the agent as well as the client, such as designated locations in a memory 104
	The scope of the first aspect is not limited to these examples	US9537403	0.86	H	A slow DAC 930, the gear shift can be made gradual. Alternatively, by using a fast DAC, the gear shift can
	The message is not received in step	US9537998	0.86	H	Step 301. Alternatively, when the message is transmitted, the mobile

patent indicators and semantic similarity of corresponding problems. We assume that the solution is more inventive when the corresponding patent is ranked among the best inventive support and corresponding problem's similarity value to the target problem is higher. This can be formulated as the function 1. It is to maximize the solution inventiveness with chosen indicators in order to rank them.

$$max f(x) = (F_1(x), F_2(x), F_3(x), F_4(x), F_5(x), F_6(x))$$
$$F_1(x) = NI(x_1, x_2, ..., x_n)$$
$$F_2(x) = CFCNF(x_1, x_2, ..., x_n)$$
$$F_3(x) = CFCF(x_1, x_2, ..., x_n) \tag{1}$$
$$F_4(x) = CBCNF(x_1, x_2, ..., x_n)$$
$$F_5(x) = CBCF(x_1, x_2, ..., x_n)$$
$$F_6(x) = SV(x_{sv1}, x_{sv2}, ..., x_{svn})$$

subjected to linear constraints:

$$x_i \geq 0, i = 1, ..., n \tag{2}$$

$$x_{sv_i} \in [0, 1], i = 1, ..., n \tag{3}$$

In (1)–(3), $NI(x)$, $CFCNF(x)$, $CFCF(x)$, $CBCNF(x)$, $CBCF(x)$, and $SV(x_{sv})$ are count of the number of inventors, cited-forward citations with no family, cited-forward citations with family, cited-backward citations with no family, cited-backward citations with family, and semantic similarity respectively. x is the vector of count variable, 0 stands for the lower bound of the i-th count variable. The linear constraints proceed from the patent cost expense (number of inventors), peer acknowledgement (number of backward citations), related knowledge reference (number of forward citations), and semantic distance (similarity value) considerations. As shown in Table 3, a real-world U.S. patent sample, the count of indicators has been listed. The higher number of inventors, the higher cost has been invested in the patent. High number of forward citations means a high technological impact of invention. High number of backward citations means that the innovation tends to refer to high number and larger range of scientific publications. High similarity value presents the short distance between solution's corresponding problem and the target problem. Thus, these six criteria are positively correlated with the corresponding values x_{ij}. PatRIS ranks the mined solutions from the most inventive to the less inventive. Different weights w_j are also assigned to criteria.

Table 3. A sample of the indicator detail

Patent number	NI	CFCNF	CFCF	CBCNF	CBCF	SV
US9535571	1	1	22	18	0	0.86
US9537403	3	3	14	15	0	0.86
US9535772	2	0	2	4	0	0.83
US9537998	2	0	2	9	0	0.86

As shown in formula 4, we normalize each value in order to set all of attributes among the same range. j-th feature $F_j = \{x_1, x_2, ...x_i\}$, $i \in \{0, n\}$ is from those six criteria, and x_{ij} is the value of the i-th solution under the j-th feature. After normalization, values x_{ij} of patent indicators reaching 1 denote that the corresponding patents are more inventive in the j-th feature F_j than other patents. Solutions might be more inventive when $x_{sv_{ij}}$ is closer to 1. Attribute weights are also applied to the corresponding values.

$$Normalization(x_{ij}, F) = \frac{x_{ij}}{sum(F)} \qquad (4)$$

TOPSIS (Technique for Order of Preference by Similarity to Ideal Solution) with other two typical types of MCDA methods, weighted sum method and weighted product method are listed as follows.

1. Weighted Sum (WS):

$$Score_i = \sum_i x_{ij} \times w_j \qquad (5)$$

Table 4. Ranking of different decision approaches

Patent number	Inventive solutions	NI	CFCNF	CFCF	CBCNF	CBCF	SV	Rank (WS)	Rank (WP)	Rank (PatRIS)
US9535571	Associated with a broadcasting application according to an exemplary embodiment of the present invention	1	1	22	18	0	0.86	3	2	3
US9537403	A slow DAC 930, the gear shift can be made gradual. Alternatively, by using a fast DAC, the gear shift can	3	3	14	15	0	0.86	4	1	4
US9535772	Are accessible to the agent as well as the client, such as designated locations in a memory 104	2	0	2	4	0	0.83	2	4	2
US9537998	Step 301. Alternatively, when the message is transmitted, the mobile	2	0	2	9	0	0.86	1	3	1

2. Weighted Product (WP):

$$Score_i = \Pi_i x_{ij}{}^{w_j} \tag{6}$$

3. TOPSIS: aims to choose alternative that has the shortest geometric distance from the positive ideal solution and the longest geometric distance from the negative ideal solution. Therefore, the best target is to close the best choice and far away from the worst choice. Major steps are as follows:

Step1: Achieve the normalized matrix.

$$\bar{x}_{ij} = \frac{x_{ij}}{\sqrt{\sum_i x_{ij}{}^2}} \tag{7}$$

Step2: Assign weights to each value to compute the weighted normalized matrix.

$$xw_{ij} = \bar{x}_{ij} \times w_j \tag{8}$$

Step3: compute the ideal best to mark attribute as having positive impact.

$$idealbest_{positive_j} = Max(xw_{ij}) \tag{9}$$

$$idealworst_{positive_j} = Min(xw_{ij}) \tag{10}$$

Step4: Compute the Euclidean distance of $idealbest_{positive}$ and $idealworst_{positive}$, respectively.

$$Best_i = \sqrt{\sum_j (xw_{ij} - idealbest_{positive_j})^2} \tag{11}$$

$$Worst_i = \sqrt{\sum_j (xw_{ij} - idealworst_{positive_j})^2} \tag{12}$$

Step5: Compute $Score_i$ and use it to rank inventive solutions.

$$Score_i = Worst_i / (Best_i + Worst_i) \tag{13}$$

Overall, as illustrated in Table 4, the final ranks of inventive solutions according to three types of MCDA methods are listed. In addition, Algorithm 1 illustrates the general idea of PatRIS based on TOPSIS towards ranking inventive solutions.

Algorithm 1. PatRIS

Input: P_{target}: target problem; P_{at}: input patents; F: patent feature; th: similarity threshold;

Output: ranking

1: $\mathcal{P} \leftarrow \text{LSTM}(P_{at}, P_{target}, th)$

2:    ▷Extract input Patents P_{at} to achieve similar problems $\mathcal{P} = \{P_1, P_2, ..., P_n\}$ with chosen similarity threshold th for the target problem P_{target}

3: $\mathcal{S} \leftarrow \text{XLNet}(P_{at}, \mathcal{P})$

4:    ▷XLNet retrieves corresponding solutions $\mathcal{S} = \{S_1, S_2, ..., S_n\}$ of $\mathcal{P}$ for P_{target} from P_{at}

5: **for each** $S_i \in \mathcal{S}$ **do**

6:    $\bar{x}_{ij} \leftarrow \dfrac{x_{ij}}{\sqrt{\sum_i x_{ij}^2}}$ normalization of each value x_{ij} under F_j

7:    $xw_{ij} \leftarrow \bar{x}_{ij} \times w_j$ assign weights w_j to achieve weighted normalized matrix

8:    $idealbest_{positive_j} \leftarrow Max(xw_{ij})$

9:    $idealworst_{positive_j} \leftarrow Min(xw_{ij})$

10:    $Best_i \leftarrow \sqrt{\sum_j (xw_{ij} - idealbest_{positive_j})^2}$

11:    $Worst_i \leftarrow \sqrt{\sum_j (xw_{ij} - idealworst_{positive_j})^2}$

12:    $Score_i \leftarrow Worst_i/(Best_i + Worst_i)$ achieve ranking $Score_i$ for S_i

13: **end for**

14: **return** ranking$(Score_i, S_i)$

---

5 Dataset and Experimental Settings

We detail the dataset and the experimental settings of PatRIS in this section. First of all, 6,161 U.S. patent documents are used as the input of LSTM neural networks to generate the problem sentence dataset for each patent. This open-source dataset[2] is issued on 03, January 2017 by the United States Patent and Trademark Office (USPTO) and it is utility patents including eight domains, human necessities (HN), performing operations (PO), textiles (T), fixed constructions (FC), mechanical engineering (ME), chemistry (C), electricity (E), and physics (P). It then generates 4,574 problems, as illustrated in Table 5. After that, 2.8 million matching pairs of different domains problems are reserved out of 10 million pairs of problem matches from 4,574 problems. Two trained identical LSTM neural networks by the labelled Quora dataset[3] are used to compute similarity value between two problem sentences. In particular, due to the shortage of open-source labelled datasets for similar patent sentence pairs, a labelled sample dataset generating by SNLI project[4] and a labelled dataset with 1,121 similar patent sentence pairs[5] from different patents, which generated by our two experts are used to evaluate the output of LSTM neural networks. LSTM neural networks eventually retrieve 327 pairs of similar problems from different domains patents

---

[2] https://bulkdata.uspto.gov/data/patent/grant/redbook/fulltext/2017/.

[3] https://www.kaggle.com/c/quora-question-pairs/data.

[4] https://nlp.stanford.edu/projects/snli/.

[5] https://drive.google.com/file/d/1JvrUuO4by_FzvyP-5gxKQAuyyQc9cadY/view.

when similarity threshold is set 0.8. After that, these chosen similar problems are first converted into 327 queries. The trained XLNet neural networks by labelled open-source Stanford Question Answering Dataset (SQuAD 2.0)[6] are then used to achieve solutions for similar problems. These solutions from different domains patents are considered as latent inventive solutions for target problems, due to their corresponding problems are similar with target problems but belonging domains are different. In addition, different weights are assigned to criteria of PatRIS in order to weight up the most valuable features like CFCNF and SV. Through several comparisons with the different weights distribution, the chosen optimized weights w_* are as {NI: 0.1, CFCNF: 0.3, CFCF: 0.1, CBCNF: 0.1, CBCF: 0.1, SV: 0.3}. PatRIS eventually ranks these latent inventive solutions according to its inventiveness, in order to let engineers achieve the most possible latent inventive solutions easily. Besides, We especially open source our dataset and codes in GitHub[7].

Table 5. Number of problems from U.S. patents

Problem	Domain							
	HN	PO	C	T	FC	ME	P	E
4,574	652	414	370	26	70	245	1,558	1,239

6   Case Study

We choose the target problem in Table 2 as an instance. The full target problem sentence is "*When the second metal layer 410 is not provided, the first metal layer 310 may be dissolved in a solvent to thus be removed.*". From the original patent US9534284, we can see that it mentions "*A metal having corrosion resistance, such as stainless steel, may have an oxide film on the surface thereof to protect the metal.*". Therefore, the solution that patent proposed for the target problem is to use stainless steel or other metals having corrosion resistance to replace the layer 310 in Fig. 2.

From Table 4, PatRIS and WeigtedSum present the same ranking. In particular, Widianta MM et al. [28] mentioned TOPSIS approach tends to be superior than other MCDA approaches. In deed, the ranking result according to PatRIS based on TOPSIS presents the stronger evidence. The first inventive solution "*Alternatively, when the message is transmitted, the mobile terminal determines whether a received or transmitted message for a call of the message exists in step 343.*" from U.S. patent US9537998 mentions a mobile terminal device. It works to determine whether the message is received. Thus, for solving the target problem, it might be a latent inventive solution to design a device like the

---

[6] https://rajpurkar.github.io/SQuAD-explorer/.
[7] https://github.com/nxnixin/PatRIS.git.

mobile terminal in patent US9537998 to detect the solvent to prevent the dissolving problem or detect the oxide film condition of the metal for the further protection. The second inventive solution *"In alternate embodiments, the client may communicate intentions to the agent via any number of WFE communication registers 108 and/or any number of other storage components that are accessible to the agent as well as the client, such as designated locations in a memory 104."* from U.S. patent US9535772 tends to use a communication register or storage components to address the issue of accessing to the agent. For the target problem, the inventive solution of the communication register is like the first inventive solution. Design an electrical device to detect the oxide film condition so that preventing the dissolving problem. Moreover, the first two latent inventive solutions belong to the same type of solution and PatRIS ranks indeed them as the neighbor solutions. The third inventive solution *"FIG. 10 is a diagram illustrating controlling display of a terminal icon associated with a broadcasting application according to an exemplary embodiment of the present invention."* from U.S. patent US9535571 mentions a controlling display of a terminal icon, as illustrated in Fig. 3. For the further details in the patent, in Fig. 3, (a) part is a screen of the terminal displaying and (b) part is a screen used to show the updated message when the graphical object change condition. Thus, we could assume that the latent inventive solution from this patent might be to add a screen device like Fig. 3 to detect the condition of layer 310 and perform then the further protecting measures. The forth inventive solution *"As mentioned above for FIG. 9, by using a slow DAC 930, the gear shift can be made gradual."* from U.S. patent US9537403 mentions a slow Digital-to-Analogue Converter (DAC) device, as shown in Fig. 3. Through details of the patent, we can see that it performs to make the gear shift be gradual by converting the counter output into an analogue signal which forms the compensation signal. Thus, we assume that we might design a device like DAC device to detect the solvent or the surface of layer 310 by sending the analogue signal or slow the dissolving speed of metal layer in the solvent and then leave the time to further protection measures. Overall, the mobile terminal, the alternate embodiment, and the controlling display mentioned in the first three inventive solutions are directly derived from the original solution sentences and could provide the inventive solving direction. The slow DAC device in the forth inventive solution converts the previous output into an analogue signal. Indeed, the analogue signal is not directly associated with the oxide film related solutions, however, it also provides an inventive solving-problem idea using a type of device to remind or prevent its related problem. Thus, the forth inventive solution is seen as a little bit less of the direct reminders to the target problem compared to the first three inventive solutions. It also matches the ranking results of PatRIS based on TOPSIS and implies that the multi-criteria optimization decision approach is suitable to rank latent inventive solutions according to the mix of corresponding patents' inventiveness indicators and the similarity indicator.

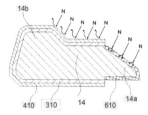

Fig. 2. The illustration figure of patent US9534284

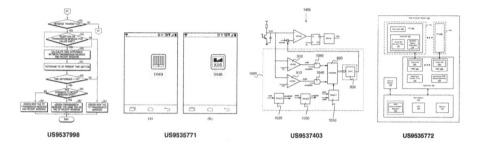

Fig. 3. The illustration figures of patents

7 Conclusion and Future Work

In this paper, an inventive solution ranking model called PatRIS based on LSTM, XLNet neural networks, and a MCDA method called TOPSIS is proposed to rank latent inventive solutions from different domains patents. In particular, it addresses the issue of ranking several latent inventive solutions with the same similarity value. We postulate that solutions' inventiveness is associated with their corresponding patents and the similarity level of corresponding problems towards the target problem. PatRIS is eventually able to rank these solutions via several chosen indicators. This work further facilitates engineers to find the most possible latent inventive solutions hidden in patent documents of different domains for the target problem in order to speed R&D activities. More importantly, it is able to avoid the failing of ranking when solutions with the same similarity value. Engineers without a broad understanding of different domains knowledge will be able to make full use of inventive knowledge from the wider range of patent documents to facilitate their inventive design inspirations. In particular, the final case study on the real-world U.S. patent sample illustrates the performance of our method and presents its latent perspective in the real world. In the future, we will explore the following direction. Combining a larger size of related knowledge will be able to further improve the accuracy of inventive solutions ranking. Thus, we will manage to explore different scientific knowledge sources, such as scientific papers, professional scientific blogs, etc. to improve and across validate the ranking results.

Acknowledgement. This work is supported by China Scholarship Council (CSC). The statements made herein are solely the responsibility of the authors.

References

1. Abrams, D.S., Akcigit, U., Grennan, J.: Patent value and citations: creative destruction or strategic disruption? Technical report, National Bureau of Economic Research (2013)
2. Ardito, L., Messeni Petruzzelli, A., Pascucci, F., Peruffo, E.: Inter-firm r&d collaborations and green innovation value: the role of family firms' involvement and the moderating effects of proximity dimensions. Bus. Strateg. Environ. **28**(1), 185–197 (2019)
3. Arora, A., Fosfuri, A., Gambardella, A.: Markets for Technology: The Economics of Innovation and Corporate Strategy. MIT Press, Cambridge (2004)
4. Block, J., Miller, D., Jaskiewicz, P., Spiegel, F.: Economic and technological importance of innovations in large family and founder firms: an analysis of patent data. Fam. Bus. Rev. **26**(2), 180–199 (2013)
5. Dernis, H., Khan, M.: Triadic patent families methodology (2004)
6. Greene, R., Devillers, R., Luther, J.E., Eddy, B.G.: GIS-based multiple-criteria decision analysis. Geogr. Compass **5**(6), 412–432 (2011)
7. Guan, J.C., Gao, X.: Exploring the h-index at patent level. J. Am. Soc. Inform. Sci. Technol. **60**(1), 35–40 (2009)
8. Guellec, D., de La Potterie, B.V.P.: The Economics of the European Patent System: IP Policy for Innovation and Competition. Oxford University Press on Demand (2007)
9. Hall, B.H., Jaffe, A., Trajtenberg, M.: Market value and patent citations. RAND J. Econ. 16–38 (2005)
10. Harhoff, D., Scherer, F.M., Vopel, K.: Citations, family size, opposition and the value of patent rights. Res. Policy **32**(8), 1343–1363 (2003)
11. Hochreiter, S., Schmidhuber, J.: Long short-term memory. Neural Comput. **9**(8), 1735–1780 (1997)
12. Huenteler, J., Schmidt, T.S., Ossenbrink, J., Hoffmann, V.H.: Technology lifecycles in the energy sector—technological characteristics and the role of deployment for innovation. Technol. Forecast. Soc. Chang. **104**, 102–121 (2016)
13. Khanna, R., Guler, I., Nerkar, A.: Fail often, fail big, and fail fast? Learning from small failures and R&D performance in the pharmaceutical industry. Acad. Manag. J. **59**(2), 436–459 (2016)
14. Lanjouw, J.O., Schankerman, M.: Patent quality and research productivity: measuring innovation with multiple indicators. Econ. J. **114**(495), 441–465 (2004)
15. Lerner, J.: The importance of patent scope: an empirical analysis. RAND J. Econ., 319–333 (1994)
16. Martínez, C.: Patent families: when do different definitions really matter? Scientometrics **86**(1), 39–63 (2011)
17. Miller, D.J., Fern, M.J., Cardinal, L.B.: The use of knowledge for technological innovation within diversified firms. Acad. Manag. J. **50**(2), 307–325 (2007)
18. Nagaoka, S., Motohashi, K., Goto, A.: Patent statistics as an innovation indicator. In: Handbook of the Economics of Innovation, vol. 2, pp. 1083–1127. Elsevier (2010)

19. Ni, X., Samet, A., Cavallucci, D.: An approach merging the IDM-related knowledge. In: Benmoussa, R., De Guio, R., Dubois, S., Koziołek, S. (eds.) TFC 2019. IAICT, vol. 572, pp. 147–158. Springer, Cham (2019). https://doi.org/10.1007/978-3-030-32497-1_13
20. Ni, X., Samet, A., Cavallucci, D.: Build links between problems and solutions in the patent. In: Cavallucci, D., Brad, S., Livotov, P. (eds.) TFC 2020. IAICT, vol. 597, pp. 64–76. Springer, Cham (2020). https://doi.org/10.1007/978-3-030-61295-5_6
21. Ni, X., Samet, A., Cavallucci, D.: Similarity-based approach for inventive design solutions assistance. J. Intell. Manuf., 1–18 (2021)
22. Noailly, J., Shestalova, V.: Knowledge spillovers from renewable energy technologies: lessons from patent citations. Environ. Innov. Soc. Trans. **22**, 1–14 (2017)
23. Park, H., Yoon, J., Kim, K.: Identification and evaluation of corporations for merger and acquisition strategies using patent information and text mining. Scientometrics **97**(3), 883–909 (2013)
24. Petruzzelli, A.M., Rotolo, D., Albino, V.: Determinants of patent citations in biotechnology: an analysis of patent influence across the industrial and organizational boundaries. Technol. Forecast. Soc. Chang. **91**, 208–221 (2015)
25. Silverberg, G., Verspagen, B.: The size distribution of innovations revisited: an application of extreme value statistics to citation and value measures of patent significance. J. Econ. **139**(2), 318–339 (2007)
26. Squicciarini, M., Dernis, H., Criscuolo, C.: Measuring patent quality (2013)
27. Van Zeebroeck, N.: The puzzle of patent value indicators. Econ. Innov. New Technol. **20**(1), 33–62 (2011)
28. Widianta, M., Rizaldi, T., Setyohadi, D., Riskiawan, H.: Comparison of multi-criteria decision support methods (AHP, TOPSIS, SAW & PROMENTHEE) for employee placement. J. Phys. Conf. Ser. **953**, 12116 (2018)
29. Yang, Z., Dai, Z., Yang, Y., Carbonell, J., Salakhutdinov, R.R., Le, Q.V.: XLNet: generalized autoregressive pretraining for language understanding. In: Advances in Neural Information Processing Systems, pp. 5754–5764 (2019)

Temporal, Spatial, and High Dimensional Databases

Shared-Memory Parallel Hash-Based Stream Join in Continuous Data Streams

Peyman Behzadnia[(⊠)]

Computer Science & Engineering Department, University of South Florida,
Tampa, FL 33620, USA
pbehzadn@usf.edu

Abstract. Stream join is known as one of the most important and computationally expensive stream operations in data stream management systems (DSMSs).
Parallelization techniques that leverage modern multi-core processor have been
proposed for stream join in literature. Equi-join is the most frequent type of join in
query workloads, and symmetric hash join (SHJ) is the most effective algorithm to
process that in data streams. In this paper, as the first research work, we propose a
shared-memory parallel symmetric hash join algorithm on multi-core processors
for equi-based stream join. Also, we introduce a novel parallel algorithm called
chunk-based pairing hash join that significantly elevates throughput and scalability. We have performed extensive experimental evaluation that demonstrates high
scalability and low latency for our proposed algorithms.

1 Introduction

Stream join, streaming counterpart of database join, is one of the most important and
computationally expensive streaming operations [1, 2]. Given the unbounded nature of
data streams, the join operation is performed only on the most recent portion of each
data stream, referred to as windows. Given high computational cost of windowed stream
joins [4], research efforts have proposed parallelization techniques for window join that
leverage modern multi-core processor architectures. In the literature, both shared-nothing
(distributed) [7–9] and shared-memory [1, 2, 4, 5, 10] parallelization techniques have
been proposed. The former targets at scaling out in multi-node and distributed systems
while the latter aims at scaling up the performance and throughput within an individual
multi-core node. The latter is the scope of this paper. Equi-join is the most frequent type of
join in query workloads and symmetric hash join (SHJ) is the most effective algorithm
to process equi-join in data streams [5]. In this paper, we propose a shared-memory
parallel symmetric hash join (PSHJ) algorithm on multi-core CPUs that achieves high-
throughput and low-latency in performing equi-based stream joins. Furthermore, we
introduce a novel algorithm called *chunk-based pairing hash join* that further boosts up
data throughput and scalability. Based on latest research surveys in literature [11], and
to best of our knowledge, we are the first research work to propose parallel hash-based
stream join algorithms on multi-core processors. We have conducted extensive set of
experiments to evaluate our proposed ideas and have compared our parallel algorithms
with the state-of-the-art parallel stream join in literature known as ScaleJoin [3]. In

© Springer Nature Switzerland AG 2021
C. Strauss et al. (Eds.): DEXA 2021, LNCS 12924, pp. 313–318, 2021.
https://doi.org/10.1007/978-3-030-86475-0_30

summary, this paper makes the following contributions: (1) As the first research work, we propose a shared-memory parallel symmetric hash join algorithm on multi-core processors that achieves high throughput, scalability, and low latency in performing equi-based stream join; (2) We introduce chunk-based pairing hash join as a novel algorithm that significantly elevates data throughout and scalability; (3) Our proposed parallel algorithms significantly outperform the state-of-the-art published result to date on parallel stream joins. The remainder of this paper is organized as follows. Section 2 introduces our proposed algorithms and illustrates on their designs, implementations, and properties. Section 3 discusses our empirical evaluation and results. We conclude in Sect. 4.

2 Proposed Parallel Hash-Based Stream Joins

In this section, we introduce our parallel hash-based stream join algorithms: parallel symmetric hash join (Sect. 2.1) and chunk-based pairing hash join (Sect. 2.2).

2.1 Parallel Symmetric Hash Join Algorithm

In this section, we present our parallelization design for symmetric hash join algorithm on multi-core CPUs. We first overview the input data stream properties. Then, we discuss distribution mechanism for delivery of input tuples to processing threads. Next, algorithmic implementation of processing threads will be explained. And last, we summarize properties of this algorithm. Equi-based stream join compares tuples received from two input streams R and S using equality condition on a common join attribute A as $A_r = A_s$ where A_r and A_s are representations for the join attributes related to stream R and S respectively. It is assumed that tuples are locally ordered based on timestamp within each data stream and globally ordered across the input data streams as well. This is called synchronous data streams which is the focus of this research. we consider round-robin method for delivering and distributing the input tuples to processing threads. We first combine tuples in R and S into a single timestamp-sorted input queue shared among all processing threads so that they can consume tuples from this shared ine of ready tuples [3, 6]. This avoids a centralized coordinator for task distribution among processing threads and ensures in-order processing of tuples. The tuples are distributed among processing threads in round-robin fashion. This helps to build resilience against skewed input streams. Once a processing thread TH receives an input tuple t from stream R (symmetrically for stream S), it performs both the *build* and *probe* phase related to the equi-join evaluation for t. First, TH adds t to the hash table for stream R in the hash bucket corresponding to join attribute value of t. Second, TH probes the hash table for stream S to find matching tuples. We integrate a near-eager tuple invalidation strategy in the build phase in our algorithm. In particular, after a newly arrived tuple is added to its hash bucket, it also invalidates the expired tuples from the hash bucket.

Since hash tables for R and S are shared data structures among all processing threads, concurrent insertion accesses to hash tables should be synchronized. Each hash bucket is protected through a latch. Proper reader-writer locks with writer-preference type are used for buckets in hash tables. Parallel symmetric hash join has the following desirable properties: (1) high throughput and low latency; (2) disjoin-parallelism in terms of independency among processing threads; (3) independent from architecture and hardware-specific optimizations.

2.2 Chunk-Based Pairing Hash Join Algorithm

In this section, we introduce our novel parallel hash-based stream join algorithm called chunk-based pairing hash join (CPHJ). Through performance analysis of our parallel symmetric hash join (PSHJ) algorithm, we aim at minimizing inter-thread synchronization overhead in CPHJ. We tackle at generalization of hashing procedure to two groups of tuples as *chunks* instead of individual tuples so that we can separate probe phase related accesses from those of build phase. In particular, we divide the tuples within each sliding window for input streams into multiple chunks. Once two new chunks of tuples arrive in sliding windows for R and S streams, the chunk-based pairing hash join algorithm performs two group-based steps on the newly arrived chunks as follows: (1) All processing threads enter to *chunk-based* build phase and populate both hash tables for R and S by inserting newly arrived tuples into buckets; (2) After synchronization through barriers, all processing threads enter the *chunk-based* probe phase and find matching tuples by probing hash table for the partner stream in a symmetric fashion. The algorithm performs the same procedure on the next two incoming chunks of tuples arriving in the sliding windows.

In the group-based build phase, as all threads need write-access to hash tables, latches are implemented for hash buckets to synchronize concurrent access. In contrast, in the group-based probe phase, as all threads need to only probe hash tables, they can freely access the shared hash tables at the same time with no synchronization. Thus, the significant portion of synchronization overhead is eliminated in our parallel design. This will minimize the overall synchronization overhead and significantly increases throughput. The only drawback is that arrival of new tuple chunks incurs a latency in join evaluation and output generation. However, as it will be show in the next section, response time for chunk-based pairing is shorter than parallel SHJ on average despite this latency while it achieves higher throughout compared to parallel SHJ.

In our implementation, we consider fixed-size chunks in terms of number of tuples. We address this parameter as λ. We have experimented wide range of values for λ to find an efficient value and have set the chunk size to 300 tuples in our implementation. Note that distribution mechanism for delivering input tuples is the round-robin schema explained in previous section. The tuple expiration policy also remains the same.

3 Empirical Evaluation

In this section, we first introduce the experimental setup including the multi-core architecture and the benchmark used in experiments. We evaluate our proposed algorithms in terms of scalability and latency. We study scalability in terms of number of tuples processed per second. We measure processing latency in terms of average end-to-end response time for input tuples. We have compared our algorithms in terms of scalability and latency with those of the state-of-the-art parallel stream join in literature, ScaleJoin.

We follow the common benchmark used by previous work on parallel two-way stream join processing for our empirical evaluation [1–4]. R tuples consists of attributes $< ts, x, y, z >$ and S tuples composed of attributes $< ts, a, b, c, d >$. Details regarding schema, attributes, and values are described in [12]. The tuple injection rate is the same for the two input streams and equals to 500 tuples per second. According to the benchmark, the tuple injection rate is steady and window size is fixed size during the course of running stream joins. Evaluation is performed on a system equipped with Intel Xeon Phi Coprocessor 5110P whose detailed specification is described in [13].

3.1 Scalability and Performance Evaluation

We assess the scalability of parallel stream joins for two streams R and S over increasing number of threads and for different widow sizes. Window sizes are 5, 10 and 15 min. The number of threads varies from 1 (non-parallel) up to 80 threads We measure the throughput in terms of number of tuples processed per second. Figure 1 [13] shows scalability results for PSHJ, CPHJ, and comparative results with ScaleJoin.

Figure 1(a) shows throughput results for parallel symmetric hash join algorithm (PSHJ). It achieves high throughput rate up to approximately 60,000 t/s, 48,000 t/s, and 33,000 for window sizes of 5, 10 and 15 min respectively. It demonstrates almost linear scalability up to 50, 45 and 30 processing threads for window sizes of 5, 10 and 15 min respectively. Figure 1(b) depicts scalability results for chunk-based pairing hash join (CPHJ). It obtains significantly high throughput rate up to approximately 71,600, 56,100, and 39,300 tuples per second for window size 5, 10 and 15 min respectively. Also, it achieves linear scalability for up to 65, 60 and 50 number of threads. Figure 1(c) depicts comparative results in terms of scalability and throughput between PSHJ, CPHJ, and ScaleJoin. The window size is 5 min in this experiment. A significant throughput difference is observed between our two algorithms when the number of threads exceeds 50 and CPHJ still keeps increasing the throughput under large number of threads up to 65. At this point, it achieves highest throughput of 71,600 *t/s*. After 65 threads, due to inevitable and high inter-thread synchronization overhead, throughput declines, but CPHJ has gradual decrease in throughput and still obtains throughput of 63,100 *t/s* for 80 threads. According to Fig. 1(c), PSHJ and CPHJ achieve up to 11 times and 12.5 times more throughput, respectively, compared to that of ScaleJoin. Also, these two algorithms provide up to around 22 times and 24.5 times more throughput, respectively, compared to that of non-parallel (sequential) stream join computation.

3.2 Latency Evaluation

Figure 2(a) [13] shows the end-to-end latency for CPHJ, PSHJ, and ScaleJoin under different number of processing threads for window size of 5 min. ScaleJoin has less latency than our proposed ideas because it performs nested loops-style join, and thus there is no shared data structures and inter-thread synchronization overhead. However, our proposed hash-based joins still have latency less than 1.5 s which is considered low and acceptable for time-sensitive stream applications. CPHJ has lower latency compared to PSHJ for large number of threads, thanks to its chunk-based design and synchronization-free probe phase. Furthermore, we have conducted an experiment to measure response

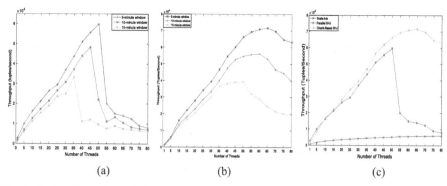

Fig. 1. Scalability: (a) Parallel symmetric hash join; (b) Chunk-based pairing hash join; (c) comparative results for PSHJ vs. CPHJ vs. ScaleJoin

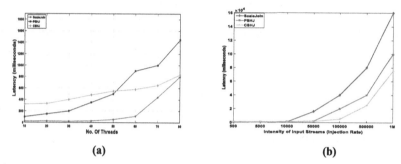

Fig. 2. Latency Evaluations for parallel schemas: CPHJ, PSHJ, ScaleJoin. (a) Under different number of threads – window size is 5 min; (b) Under different input tuple intensities

time under intense input data streams where the tuple injection rate grows very large. Figure 2(b) [13] shows the results for this type of latency evaluation for the three parallel join schemas. Our algorithms show significantly better latency for intense workloads compared to ScaleJoin which shows exponential jump in latency for input rates more than 10,000. According to Fig. 2(b), CPHJ can scale-up to intense workload in terms of latency better than parallel SHJ.

Stream join is a *dynamic* and costly operation and potentially consume more energy compared to other database operations. According to research analyzing energy efficiency in database servers [14], there is a strong linear correlation between energy-efficiency and performance of in-memory parallel hash joins in database servers. Thus, our proposed in-memory parallel hash joins also achieve significant energy-efficiency for the systems running stream joins [14] and thus has contribution to research area on dynamic energy-aware databases as well [15, 16].

4 Related Work and Conclusion

The existing research work on parallel stream join can be classified into two categories: shared-nothing (distributed) [7–9] and shared-memory [1–4, 10]. A detailed discussion on the related work is presented in [13]. In this paper, we proposed two in-memory parallel hash-based stream join algorithms. Our extensive experimental evaluation showed high scalability and low latency for our proposed ideas. Also, they significantly outperformed the best existing work in literature in terms of scalability.

References

1. Teubner, J., Mueller, R.:How soccer players would do stream joins. In: Proceedings of the 2011 ACM SIGMOD International Conference on Management of data (2011)
2. Roy, P., Teubner, J., Gemulla, R.: Low-latency handshake join. In: Proceedings of the VLDB Endowment (2014)
3. Gulisano, V., Nikolakopoulos, Y., Papatriantafilou, M., Tsigas, P.: Scalejoin: a deterministic, disjoint-parallel and skew-resilient stream join. In: Proceedings of IEEE International Conference on Big Data (2015)
4. Gedik, B., Bordawekar, R.R., Philip. S.Y.: CellJoin: a parallel stream join operator for the cell processor. VLDB J. **18**, 501–519 (2009)
5. Xie, J., Yang, J.: A survey of join processing in data streams. In: Aggarwal, C.C. (ed.) Data streams. Advances in Database Systems, vol. 31, pp. 209–236. Springer, Boston (2007). https://doi.org/10.1007/978-0-387-47534-9_10
6. Cederman, D., Gulisano, V., Nikolakopoulos, Y., Papatriantafilou, M., Tsigas. P.: Concurrent data structures for efficient streaming aggregation. In: 26th ACM SPAA (2014)
7. Qiu, Y., Papadias, S., Yi, K.: Streaming Hypercube: a massively parallel stream join algorithm. In: 22nd International Conference on Extending Database Technology (EDBT) (2019)
8. Lin, Q., Ooi, B.C., Wang, Z., Yu, C.: Scalable distributed stream join processing. In: Proceedings of the 2015 ACM SIGMOD (2015)
9. Zhang, F., Chen, H., Jin, H.: Simois: a scalable distributed stream join system with skewed workloads. IEEE ICDCS **2019**, 176–185 (2019)
10. Najafi, A.M., Sadoghi, M., Jacobsen, H.: Splitjoin: a scalable, low-latency stream join architecture with adjustable ordering precision. In: Proceedings of USENIX ATC (2016)
11. Zhang, A.S., Zhang, F., Wu, Y., He, B., Johns, P.: Hardware-conscious stream processing: a survey. SIGMOD Rec. **48**(4), 44–53 (2020)
12. Kritikakis, C., Chrysos, G., Dollas, A., Pnevmatikatos, D.N.: An FPGA-based high-throughput stream join architecture. In: 26th International Conference on FPL (2016)
13. Behzadnia, P.: Dynamic energy-aware database storage and operations. Ph.D. Dissertation, Graduate Thesis and Dissertations – Scholar Commons, University of South Florida (2018)
14. Tsirogiannis, D., Harizopoulos, S., Shah, M.: Analyzing the energy efficiency of a database server. In: Proceedings of ACM SIGMOD (2010)
15. Behzadnia, P., Yuan, W., Zeng, B., Tu, Y.-C., Wang, X.: Dynamic power-aware disk storage management in database servers. In: Hartmann, S., Ma, H. (eds.) DEXA 2016. LNCS, vol. 9828, pp. 315–325. Springer, Cham (2016). https://doi.org/10.1007/978-3-319-44406-2_25
16. Behzadnia, P., Tu, Y., Zeng, B., Yuan, W.: Energy-aware disk storage management: online approach with application in DBMS. International Journal of Database Management Systems **9**(1), 1–22 (2017)

Event Related Data Collection from Microblog Streams

Manoj K. Agarwal[1(✉)], Animesh Baranawal[2], Yogesh Simmhan[2], and Manish Gupta[1]

[1] Search Technology Center, Microsoft India, Hyderabad 500032, India
{agarwalm,gmanish}@microsoft.com
[2] Indian Institute of Science, Bangalore 560012, India
{animeshb,simmhan}@iisc.ac.in

Abstract. Many studies have established that microblog streams, e.g., Twitter and Weibo, are leading indicators of emerging events. However, to statistically analyze and discover the emerging trends around these events in microblog message streams, e.g., popularity, sentiments, or aspects, one must identify messages related to an event with high precision and recall. In this paper, we propose a novel problem of automatically discovering meaningful *keyword rules*, which help identify the most relevant messages in the context of a given event from fast moving and high-volume social media streams. For the specified event, such as {#trump} or {#coronavirus}, our technique automatically extracts the most relevant keyword rules to collect related messages with high precision and recall. The rule set is dynamic, and we continuously identify new rules that capture the event evolution. Experiments with millions of tweets show that the proposed rule extraction method is highly effective for event-related data collection and has precision up to 99% and up to 4.5X recall over the baseline system.

Keywords: Rules extraction · Event analysis · Streaming data · Algorithm

1 Introduction

The real-time and high-volume nature of microblog streams such as Twitter makes it a great source of information about recent or emerging events. In many existing works [22, 23], an event has been defined as 5W1H tuple – "What?", "Where?", "When?", "Who?", "Why?" and "How?". However, we use a more relaxed definition of an event as *"messages, posted by multiple users, in the same context, within a bounded time window"* [3, 10, 18, 24]. The *"context"* can be a real-world event, such as an earthquake or a football match or an abstract topic (e.g., a group of people discussing "climate change" on a particular day). Given an event, identifying the messages (e.g., tweets) related to the event can be useful for multiple applications such as understanding the sentiments towards the event [13], knowing the current hype around the event and predicting its future popularity [3], summarizing the event to obtain social news [4], predicting event location [5, 9], predicting election outcomes [12], detecting disaster events [24] and sending warnings [10], etc. For such applications, identifying the tweets or messages

© Springer Nature Switzerland AG 2021
C. Strauss et al. (Eds.): DEXA 2021, LNCS 12924, pp. 319–331, 2021.
https://doi.org/10.1007/978-3-030-86475-0_31

with a high recall is of utmost importance. Thus, a high-precision and high-recall system to collect event related tweets is a pre-requisite for such applications [30, 31].

Given an incoming Twitter message stream, a naïve way to obtain related tweets for an event is to query the stream using the event words or hashtags. For example, "*#brexit*" could be used to extract tweets for the event "United Kingdom withdrawal from the European Union". In many recent works [12, 24, 25], the tweets related to the events are collected using a small number of known hashtags. Similarly, most of these illustrative applications rely on hand crafted keyword rules [3, 12] (we refer to words and hashtags collectively as keywords). However, such a system would clearly have a poor recall, especially for events that are geographically and culturally diverse since events can have multiple popular representative words and/or hashtags used by potentially disjoint sections of Twitter users. Hence, such approaches are not only susceptible to bias, but also inadequate due to the dynamicity of real-world events.

Another method to collect event related tweets is to identify other keywords that co-occur frequently with the main event keyword [26, 30, 31], and fetch the tweets containing them. However, this results in low precision as frequently co-occurring words are also often part of other events. For example, in [26], the most frequently co-occurring words for the main event words {'thanksgiving, turkey'} were {'hope', 'happy'}. However, the tweets fetched based on just the keywords 'happy' or 'hope' will lead to low precision. Similarly, in [26], for main event words {'tiger', 'woods'}, the frequently co-occurring words found were 'accident' and 'crash' for a Tweet corpus from November 2009, when Tiger Woods had met an accident. Clearly, tweets fetched on just 'accident' will lead to low precision related to tweets about the event 'tiger wood car accident'. But, if we search for tweets containing the words {'tiger', 'accident'}, the collected tweets will be highly relevant. In this paper, a set of such keywords, used to fetch tweets related to an event with high precision, is called a rule. If we also use an additional rule {'tiger', 'crash'}, we get further tweets related to the main event.

If such rules related to an event can be identified automatically, the tweets related to the event can be collected with high precision and recall. Naturally, as the event evolves, the rules must also evolve accordingly.

In this paper, we propose a novel problem: For a given event, automatically identify a set of keyword rules such that these rules can be used to collect the event related tweets with high precision and high recall. The rule set is dynamic, and rules are added and removed from this rule set as the event evolves. It is natural that the keywords in the discovered rules also act as an event word cloud, and hence can be used to track the event; but the reverse is not true. Such a system is particularly useful for many applications to gauge the intensity, popularity, sentiments of an event or event summarization.

In summary, an event is represented using a disjunction of rules where each rule is a conjunction of multiple keywords. The characteristics of the rules are:

– The keywords must have high contextual frequency, i.e., they must occur in event related tweets with a high frequency.
– Keywords within a rule must be cohesive and representative of the event of interest.
– Rules should be able to capture the event evolution and the event drift, and hence must be dynamically updated.

The organization of this paper is as follows: In Sect. 2, we present the related work. In Sect. 3, we present our methodology, followed by description of the baseline system in Sect. 4. Next, we present the experimental results in Sect. 5. Finally, we present the conclusion in Sect. 6.

2 Related Work

In one of the earliest works, Li et al. [30], highlighted the need of a search system with high precision and high recall in the context of medical search, legal search, and social search, etc., and proposed a double-loop human supervised method. Similarly, in [31], authors presented a high precision, high recall system to extract tweets related to a given event and proposed a semi-supervised approach to identify the relevant keywords based on a word importance score. Finally, in a recent work [21], the authors proposed a supervised method to maximize the information coverage (i.e., high recall) for long running events and propose a human assisted method.

In summary, there exists a significant body of work to build a high precision and high recall system, to extract related traffic for the event of interest. However, the existing methods 1) mostly rely on supervised mechanism to come up with set of event related keywords, and 2) more importantly, each single keyword in this set is used as Boolean filter to extract event related traffic. Such systems either result in a low precision if a keyword is too generic, or a low recall if *only* highly informative keywords are discovered. In this paper, we propose to discover the *rules* comprising multiple keywords.

There has been significant work on tracking user specified or automatically discovered events. This work is collectively termed 'Event Tracking' [15–17, 27]. The objective of such systems is to monitor the evolving events. However, all these systems invariably identify and track the word cloud related to the source event, but their objective is not to collect event related tweets with high precision and recall. Although the word cloud evolves as the event progresses, it cannot be used to automatically create the rules unless a principled mechanism is defined to convert them into such rules. If a keyword rule is too complex (e.g., contains most of the keywords in an event word cloud), it will have a poor recall and if is it too simple, e.g., every keyword in the event word cloud becomes a rule by itself, it will have a poor precision.

Another relevant and well-studied problem is Query Expansion Techniques [7, 11, 19]. Existing works on query expansion techniques use keywords [1, 10, 13, 20], topic words [1, 8, 10], social factors [2], or visual contents [2]. Though query expansions methods over Microblog streams [6, 20, 21] appear close to our goal, the objective of such methods is to expand a query to retrieve additional semantically relevant tweets *only in the context of the given query*. On the other hand, our objective is to *continue* to retrieve the keyword rules for the entire event lifecycle. Thus, the query expansion methods cannot be trivially adapted to an event data collection system. Further, the objective of query expansion methods is to identify *semantically similar* alternate queries for improving the recall, whereas the keywords rules can be semantically different, covering different event aspects.

3 Methodology

The rule extraction problem is defined as follows.

Input: A set of one or more rules R_0 representing an event E that needs to be monitored.

Output: A dynamic set of rules R for gathering the tweets related to the event E with high precision and recall during the event lifetime.

We consider N tweets at a time (called batch) over the input tweet stream.

Definition: Given a rule set R, a tweet t is related to an event E, if $\exists r \in R$ **such that** $\{r\} \subseteq \{t\}$. $\{r\}$ is the set of keywords in the rule r and $\{t\}$ is the set of keywords in tweet t.

Let T_{R_i} be the set of tweets filtered based on the rule set R_i in batch i. Note that, the rule set R_0 (e.g., $R_0|\text{'}r = \text{\#brexit'}$) is defined by the user (specifying the event). Thus, the tweets in the set T_{R_0} represents the best estimate of ground truth. Similarly, we assume, tweets in batch i, matching the current rule set R_i represent the best estimate of ground truth for batch i.

We propose a two-step approach. In first step, we identify a set of candidate rules C_R. In second step, a rule $r \in C_R$ is added as the final rule if $(benefit_r/cost_r) \geq \alpha$, where the benefit and cost of a rule r is calculated over the tweets collected based on this rule. Our premise is that the keyword distribution in set T_{R_i} is the representative distribution of the event in iteration i. For a new rule r to be admitted into the rule set in iteration $i + 1$, the distribution of the keywords in the tweet set T_r, filtered based on rule r, should be *similar* to this distribution. We next explain our detailed methodology.

First, we extract the set of frequent words W_i and hashtags H_i from tweets in T_{R_i} in batch i. For a given fraction threshold f, the keywords $K_{R_i} = \{W_i \cup H_i | f\}$ are considered where K_{R_i} is set of top f fraction of keywords in the batch (if $f = 0.05$, we consider top 5% most frequent keywords in set K_{R_i}). W_i represents the set of words and H_i is set of hashtags in K_{R_i}. A new rule is subset of keywords in K_{R_i}, $r = \{w | w \in K_{R_i}\}$.

Partitioning Keywords Based on Their Frequency: Next, we partition the set of keywords into buckets such that approximately *similar* frequency keywords are in the same bucket. Towards that, we sort the list of keywords in set K_{R_i} in non-increasing order of their frequency. If the frequency of two keywords is vastly different, one of them cannot be frequently co-occurring with the other in the tweets. Therefore, to identify the set of keywords with similar frequencies, we identify the change points on the frequency distribution curve of the keywords in K_{R_i}. A change point over a frequency distribution is a point where the mean frequency within a small window around it changes significantly. We use the classical CUSUM algorithm [28] to detect the change points over the frequency distribution. Keywords, corresponding to frequencies between two change points are put in the same bucket, representing a set of keywords with similar frequency.

Building Co-occurrence Graph: For each bucket of keywords, we induce a graph $G(V, E)$ over the keywords in the bucket such that each keyword is a node in G. Two keywords A and B have an edge between them *if* their co-occurrence score $\frac{|T_A \cap T_B|}{|T_A \cup T_B|} \geq \gamma$; γ ($0 < \gamma \leq 1$) is the user specified threshold. Without loss of generality, T_A is set of tweets containing keyword A in batch i.

We identify all the cliques of size greater than one in the graph $G(V, E)$, such that the frequency of the clique is above a threshold t_h:

$$t_h = \frac{2fN\gamma}{1+\gamma} \qquad (1)$$

Thus, each pair of keywords in the clique must appears in $\geq t_h$ tweets together. Note, $\frac{2\gamma}{1+\gamma} \leq 1$, and for a given fraction f, and a batch size of N, the minimum number of tweets in which a keyword appears is $f.N$. Thus, the threshold t_h ensures that the words in a clique are mostly co-appearing in the tweets. There is no limit on the clique size, however, due to limit on the tweet size, we find that cliques are no larger than six keywords.

The keywords in a clique become the candidate rules. Hence, a rule contains at least two keywords. However, if a clique of size one, with frequency greater than $2t_h$ (i.e., $2fN$) is a hashtag, it is also a rule. Note, popular hashtags for an event are likely to appear as single keyword rules (as they co-appear with many keywords).

Keyword Quality under a Rule Set: Given the ith batch of N tweets, a keyword k and a rule set R_i, we define the quality of keyword k under R_i as

$$Q(k|R_i) = \frac{|T(k|R_i)|}{|T_k|} \qquad (2)$$

T_k is the set of tweets containing keyword k in batch i (of N tweets) and $T(k|R_i)$ is set of tweets containing keyword k in set T_{R_i}.

Algorithm 1: Pseudo Code for Rule Extraction

```
Data: Incoming Tweets
Result: Extracted Rules
while True do
    for batch i do
        for tweet t in batch i do
            keywordList = GETHASHTAGS(t) ∪ GETWORDS(t)
            /* update keywords profile */
            UPDATEKEYWORDSPROFILE(t, keywordList)        // Eqs. 2,3
            /* identify matching rules */
            rIDs = GETRULEIDS(keyWordList)
            for r ∈ rIDs do
                /* update candidate, existing Rule Profile */
                if r ∈ candidateRules then
                    /* compute cost and benefit */
                    UPDATECANDIDATERULEPROFILE(r, t, keywordList)

        /* cost-benefit analysis on candidate rules */
        ANALYZECANDIDATERULES()
        /* create new rules */
        CREATENEWCANDIDATERULES()
```

Tweet Quality under a Rule Set: Given the ith batch of N tweets, a tweet t and a rule set R_i, we define the quality of the tweet t under the rule set R_i as

$$Q(t|R_i) = \frac{\sum_{k \in K_{R_i} \cap k \in t} Q(k|R_i)}{|k \in K_{R_i} \cap k \in t|} \qquad (3)$$

A candidate rule r is accepted as the final rule if in $(i + 1)^{th}$ batch, (a) it adds many novel tweets containing hashtags and words in set K_{R_i} (high benefit). Benefit is calculated as $\sum_{t \in T_r} Q(t|R_i)$ and (b) if the increase in frequency of hashtags/words in set K_{R_i} is more than the increase in frequency of hashtags not in K_{R_i} (low cost). Cost is calculated as $\left| T_{R_{i+1 \cup r}} - T_{R_{i+1}} \right|$. We define α as the ratio of benefit and cost. Thus, a rule is discovered in batch i, and if its impact is positive in batch $i + 1$, it is accepted. The updated rule set is used to collect the matching tweets from the next batch of N tweets and the process continues. The overall method is illustrated in Algorithm 1.

4 Baseline

We choose the system presented in [19] as the baseline, as it comes closest to the objective of our system. The objective in [19] is to search about an event *retrospectively*, given query term(s). The twitter corpus is divided into batches of fixed time length, called timespan. A burstiness score of a word w, $b(w|T_i) = \frac{P(w|T_i)}{P(w)}$ is calculated for each word appearing in a timespan T_i where $P(w|T_i) = \frac{f_{w|T_i} + \mu \frac{f_w}{N}}{|T_i| + \mu}$ and $P(w) = \frac{f_w + K}{N + K|V|}$. f_w is the frequency of word w in the entire corpus, and $f_{w|Ti}$ is the frequency of the word w in Ti. N is the total number of terms in the entire corpus, $|Ti|$ is the number of terms in timespan Ti. $|V|$ is the total size of the vocabulary. μ and K are the smoothing parameters. *Top-k* words are used to expand the initial

They divide the twitter timeline into batches, and consider keyword co-occurrence scores to identify correlated keywords, similar to our system. The key difference is, they consider all temporally correlated keywords. The assumption that temporally correlated keywords are related to the *same* event may not hold for disjoint events which, nonetheless, occur together. On the other hand, we consider temporally and spatially related terms (i.e., appearing in the same tweet) in the *context* of the event specific rule set. Further, unlike our system, where we dynamically discover rules, their system works on *entire* tweet corpus, and it is not a *live* system.

5 Experiments

5.1 Dataset

We consider two datasets: **DS 1)** 71.7 million tweets, from 17th Oct 2016 to 16th Dec 2016. **DS 2)** 61.5 million tweets from 5th March 2020 to 30th April 2020. Both these sets were collected using Twitter 1% traffic API. The two time periods were selected carefully, to represent tweets from two different periods, before Twitter changed its character limit from 140 to 280 characters per tweet in 2017. We used a small set of stop-words, to discard certain frequently occurring keywords.

Given a rule set R_i in i^{th} batch of tweets, the precision of a rule is defined as

$$prec(r|R_i) = \frac{\sum_{t \in T_r} Q(t|R_i)}{|T_r|} \tag{4}$$

Thus, each tweet contributes to the precision proportional to the quality of keywords it contains, w.r.t. rule set R_i. Recall of rule r is computed as

$$rec(r|R_i) = \frac{\sum_{t \in T_r} Q(t|R_i)}{\sum_{t \in T_{R_i \cup r}} Q(t|R_i)} \tag{5}$$

Thus, we compare the quality of the tweets collected due to a new rule r with the quality of all the tweets collected due to combined rule set of r and existing rules. As defined earlier, we consider the current rule set as the ground truth. Note, all tweets at R_0 (e.g., R_0"$r = \#brexit$") are considered related to Brexit event. We considered three types of events, 1) Point events with lifespan of a day; 2) Medium term events, with lifespan of approximately a week; 3) Long term events, spread over multiple weeks.

5.2 System Configuration

We Apache Flink [14], a stream processing framework and deployed in local mode, supporting Java 11 to implement the described rule extraction framework. For each of our experiments, we set the following parameters, unless specified otherwise: $\gamma = 0.2$, $\alpha = 0.1, f = 2.5\%$ (top 2.5% keywords sorted on frequency are selected, in a batch). Note, the number of tweets collected by our system and base-line are different. Since we have a fixed batch size, the two systems may cover different number of batches (x-axis in Figs. 1, 2, 3 and 4).

5.3 Short Term Events

We evaluate a point event, #earthday (**DS2**). We set batch size $N = 800$ and $\gamma = 0.2$.

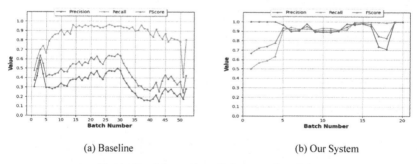

(a) Baseline   (b) Our System

Fig. 1. Precision and recall for event #earthday

In Fig. 1, we plot the precision-recall of our system w.r.t. baseline. Note that, the first batch of tweets will always have a precision of 100% because we start with the event hashtag as the only seed rule. We see, the recall of our system improves quickly. The precision for our system drops due to addition of spurious rules, but then improves subsequently with addition of better rules. The baseline precision is low because of the presence of many unrelated keywords.

5.4   Medium Term Events

We selected the event with medium term impact, of up to seven days, namely *#blackfriday* (**DS1**). Figure 2 shows the trends due to *#blackfriday* event.

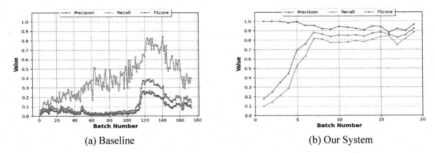

(a) Baseline   (b) Our System

Fig. 2. Precision and recall for event *# blackfriday*

We see, the baseline trends are because of many keywords discovered as rules related to another event, namely, *#thanksgiving*. This leads to a significant drop in precision as well as recall. our system starts with a low recall but steadily improves. The precision of our system remains high throughout.

5.5   Long Term Events

We cover two long term events, *#trump* (during the trump election campaign) in **DS1** (Fig. 3), and Covid-19 Virus (*#coronavirus*), in **DS2** (Fig. 4).

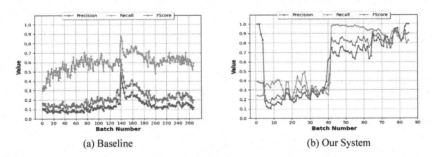

(a) Baseline   (b) Our System

Fig. 3. Precision and recall for event *# trump*

We see, our system starts with low F1Score for *#coronavirus*, which improves slowly initially but subsequently adds many relevant rules.

For such long running events, determining the ground truth rules is difficult (as events change fundamentally multiple times over its lifespan), therefore, it is difficult to compute the precision-recall curve for the whole event. We, thus, zoom in around the time where they were most popular for a period of one week. Assuming that these long running events are relatively stable during the zoomed in period, we compare the baseline with our system in this time period.

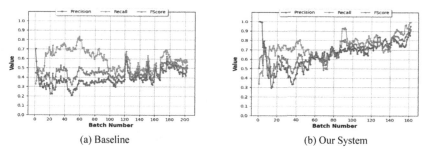

(a) Baseline   (b) Our System

Fig. 4. Precision and recall for event *# coronavirus*

5.6 System Performance

We have primarily two tunable parameters, Score threshold α, and similarity threshold γ. The batch size variation did not impact the performance much, as in a short span, the event characteristics do not change much. A smaller value of γ and α implies more rules are added in the system. In the heatmap plot, in Fig. 5, we see, as α and/or γ reduce, the recall of our system improves, while reverse is true in case of precision.

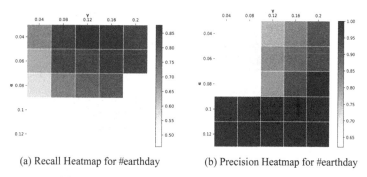

(a) Recall Heatmap for #earthday   (b) Precision Heatmap for #earthday

Fig. 5. System Performance under Parameter Tuning

Our system is very efficient: we could process the popular event *#trump* in 40 min. Since, the event is spread over the entire dataset, comprising ~71.7 million tweets, we get the processing speed of ≈30k tweets per second.

5.7 Human Judgements

In this experiment, the tweets retrieved by our system and the baseline system are judged by the human judges. We analyzed the rules over one day of peak period for each event. For an event, the systems discover multiple rules. We sample the rules and for each sampled rule, we uniformly randomly sampled m tweets, out of all the tweets filtered based on that rule, which were then presented to the judges. Thus, if an event has k rules discovered, up to $k.m$ tweets were judged (some rules filtered less than m tweets). The judges were trained in the task, i.e., they were briefed about the event, and asked to judge if a tweet is related to the event or not. m is set to 20. We considered a rule relevant, if 80% or more tweets filtered based on the rule are considered relevant to the event by the judge. For relevant rules, we multiply the relevance score (for 80% threshold, it ranges from 0.8 to 1.0) with the numbers of tweets filtered based on that rule, to compute the final number of relevant tweets. The results are presented in Table 1 for 80% relevance threshold. We show relevant tweets vs. all tweets by two systems.

Table 1. Precision and recall (baseline vs. ours)

Event name	Rule precision (our system)	Rule precision (baseline)	Relevant/All (our system)	Relevant/All (baseline)
#earthday	45/58	2/11	5696/7148	2040/4722
#coronavirus	73/74	6/11	45216/47167	10334/14081
#trump	80/80	7/10	65043/65534	97901/121277
#blackfriday	14/16	3/11	6554/7772	3570/17616

For events #*trump*, our recall was inferior to the baseline. However, our precision at 99% was better. For #*earthday* and #*coronavirus* both the recall as well as precision are improved over baseline. For event #*coronavirus,* the recall was 4.5X over the baseline. However, as we would analyze next, even for #*trump* and #*blackfriday,* the quality of our rules is significantly better.

Table 2. Rules discovered by our system and baseline (rules in bold are relevant)

Event name	Rules – baseline	Rules – our system
#earthday	@adamcvean, @coco_who, @lowkeyel, @_mthegem, **#earthday, #earthday2020**	'earth, mother', 'planet, protect', 'pledge, save, water', 'first, years', **#earthday, #earthday2020**, '50th, anniversary', 'come, every', **'plant, trees'**
#coronavirus	**italians, cancellations, sanitizer, panicking**, #bestlyrics, **#coronavirus**, #kca	'actually, people', 'house, president', 'declares, emergency', **'got, shit'**, 'know, one', **'negative, tested, trump', 'covid19, news', #coronavirus, '#covid19, #covid'**
#trump	**elected, bernie**, protest, **obama, michelle, #trump**, #mannequinchallenge	**'president, states, united'**, 'racism, win', 'hope, win', 'make, today', **'lives, matter'**, 'america, great, make', **#trump, #notmypresident, '#electionnight, #election2016'**
#blackfriday	moana, inner, **friday**, recount, **deals, #blackfriday**, #pizzagate, #thanksgiving	'chance, giveaway', 'like, people', **'black, friday'**, 'new, one' **'free, shipping', '#giveaway, #win', #blackfriday, '#amazongiveaway'**

In Table 2, we show some of the rules discovered by the two systems. For event, #*trump,* tweets were collected over Nov 9th and 10th, 2016. The baseline system collected more tweets than our system (as shown in Table 1), however, if we analyze the rules generated by the two systems, we see, most of the rules by the baseline system were *Obama, Michelle, and Bernie*. Even though, they were considered related to the event

'*Election of Donald Trump*', but not precisely to the main intent #*trump*. Further, many unrelated hashtags such as # *mannequinchallenge*, which were co-occurring, were also identified. #*protest* was a mixed rule, with less than half of all the tweets related to election of Trump. On the other hand, the rules discovered by our system never deviated from the main intent. We identified rules related to all the major intents for this event, such as '*racism, win*' or '*hope, win*'. Even seemingly unrelated rules '*make, today*' were found to be related to the primary intent of the event.

Similarly, for #*blackfriday* (26–27 Nov 2016) many unrelated rules such as '*moana*' (about a movie), or #*pizzagate* were found by the baseline system. Similarly, unrelated rule #*thanksgiving* resulted into a huge loss of precision (Table 1). On the other hand, except two rules, all the rules by our system were related to the main event intent. Further, our system was able to remove co-occurring tweets related to #*thanksgiving*.

Finally, for #*coronavirus*, the rules discovered by our system covered varied intents. For instance, rule '*actually, people*' was about denying that anything serious called Coronavirus exists. Even a rule like '*got, shit*' that added 232 tweets was highly precise rule. '*house, president*' was a rule about 'Coronavirus bill passed by the house', etc.

6 Conclusion

We presented a novel system to automatically extract the meaningful keyword rules from live data streams, in the context of a given event with the objective of collecting event data with high precision and high recall. Ours is a first system that identifies conjunction of multiple keywords as rules. Our system could identify meaningful rules for events with varying dynamicity. The number of rules remained bounded. For long running events, the rules automatically captured the event evolution with high precision. Further, our system was highly efficient while computing the rules. Our future work direction is to automatically identify the different event facets using the rules.

References

1. Chen, C., Li, F., Ooi, B.C.: TI: an efficient indexing mechanism for real-time search on tweets. In: Proceedings of the 2011 ACM SIGMOD International Conference on Management of data, pp. 649–660 (2011)
2. Gao, Y., Wang, F., Luan, H.: Brand data gathering from live social media streams. In: Proceedings of International Conference on Multimedia Retrieval, pp. 169–176 (2014)
3. Gupta, M., Gao, J., Zhai, C.: Predicting future popularity trend of events in microblogging platforms. Proc. Am. Soc. Inf. Sci. Technol. **49**(1), 1–10 (2012)
4. Kwak, H., Lee, C., Park, H.: What is Twitter, a social network or a news media?. In: Proceedings of the 19th International Conference on World Wide Web, pp. 591–600 (2010)
5. Li, R., Lei, KH., Khadiwala, R.: TEDAS: a Twitter-based event detection and analysis system. In: 2012 IEEE 28th International Conference on Data Engineering, pp. 1273–1276 IEEE (2012)
6. Massoudi, K., Tsagkias, M., de Rijke, M., Weerkamp, W.: Incorporating query expansion and quality indicators in searching microblog posts. In: Clough, P., et al. (eds.) ECIR 2011. LNCS, vol. 6611, pp. 362–367. Springer, Heidelberg (2011). https://doi.org/10.1007/978-3-642-20161-5_36

7. Nagmoti, R., Teredesai, A., De Cock, M.: Ranking approaches for microblog search. In: 2010 IEEE/WIC/ACM WI-IAT, vol. 1, pp. 153–157. IEEE (2010)

8. O'Connor, B., Krieger, M., Ahn, D.: TweetMotif: exploratory search and topic summarization for Twitter. In: Proceedings of the ICWSM, vol. 4, no. 1 (2010)

9. Sadilek, A., Kautz, H., Bigham, JP.: Finding your friends and following them to where you are. In: Proceedings of the Fifth ACM WSDM, pp. 723–732 (2012)

10. Sakaki, T., Okazaki, M., Matsuo, Y.: Earthquake shakes twitter users: real-time event detection by social sensors. In: Proceedings of the 19th WWW, pp. 851–860 (2010)

11. Teevan, J., Ramage, D., Morris, MR.: #TwitterSearch: a comparison of microblog search and web search. In: Proceedings of the Fourth ACM WSDM, pp. 35–44 (2011)

12. Tumasjan, A., Sprenger, T., Sandner, P.: Predicting elections with Twitter: what 140 characters reveal about political sentiment. In: Proceedings of ICWSM, vol. 4, no. 1 (2010)

13. Wang, H., Can, D., Kazemzadeh, A.: A system for real-time twitter sentiment analysis of 2012 us presidential election cycle. In: Proceedings of the ACL 2012 System Demonstrations, pp. 115–120 (2012)

14. Carbone, P., Katsifodimos, A., Ewen, S.: Apache flink: stream and batch processing in a single engine. In: Bulletin of the IEEE TCDE, vol. 36, no. 4 (2015)

15. Osborne, M., Moran, S., McCreadie, R.: Real-time detection, tracking, and monitoring of automatically discovered events in social media. In: Proceedings of 52nd ACL, pp. 37–42 (2014)

16. Lin, CX., Zhao, B., Mei, Q.: Pet: a statistical model for popular events tracking in social communities. In: Proceedings of the 16th ACM SIGKDD, pp. 929–938 (2010)

17. Weiler, A., Grossniklaus, M., Scholl, MH.: Event identification and tracking in social media streaming data. In: EDBT/ICDT, pp. 282–287 (2014)

18. Agarwal, MK., Ramamritham, K.: Real time contextual summarization of highly dynamic data streams. In: EDBT, pp. 168–179 (2017)

19. Metzler, D., Cai, C., Hovy, E.: Structured event retrieval over microblog archives. In: Proceedings of the Conference of NAACL-HLT, pp. 646–655 (2012)

20. Wang, Y., Huang, H., Feng, C.: Query expansion based on a feedback concept model for microblog retrieval. In: Proceedings of the 26th WWW, pp. 559–568 (2017)

21. Srikanth, M., Liu, A., Adams-Cohen, N.: Dynamic social media monitoring for fast-evolving online discussions. arXiv preprint arXiv:2102.12596 (2021)

22. Mu, L., Jin, P., Zheng, L.: Lifecycle-based event detection from microblogs. In: Companion Proceedings of the the Web Conference 2018, pp. 283–290 (2018)

23. Jin, P., Mu, L., Zheng, L.: News feature extraction for events on social network platforms. In: Proceedings of the 26th International Conference on WWW Companion, pp. 69–78 (2017)

24. Rudra, K., Goyal, P., Ganguly, N.: Identifying sub-events and summarizing disaster-related information from microblogs. In: The 41st International ACM SIGIR, pp. 265–274 (2018)

25. Phuvipadawat, S., Murata, T.: Breaking news detection and tracking in Twitter. In: 2010 IEEE/WIC/ACM International Conference on Web Intelligence and Intelligent Agent Technology, vol. 3, pp. 120–123. IEEE (2010)

26. Guille, A., Favre, C.: Event detection, tracking, and visualization in twitter: a mention-anomaly-based approach. In: Proceedings of SNAM, vol. 5, no. 1 (2015)

27. Lee, P., Lakshmanan, LV., Milios, EE.: Event evolution tracking from streaming social posts. arXiv preprint arXiv:1311.5978 (2013)

28. Agarwal, MK., Gupta, M., Mann, V.: Problem determination in enterprise middleware systems using change point correlation of time series data. In: Proceedings of NOMS, pp 471–482, April 2006

29. Magdy, A., Abdelhafeez, L., Kang, Y.: Microblogs data management: a survey. VLDB J. **29**(1), 177–216 (2020)

30. Li, C., Wang, Y., Resnick, P.: ReQ-ReC: high recall retrieval with query pooling and interactive classification. In: Proceedings of the 37th International ACM SIGIR Conference on Research & Development in Information Retrieval, pp. 163–172 (2014)
31. Zheng, X., Sun, A., Wang, S.: Semi-supervised event-related tweet identification with dynamic keyword generation. In: Proceedings of the 2017 ACM on Conference on Information and Knowledge Management, pp. 1619–1628 (2017)

GACE: Graph-Attention-Network-Based Cardinality Estimator

Daobing Zhu, Dongsheng He, Shuhuan Fan, Jianming Liao,
and Mengshu Hou[✉]

University of Electronic Science and Technology of China, No. 2006, Xiyuan Ave,
West Hi-Tech Zone, Chengdu 611731, Sichuan, China
{dbzhu,ehds}@std.uestc.edu.cn, {fansh,liaojm,mshou}@uestc.edu.cn

Abstract. Cardinality estimation plays a vital role in query optimizer, the key factors challenge its accuracy are join-crossing correlations between different attributes. Traditional estimation techniques provide poor estimation quality for complex queries with many joins for the lack of correlations. Recently, much work has shown that machine learning based methods overcome the challenge to a certain extent. However, the existing learning-based approaches have two major downsides. First, lack of explicit utilization of the effective and explainable feature: conjunctive predicates between different attributes. Second, the dynamic information associated with the dataset status is not used to encode the static query structure. In this paper, we propose GACE, a cardinality estimator based on Graph Attention Network (GAT) to capture join-crossing correlations between attributes from join graphs constructed on conjunctive predicates. GACE leverages the reliable features Base Table Selectivity and Join Selectivity obtained through traditional techniques to enrich the encoding of query structure and track the status alterations of dataset. Taking the regularity of join graphs into account, a GAT model with batch training supporting is implemented to sufficiently decrease the overhead of training. The results of our empirical evaluation on real-world dataset (IMDb) demonstrate that GACE achieves improvement in quality of cardinality estimation, especially for more joins.

Keywords: Query optimization · Cardinality estimation · Join-crossing correlations · Graph Attention Network · Feature encoding

1 Introduction

Choosing an optimal execution plan with minimum cost for an SQL query is the ultimate goal of query optimization. Most classical query optimizers belong to cost-based optimization (COB) category, which picks a plan to execute according to estimated cost. COB is composed mainly of cardinality estimation and

Supported by the National Key R&D Program of China (grant No. 2019YFB1705601) and the Natural Science Foundation of China (grant No. 62072075).

C. Strauss et al. (Eds.): DEXA 2021, LNCS 12924, pp. 332–345, 2021.
https://doi.org/10.1007/978-3-030-86475-0_32

cost model. Cost model estimates the overhead of different execution plans in the search space to choose the best execution plan grounding on the estimated cardinalities of intermediate results. As shown by prior work, Cardinality estimation has much more influence on performance of the chosen plan than the cost model [11]. However, query optimizers implemented by traditional cardinality estimation methods like histograms often choose sub-optimal plans because of the ignorance of join-crossing correlations between attributes, which are the key factors that influence the accuracy of cardinality estimation on real-world datasets [11,12]. They simplify computational model by making assumptions such as uniformity and independence etc., leading to the miss of correlations.

Recently, many machine learning techniques have been applied to improve the accuracy of cardinality estimation and reduce the consumption of cardinality estimation. MSCN [10] leverages query structure involved in every query as the input of its multi-set convolutional network to capture the correlations. E2E [19] is also based on query structure, which further encodes the features of query structure at the deeper level of tree-like physical query plan, and employs a two-level tree LSTM model to capture join-crossing correlations between attributes. However, inadequate feature extraction and encoding limit their further enhancements in the accuracy and scalability, specifically on workloads with more joins. First, conjunctive predicates that imply correlations between different attributes are not fully utilized. The cardinalities of the conjunctive operation results are related to the correlations between the participating join attributes, the more correlations, the larger the cardinalities. Second, Their one-hot encoding for query structure is static and not associated with the status of the dataset. If the status of the dataset changes, the accuracy of the model may fluctuate.

To break through these limitations, we first propose a Graph-Attention-Network-Based cardinality estimator (GACE) with more robust encoding methods for query structure that captures the correlations between attributes from joins (conjunctive predicates) of query workloads through GAT to the best of our knowledge. Experiments on real-world dataset (IMDb) and query workload show that GACE is effective and feasible to catch join-crossing correlations between attributes from query workload.

In summary, we make the following contributions.

(1) A cardinality estimator based on GAT that takes advantage of the features extracted from join graphs. GAT helps to extra from join graphs with the identical shape form by conjunctive predicates of queries, and convert the correlations between join attributes into attention weights between nodes.

(2) A more feature-rich encoding approach that combines the weaker one-hot encoding with dynamic features: Base Table Selectivity σ and Join Selectivity $\bowtie$ associated with the status of dataset, which are fetched through traditional technologies.

(3) A GAT implementation that supports batch training, which reduce the overhead of training. It is workable because the shapes of join graphs formed by attributes on the dataset are regular.

2 Related Work

2.1 Traditional Cardinality Estimation Methods

Cormode et al. [2] gave a comprehensive surveys about traditional cardinality estimation methods: sampling, histogram, sketching and wavelet. i) Sampling has become the most widely implemented traditional methods [12,13,23] because of its simplicity and effectiveness. There are three main challenges with sampling: vanishing problem, 0-tuple problem and memory and time overhead. Leis et al. [12] optimizes the problems by using indexes and limiting the total sampling overhead during enumeration, however this method is restricted by the existence of the index. The idea of sampling is also widely combined by other traditional methods and learning-based methods. ii) Histogram [9] summarizes a dataset by grouping the data values into buckets. The single-attribute histogram approaches work well when dealing with queries involving only single table. When it comes to queries with joins between multiple tables, the assumption of independence leads to a general underestimation. [7,16–18] use multi-dimensional histogram approaches to capture the statistical relationships between different attributes. What follows is more memory and time overhead caused by choosing bucket boundaries from more degrees. iii) Sketching (FM sketch [5], Count-Min sketch [3], LinearCount [22], HyperLogLog [4]) are widely used in cardinality estimation in the context of big data field, they use hash functions and bitmap like data structure to appropriately count distinct values. They are used directly in point query case, extra steps are required to fit range queries. iv) Wavelet-based work [1,15] use wavelets to approximate the underlying correlations between multi attributes. They built wavelet coefficient synopses of data and use these synopses to provide approximate answers to queries. Compared with previous methods, wavelets have not yet been adopted for use in commercial database systems.

2.2 Learning-Based Cardinality Estimation

Learning-based methods can be divided into two categories according to the source of feature encoding: learning from query structure and learning from data distribution. i) Tanu et al. [14] first use queries' attributes, operators, etc. to perform cardinality estimates without knowledge of query execution plans and data distribution. Recently, MSCN [10] separately encodes tables, joins, and predicates from query structure, which achieves considerable cardinality estimation accuracy on IMDb dataset using multi-set convolutional network. E2E [19] extracts query structure feature from deeper layer: physical query plan by a two-level tree LSTM model. What's more, it proposes a set of word vector models to embed string values for JOB query workload. ii) Probabilistic relational models (PRMs) based work [6,20] have made improvements in factoring the joint probability distribution of all the attributes in the database into materialized tables. Deepdb [8] is also a work of this data-driven family, using Sum-Product Networks (SPNs) to learn correlations from data rather than queries. Yang et al. [25] also uses unsupervised autoregressive model and Monte Carlo integration to learn the correlations between attributes from data in a single table. Later Yang et al. proposed NeuroCard [24], a join cardinality estimator supporting multi-tables based on Naru.

3 Overview

Our work absorbs the strengths of MSCN [10], and adds Graph Attention Network to capture join-crossing correlations between attributes from join graphs. Figure 1 shows the overview of our cardinality estimator, GACE.

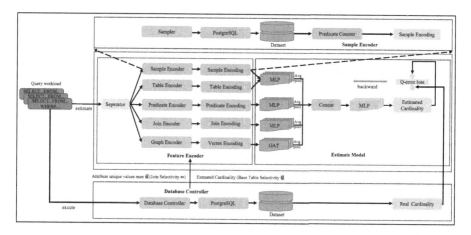

Fig. 1. Overview of GACE, GACE is composed of four modules: Database Controller, Feature Encoder, Sample Encoder, and Estimate Model.

Database Controller. Database Controller is a component that helps to handle the database backend (PostgreSQL version 10.3). It is used for three main purposes: i) executing query workload for real cardinalities, ii) estimating the cardinalities of base table for Base Table Selectivity σ, iii) fetching unique values number of attributes for Join Selectivity $\bowtie$. We normalized the real and estimated cardinalities with min-max normalization in log scale

$$C_{norm} = \frac{\ln(c) - \ln(c_{\min})}{ln(c_{\max}) - ln(c_{\min})}. \tag{1}$$

Feature Encoder and Sample Encoder. Separator in Feature Encoder component divides a query $q \in Q$ into different parts: tables $T_q \in T$, predicates $P_q \in P$, joins $J_q \in J$. Each T_q has a one-to-one correspondence with sample bitmap $S_q \in S$, that is, $s \in S_q$ represents a bitmap that stores boolean values of the corresponding $t \in T_q$'s sample tuples after applying predicates $p \in P_q$. This part of work is finished by the Sample Encoder component on the top of Fig. 1.

Graph Encoder uses j_q of every query q to encode vertexes features $V_q \in V$. T_q, P_q, J_q, S_q, V_q specifically refer to the elements participating in query q. T, J, P, S, V describe the sets of all available tables, predicates, joins, samples, vertexes in query workload respectively. More encoding detail is discussed in Sect. 4.

Estimate Model. We used four modules to learn features from Table Encoding $\{T_q, S_q\}$, Predicate Encoding P_q, Join Encoding J_q, Vertex Encoding V_q respectively, the first three are Multilayer Perceptrons (MLP), and the last one is a Graph Attention Network. We use the commonly used metric q-error as the loss of our model defined in Formula (2)

$$q\text{-}error_{max} = \max(\frac{\hat{C}}{C}, \frac{C}{\hat{C}}), \tag{2}$$

where C and $\hat{C}$ stand for real and estimated cardinality normalized by Formula (1) respectively.

4   Encoding and Model

4.1   Encoding

In this section, we use a specific query q as an example to describe our encoding process in detail, focusing on our strategy to generate dynamic features: Base Table Selectivity σ and Join Selectivity $\bowtie$. As shown in Fig. 2, we sort all tables, all attributes, all operations, and all joins separately for one-hot like encoding, and fetch sample rows for sample bitmap encoding. Operation Encoding uses pure one-hot encoding, setting the corresponding bit position with 1.

Sample Encoding. The Sampler Encoder Component in our system takes out a certain number of sample tuples (1000 tuples) from the database. For each q, and it will apply predicates P_q of T_q correspondingly, the bits in sample bitmap S_q of tuples is set to 1 if meet predicates P_q. Sample Encoding uses the idea of sampling to reflect the cardinality of base table after the pushdown of predicates.

Base Table Selectivity σ. We propose the Base Table Selectivity to enrich Table Encoding and Attribute Encoding. Compared to the traditional one-hot like encoding, we set the bit position with Base Table Selectivity $\sigma_q \in \sigma_{all}$ rather than 1. A Base Table Selectivity $\sigma \in \sigma_q$ corresponds to a table $t \in Tq$ is a decimal $\in [0, 1]$. Formula (3) shows the definition

$$\sigma = \frac{\hat{C}_{tp}}{Rows(t)}, \tag{3}$$

Table Encoding		Attribute Encoding	
cast_info ci	...σ00000...	ci.person_id	...σ00000...
movie_companies mc	...0σ0000...	ci.role_id	...0σ0000...
movie_info mi	...00σ000...	mc.company_id	...00σ000...
movie_info_idx mi_idx	...000σ00...	mc.company_type_id	...000σ00...
movie_keyword mk	...0000σ0...	t.kind_id	...0000σ0...
title t...	...00000σ...	t.production_year...	...00000σ...

Join Encoding		Operation Encoding		Sample Rows	
no join case	...⋈00000...	<	...100...	1	row1
t.id=ci.movie_id	...0⋈0000...	=	...010...	2	row2
t.id=mc.movie_id	...00⋈000...	>	...001...	...	...
t.id=mi.movie_id	...000⋈00...			1000	row1000
t.id=mi_idx.movie_id	...0000⋈0...				
t.id=mk.movie_id...	...00000⋈...	Value Encoding $V_{normalized} = \frac{V - V_{min}}{V_{max} - V_{min}}$			

```
SELECT COUNT(*)
FROM title t, movie_companies mc
WHERE t.id = mc.movie_id AND
t.production_year > 2010 AND mc.company_id = 5
```
$T_q[$ t_0 {[...00000σ_{t_0}...],[...010...1]}, t_1 {[...0σ_{t_1}0000...],[...100...0]}]
$J_q[J_0[...00\bowtie_{t_0 \cdot t_1} 000...]]$
$P_q[$ p_0 {[...00000σ_{t_0}...],[...001...],[0.72]}, p_1 {[...000σ_{t_0}00...],[...010...],[0.14]}]

Fig. 2. Query encoding example, σ_t and $\bowtie_{t0*t1}$ represent Base Table Selectivity and Join Selectivity respectively, they are used for replacing 1 in one-hot encoding.

where $Rows(t)$ stands for the total rows of base table t, $\hat{C}_{tp}$ represents the estimated cardinality of base table t applying predicates p got through traditional cardinality estimation technologies. If a table t has no predicates, the Base Table Selectivity σ of it is 1. For each base table in a query, Feature Encoder uses its estimated cardinality $\hat{C}_{tp}$ fetched by Database Controller with $EXPLAIN$ command and persistent $Rows(t)$ to calculate its Base Table Selectivity σ. As for attributes in predicates, they use the Base Table Selectivity σ of corresponding base tables.

Join Selectivity $\bowtie$ and Vertex Encoding. Join Encoding and Vertex Encoding are different from Tables Encoding and Attribute Encoding, they involve conjunctive operations between different attributes. We propose Join Selectivity for Join Encoding and Vertex Encoding. Formula (4) shows its definition

$$\bowtie = \begin{cases} \min(\frac{1}{\mu_{inner}}, \frac{1}{\mu_{outer}}), & \text{join case} \\ 1, & \text{no join case} \end{cases} \quad (4)$$

where μ_{inner} and μ_{outer} are the number of unique values of inner table and outer table's conjunctive attributes respectively. Figure 3 shows our Vertex Encoding for the example query, where table title join table *movie_companies* on $t.id = mc.movei\_id$, we set corresponding attributes to $\bowtie_{t*mc}$, which riches the feature of vertexes. Base Table Selectivity σ and Join Selectivity $\bowtie$ can not only enrich the one-hot like encoding, but also better adapt to the status shifts of the dataset, such as the addition and deletion of tuples, their contribution to estimation quality is discussed in Sect. 5.

Generated S_q and T_q are spliced into Table Set as the input of Table Set MLP. The input of Predicate Set MLP are composed of Attribute Encoding, Operations Encoding and value normalized by the same min-max normalization method as cardinalities defined in Formula (1). Join Encoding J_q and Vertex Encoding V_q construct the input of Join Set MLP's and GAT's respectively.

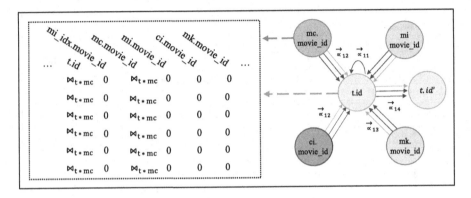

Fig. 3. Vertex encoding example of graph attention network

4.2 Model

Graph Attention Network [21] GAT belongs to Graph Convolutional Network family, it can effectively aggregate the features of different vertexes in the graph. By introducing the attention mechanism, the graph attention network peels off the correlation between vertex features and graph structure, so GAT does not need to obtain information of the entire graph, which improves the generalization ability of model. GAT parameterizes the adjacency matrix of graph, and allows the model to learn the degrees of contribution between different nodes through self-learning, that is the attention weights

$$\alpha_{i,j} = \frac{\exp(LeakyReLU(\overrightarrow{a}^T[W\overrightarrow{h_i}||W\overrightarrow{h_j}]))}{\sum_{k \in N_i} \exp(LeakyReLU(\overrightarrow{a}^T[W\overrightarrow{h_i}||W\overrightarrow{h_j}]))}. \tag{5}$$

Formula (6) describes the computing layer of GAT

$$\overrightarrow{h_i'} = \prod_{k=1}^{K} \sigma(\sum_{j \in N_i} \alpha_{ij}^k W^k \overrightarrow{h}_j), \tag{6}$$

where α_{ij}^k are attention weights, W^k is weight matrix, and $\overrightarrow{h}_j$ are features of nodes.

In the context of query optimization, conjunctive predicates between attributes can be seen as edges of join graph. As shown on the right of Fig. 3, queries involve the joins of *id* of table *t(title)* and *movie_id* of table *mc(movie_companies)* and table *mi(movei_info)* form join-crossing edges between them. This kind of conjunctive predicates between primary keys and foreign keys is very common in real-world query workload. The cardinalities of results applying the join predicates are closely related to both the cardinalities of base tables (Base Table Selectivity σ) and the join-crossing correlations (Join Selectivity $\bowtie$) between the join base tables. The more the correlations are captured, the more accurate the cardinalities predicted by the model are. We use GAT to learn the attention weights between different vertexes of join graph from query workload including join-crossing correlations between different attributes.

Three MLPs and GAT are used to learn features from query structure and query workload. Formulas below show the deep detail of our model in Estimate Model in Fig. 1

$$Table module : w_T = LN(\frac{1}{|T_q|}\sum_t t \in T_q MLP_T(v_t)),$$

$$Join module : w_J = LN(\frac{1}{|J_q|}\sum_t j \in J_q MLP_J(v_j)),$$

$$Predicate module : w_P = LN(\frac{1}{|P_q|}\sum_t p \in P_q MLP_P(v_p)),$$

$$GCN module : w_G = LN(\frac{1}{|V_q|}\sum_t v \in V_q GAN(v_v)),$$

$$Merge\&predicate : w_{out} = MLP_{out}([w_T, w_J, w_P, w_G]).$$

Table, Join, Predicate MLP modules extract basic features from query structure: Table Encoding (including Sample Encoding S_q) T_q, Join Encoding J_q, Predicate Encoding P_q, respectively. The most important feature of them are Base Table Selectivity and Join Selectivity associated with the status of dataset. GAT learns the attention weights between attributes from join graphs and Vertex Encoding with Join Selectivity V_q, including the join-crossing correlations. We put LayerNorm layer at the end of each model to normalize the distributions of intermediate layers. The features of four models are concatenated by the last layer of MLP, estimated cardinality is a value between 0 and 1 after activation by sigmoid function and the estimated cardinalities can be recovered via denormalization function defined in Formula (7).

$$\hat{C}_{unnorm} = e^{\hat{C}_{norm} \times (C_{norm(max)} - C_{norm(min)}) + C_{norm(min)}}, \tag{7}$$

where $\hat{C}_{norm}$ is cardinality estimated by model, $C_{norm(max)}$ and $C_{norm(min)}$ represent the maximum and minimum real cardinalities after normalization.

5   Evaluation

5.1   Experimental Setup

Query Workloads. We use the IMDb dataset used in JOB[11] and query workloads proposed by MSCN [10]. Table 1 shows the join number distribution of each query workload: i) Train query workload contains 100k queries, the max number of joins is 2. ii) Synthetic workload is generated by the same query generator as train workload, containing both (conjunctive) equality and range predicates on non-key attributes, the maximum number of joins is 2. iii) Scale is a comprehensive workload with join number uniformly distributed from 0 to 4. iv) JOB-light contains 70 queries derived from 113 queries in JOB, the maximum join number is 4. Synthetic, Scale and JOB-light query workloads are used as test workloads to evaluate the performance of models.

Batch Training. Different from the normal tasks using GAT, which treat the input of the entire dataset as a complete graph, our task regards each query in the workload as a small graph. Therefore, the traditional GAT cannot be directly transferred to our task for batch training. In the scenario of cardinality estimation, not only can we obtain the max number of attributes before training, but also the shape of join graphs formed between attributes are identical the same. So it is practicable to consider multiple join graphs of queries as a batch, sharing parameters in the same GAT. Based on the fact, we have implemented a GAT that supports batch training. Figure 4 compares the time overhead of training in epochs with $batch\_size = 8$ on machine with Intel(R) Xeon(R) E5-2678 v3, 128 GB Memory, and NVIDIA TITAN RTX, BatchGAT achieves 3x acceleration.

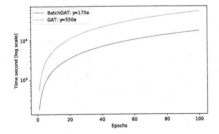
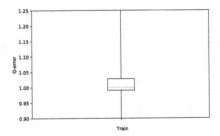

Fig. 4. BatchGAT vs. GAT   **Fig. 5.** Base table estimation quality

Metrics. $Q\text{-}error_{max}$ with max operation defined in Formula (2) is adopted as a metric to evaluate the results. Both real and estimated cardinalities are lower bounded by 1, and max operation is used in the calculation, so the minimum attainable $q\text{-}error_{max}$ is 1. We give $q\text{-}error_{max}$ at median, 90th, 95th, 99th, max and mean percentiles of each query workload in the form of table to show the estimation quality synthetically. On box plots, we replace $q\text{-}error_{max}$ with *q-error* without max operation defined in Formula (8) to illustrate the quality more intuitively

$$q\text{-}error = \frac{\hat{C}}{C}. \tag{8}$$

Q-error preserves the estimated trend of models: overestimation if greater than 1 and vice versa is underestimation. The lower whisker, lower quartile, median, upper quartile, upper whisker of box plots represent q-error at $5\%, 25\%, 50\%, 75\%, 95\%$ after sorted, respectively.

Table 1. Join distribution of workloads

Number of joins	0	1	2	3	4	Overall
Train	26818	29888	43294	0	0	10000
Synthetic	1636	1407	1957	0	0	5000
Scale	100	100	100	100	100	500
JOB-light	0	3	32	23	12	70

Table 2. $q\text{-}error_{max}$ on synthetic

	Median	90th	95th	99th	Max	Mean
PostgreSQL	1.69	9.57	23.9	465	373901	154
MSCN	1.19	3.32	6.84	30.51	1322	2.89
New-Enc MSCN	1.18	3.29	6.73	31	1200.1	2.85
E2E	**1.18**	**3.19**	**6.05**	**24.5**	**323**	**2.81**
GACE	**1.18**	3.28	6.8	29.61	1008.6	2.83

Table 3. $q\text{-}error_{max}$ on Scale

	Median	90th	95th	99th	Max	Mean
Postgre SQL	2.59	200	540	1816	233863	568
MSCN	1.42	37.4	140	793	3666	35.1
New-Enc MSCN	1.35	37.38	108.50	503.57	2799.54	26.41
E2E	1.42	37.3	125	345	**1813**	26.3
GACE	**1.35**	**24.93**	**58.71**	**253.09**	2813.74	**19.19**

Table 4. $q\text{-}error_{max}$ on JOB-light

	Median	90th	95th	99th	Max	Mean
Postgre SQL	7.93	164	1104	2912	3477	174
MSCN	3.82	78.4	362	927	1110	57.9
New-Enc MSCN	2.44	17.00	41.24	135.64	269.31	11.59
E2E	3.51	48.6	139	244	272	24.3
GACE	**2.11**	**10.92**	**19.12**	**103.13**	**219.59**	**8.16**

5.2 Dynamic Features for Query Structure

Base Table Selectivity σ. We construct base table queries (216k) with 0 join by extracting all tables and corresponding predicates from the train query workload. An example query may look like

```
SELECT * FROM title AS t WHERE t.production_year>2010;
```

Figure 5 displays the distribution of *q-error* via *EXPLAIN* (PostgreSQL version 10.3) command on our man-made queries, the vast majority of *q-error* is around 1. First subplot of Fig. 7 shows the 0 join case of all test query workloads, the estimation error is also relatively small.

Join Selectivity $\bowtie$. We execute all conjunctive queries with 1 join between two tables to calculate the $q\text{-}error_{max}$ between real cardinalities and cardinalities estimated by $\bowtie \times Rows(t_{inner}) \times Rows(t_{outer})$ in order to evaluate the reliability of Join Selectivity. The min, max, mean $q\text{-}error_{max}$ are 1.65, 1.90, and 1.74 respectively.

Our experiments demonstrate that Base Table Selectivity and Join Selectivity are reliable features, the estimation quality of new-enc MSCN (MSCN with Base Table Selectivity and Join Selectivity embedded) on test workloads in Sect. 5.3 illustrates their effectiveness.

5.3   Estimation Accuracy

We compare the estimation quality of MSCN, new-enc MSCN and GACE (with dynamic features and GAT) from two perspectives: query workloads and number of joins. Quality of PostgreSQL is generally underestimated poorly because its traditional cardinality estimation technologies miss join-crossing correlations.

Query Workloads. Table 2, Table 3, and Table 4 give detail $q\text{-}error_{max}$ on different test workloads. On Synthetic workload, new-enc MSCN perform slightly better than MSCN [10] at all percentiles (1% at mean) other than 99th, GACE performs similarly as new-enc MSCN and both inferior to E2E [19], the overall estimated cardinalities of them are very close to the real cardinalities, limiting the room for improvement. On scale workload, the overall estimated quality is also comparatively accurate. New-enc MSCN outperforms MSCN slightly at all percentiles (20% at mean), the improvement of GACE is considerable, it outperforms E2E in all cases by 5%–68% except max percentile. On JOB-light workload, which mainly contains queries with 2–4 joins, the performance of MSCN and E2E is not impressive. After adding more dynamic features, new-enc MSCN performs better than E2E in all cases, by 2.09x on mean error. As for the GACE with new feature encoding and GAT module, the estimation quality has exciting advancement at all percentiles, better than E2E by almost 3x on mean error. Figure 6 shows the q-errors box plots of models on three test workloads, from the angle of q-error, the overall conclusion is basically the same as $q\text{-}error_{max}$, it can be found that all cardinality estimators have a tendency of underestimation. Figure 7 illustrates the estimation quality from the view of the join number on different workloads.

Number of Joins. It can be concluded that when the join number is 0–2, the underestimation and overestimation q-error of new-enc MSCN, and GACE are almost the same, slightly better than MSCN, which is consistent with their performance on the Synthetic query workload. On the case of 3 join, the estimation quality of new-enc MSCN is slightly better than MSCN, GACE achieves considerable improvements. This is because when the number of joins is 0–3, the overall estimation quality is relatively high, leaving little space for promotion. When join number reaches 4, the performance of previous work is poor, the introduction of new dynamic features and the GAT network bring impressive improvements.

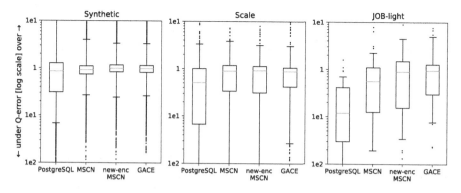

Fig. 6. Q-error distribution of PostgreSQL, MSCN, New-enc MSCN, GACE on different query workloads.

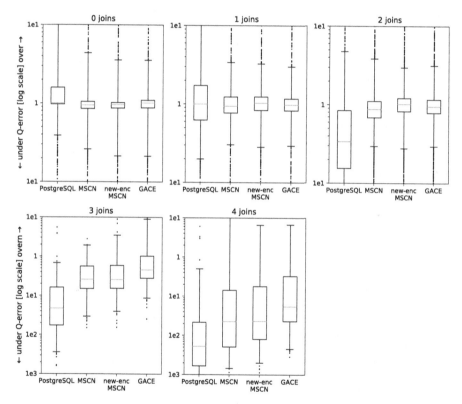

Fig. 7. Q-error distribution of PostgreSQL, MSCN, New-enc MSCN, GACE on different number of joins

These observations verify the performance of new-enc MSCN and GACE in three different test workloads from another perspective. On Synthetic workload with up to 2 joins, the overall estimation quality of all models are relatively reliable, new-enc MSCN and GACE perform similarly with MSCN and inferior to E2E. On Scale workload, new-enc MSCN and GACE improve slightly due to the even distribution of join number from 0 to 4. New-enc MSCN and GACE have each accomplished improvements on JOB-light mainly containing queries with 2–4 joins, the latter is more significant.

6   Conclusion

In this paper, we propose a Graph-Attention-Network-based cardinality estimator (GACE) that effectively catches join-crossing correlations from join graphs extracted on query workloads in addition to query structure. GACE leverages the Graph Attention Network to learn attention weights between different attributes across tables, including the key factors that challenge the accuracy of cardinality estimation: join-crossing correlations. GACE combines one-hot encoding with dynamic features with dataset status embedded (Base Table Selectivity and Join Selectivity) to enrich the feature encoding of query structure. Exploiting the regularity of join graphs, a GAT with batch training support is implemented to mitigate the training overhead. Experiments on IMDb dataset for query workloads with different distribution of joins demonstrate that the combination of one-hot encoding with Base Table Selectivity and Join Selectivity is helpful in improving the quality of cardinality estimation and GAT can effectively capture the join-crossing correlations between attributes from join graphs especially on workloads with more joins. It inspires us to explore the possibility of combining graph neural networks with methods learning from data in our next work.

References

1. Chakrabarti, K., Garofalakis, M., Rastogi, R., Shim, K.: Approximate query processing using wavelets. VLDB J. **10**(2), 199–223 (2001)
2. Cormode, G., Garofalakis, M., Haas, P.J., Jermaine, C.: Synopses for massive data: samples, histograms, wavelets, sketches. Found. Trends Databases **4**(1–3), 1–294 (2012)
3. Cormode, G., Muthukrishnan, S.: An improved data stream summary: the count-min sketch and its applications. J. Algorithms **55**(1), 58–75 (2005)
4. Flajolet, P., Fusy, É., Gandouet, O., Meunier, F.: HyperLogLog: the analysis of a near-optimal cardinality estimation algorithm. In: Discrete Mathematics and Theoretical Computer Science, pp. 137–156. Discrete Mathematics and Theoretical Computer Science (2007)
5. Flajolet, P., Martin, G.N.: Probabilistic counting algorithms for data base applications. J. Comput. Syst. Sci. **31**(2), 182–209 (1985)
6. Getoor, L., Taskar, B., Koller, D.: Selectivity estimation using probabilistic models. In: Proceedings of the 2001 ACM SIGMOD International Conference on Management of Data, pp. 461–472 (2001)
7. Gunopulos, D., Kollios, G., Tsotras, V.J., Domeniconi, C.: Selectivity estimators for multidimensional range queries over real attributes. VLDB J. **14**(2), 137–154 (2005)

8. Hilprecht, B., Schmidt, A., Kulessa, M., Molina, A., Kersting, K., Binnig, C.: DeepDB: learn from data, not from queries! arXiv preprint arXiv:1909.00607 (2019)
9. Ioannidis, Y.: The history of histograms (abridged). In: Proceedings of the 29th International Conference on Very Large Data Bases, VLDB 2003, vol. 29. p. 19–30. VLDB Endowment (2003)
10. Kipf, A., Kipf, T., Radke, B., Leis, V., Boncz, P., Kemper, A.: Learned cardinalities: estimating correlated joins with deep learning. arXiv preprint arXiv:1809.00677 (2018)
11. Leis, V., Gubichev, A., Mirchev, A., Boncz, P., Kemper, A., Neumann, T.: How good are query optimizers, really? Proc. VLDB Endow. **9**(3), 204–215 (2015)
12. Leis, V., Radke, B., Gubichev, A., Kemper, A., Neumann, T.: Cardinality estimation done right: index-based join sampling. In: CIDR (2017)
13. Lipton, R.J., Naughton, J.F., Schneider, D.A.: Practical selectivity estimation through adaptive sampling. In: Proceedings of the 1990 ACM SIGMOD International Conference on Management of Data, pp. 1–11 (1990)
14. Malik, T., Burns, R.C., Chawla, N.V.: A black-box approach to query cardinality estimation. In: CIDR, pp. 56–67. Citeseer (2007)
15. Matias, Y., Vitter, J.S., Wang, M.: Wavelet-based histograms for selectivity estimation. In: Proceedings of the 1998 ACM SIGMOD International Conference on Management of Data, pp. 448–459 (1998)
16. Muralikrishna, M., DeWitt, D.J.: Equi-depth multidimensional histograms. In: Proceedings of the 1988 ACM SIGMOD International Conference on Management of Data, pp. 28–36 (1988)
17. Poosala, V., Haas, P.J., Ioannidis, Y.E., Shekita, E.J.: Improved histograms for selectivity estimation of range predicates. ACM SIGMOD Rec. **25**(2), 294–305 (1996)
18. Poosala, V., Ioannidis, Y.E.: Selectivity estimation without the attribute value independence assumption. In: VLDB, vol. 97, pp. 486–495. Citeseer (1997)
19. Sun, J., Li, G.: An end-to-end learning-based cost estimator. PVLDB **13**(3), 307–319 (2019)
20. Tzoumas, K., Deshpande, A., Jensen, C.S.: Lightweight graphical models for selectivity estimation without independence assumptions. Proc. VLDB Endow. **4**(11), 852–863 (2011)
21. Veličković, P., Cucurull, G., Casanova, A., Romero, A., Lio, P., Bengio, Y.: Graph attention networks. arXiv preprint arXiv:1710.10903 (2017)
22. Whang, K.Y., Vander-Zanden, B.T., Taylor, H.M.: A linear-time probabilistic counting algorithm for database applications. ACM Trans. Database Syst. (TODS) **15**(2), 208–229 (1990)
23. Wu, W., Naughton, J.F., Singh, H.: Sampling-based query re-optimization. In: Proceedings of the 2016 International Conference on Management of Data, pp. 1721–1736 (2016)
24. Yang, Z., et al.: NeuroCard: one cardinality estimator for all tables. arXiv preprint arXiv:2006.08109 (2020)
25. Yang, Z., et al.: Deep unsupervised cardinality estimation. arXiv preprint arXiv:1905.04278 (2019)

A Two-Phase Approach for Enumeration of Maximal (**Δ, γ**)-Cliques of a Temporal Network

Suman Banerjee[1(✉)] and Bithika Pal[2]

[1] Indian Institute of Technology, Jammu, India
suman.banerjee@iitjammu.ac.in
[2] Indian Institute of Technology, Kharagpur, India
bithikapal@iitkgp.ac.in

Abstract. A *Temporal Network* is a graph whose topology is changing over time and represented as a collection of triplets of the form (u, v, t) that denotes the interaction between the agents u and v at time t. Analyzing and enumerating different structural patterns of such networks are important in different domains including social network analysis, computational biology, etc. In this paper, we study the problem of enumerating one such pattern: maximal (Δ, γ)-Clique. Given a temporal network $\mathcal{G}(V, E, \mathcal{T})$, a (Δ, γ)-Clique is a vertex subset, time interval pair $(\mathcal{X}, [t_a, t_b])$ such that between every pair of vertices of $\mathcal{X}$, there exist at least γ links in each Δ duration in $[t_a, t_b]$. The proposed methodology is broadly divided into two phases. In the first phase, each temporal link is processed for constructing (Δ, γ)-Clique(s) with maximum duration. In the second phase, these initial cliques are expanded by vertex addition to form the maximal cliques. We show that the proposed methodology is correct, and running time, space requirement analysis has been done. From the experimentation on three real datasets, we observe that the proposed methodology enumerates all the maximal (Δ, γ)-Cliques efficiently, particularly when the dataset is sparse. As a special case $(\gamma = 1)$, the proposed methodology is also able to enumerate $(\Delta, 1) \equiv \Delta$-cliques with much less time compared to the existing methods.

Keywords: Temporal Network · (Δ, γ)-Clique · Enumeration algorithm

1 Introduction

Network (also called graph) is a mathematical object which has been used extensively to represent a *binary relation* among a group of agents. Analyzing such networks for different structural patterns remains an active area of study in different domains including *Computational Biology* [11], *Social Network Analysis* [19], *Computational Epidemiology* [14], and many more. Among many one such

---

Both the authors have contributed equally in this work and they are joint first authors.

C. Strauss et al. (Eds.): DEXA 2021, LNCS 12924, pp. 346–357, 2021.
https://doi.org/10.1007/978-3-030-86475-0_33

structural pattern is the maximally connected subgraphs, which are popularly called as *Clique*. Finding the maximum cardinality clique in a given network is a well-known *NP-Complete* Problem [7].

Real-world networks are *time-varying*, which means that the existence of an edge changes with time. Such networks are represented as temporal networks [10] (also known as *time-varying networks*). For these types of networks, a natural supplement of clique is the *temporal clique* which consists of two things: a subset of the vertices, and a time interval. Viard et al. [20] put forward the notion of Δ-Clique. A vertex subset along with a time interval is said to be a Δ-Clique if every vertex pair from that set have at least a single edge in every Δ duration within the time interval. However, for practical applications such as community structure in temporal networks etc. Δ-Clique appears to be a too sparse structure. Hence, the notion of Δ-Clique has been extended to (Δ, γ)-Clique by incorporating frequency along with the duration [3].

The problem of maximal clique enumeration is a classic computational problem that has been extensively studied on static graphs. Akkoyunlu [1] was the first to propose an algorithm for this problem. Later, Bron and Kerbosch[5] introduced a recursive approach for the maximal clique enumeration problem. These two studies are the foundations on maximal clique enumeration and trigger a huge amount of research due to many practical applications from computational biology to spatial data analytics [2,4]. In the past two decades, several methodologies have been developed for enumerating maximal cliques in different computational paradigms, and different kinds of networks, such as in sparse graphs [6], in large networks [18], in uncertain graphs [16], and many more. However, the literature on temporal clique enumeration is limited. To the best of our knowledge other than the enumeration of Δ-Clique [8,19], (Δ, γ)-Clique [3], isolated cliques [15] there are no other studies.

A maximal (Δ, γ)-Clique signifies different meanings in different contexts. In case of a time-varying social network, a (Δ, γ)-Clique may signify a temporal community in the network. Similarly, in case of a protein-protein interaction network, a (Δ, γ)-Clique signifies a set of proteins that frequently dock and form a new protein. Hence, enumerating maximal (Δ, γ)-Clique is an important problem in temporal network analysis. The key contributions of this paper are as follows:

- We propose an efficient two phase approach for enumerating maximal (Δ, γ)-Cliques of a temporal network.
- We prove that the proposed methodology is correct and do an analysis for understanding its time and space requirement.
- We perform an extensive set of experiments to show the effectiveness of the proposed methodology.

The rest of the paper is organized as follows. Section 2 defines the problem. Section 3 discusses the proposed solution methodology. Section 4 contains the experimental evaluation, and finally, Sect. 5 concludes our study.

2   Problem Definition

In this section we present some preliminary concepts and defines our problem.

Definition 1 (Temporal Network) [10]. *A temporal network is defined as* $\mathcal{G}(V, E, \mathcal{T})$, *where* $V(\mathcal{G})$ *is the set of vertices of the network and* $E(\mathcal{G})$ *is the set of edges among them.* $\mathcal{T}$ *is the mapping that maps each edge of the graph to its occurrence time stamp(s).*

The figure beside shows the links of a temporal network. In a static network, a subset of vertices, where every pair of them is adjacent is known as a *clique*. Recently, Banerjee and Pal [3] generalized the notion of clique and introduced the notion of (Δ, γ)-Clique which is stated in Definition 2.

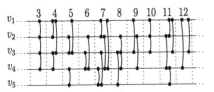

Fig. 1. Links of a Temporal Network

Definition 2 ((Δ, γ)-Clique). *Given a temporal network* $\mathcal{G}(V, E, \mathcal{T})$, *time duration* Δ, *and a frequency threshold* $\gamma \in \mathbb{Z}^+$, *a* (Δ, γ)-*Clique of* $\mathcal{G}$ *is a tuple consisting of vertex subset, and time interval, i.e.,* $(\mathcal{X}, [t_a, t_b])$ *where* $\mathcal{X} \subseteq V(\mathcal{G})$, $|\mathcal{X}| \geq 2$, *and* $[t_a, t_b] \subseteq \mathbb{T}$. *Here* $\forall v_i, v_j \in \mathcal{X}$ *and* $\tau \in [t_a, max(t_b - \Delta, t_a)]$, *there must exist at least* γ *number of edges, i.e.,* $(u, v, t_{ij}) \in E(\mathcal{G})$ *and* $f_{(u,v)} \geq \gamma$ *with* $t_{ij} \in [\tau, min(\tau + \Delta, t_b)]$. $f_{(u,v)}$ *denotes the frequency of the static edge* (u, v).

Let, All the links of $\mathcal{G}$ are within the time interval $[t^{min}, t^{max}]$. Next, we define the Maximal (Δ, γ)-Clique.

Definition 3 (Maximal (Δ, γ)-Clique). *Given a temporal network* $\mathcal{G}(V, E, \mathcal{T})$ *and a* (Δ, γ)-*Clique* $(\mathcal{X}, [t_a, t_b])$ *of* $\mathcal{G}$, $(\mathcal{X}, [t_a, t_b])$ *will be maximal if none of the following is true.*

- $\exists v \in V(\mathcal{G}) \setminus \mathcal{X}$ *such that* $(\mathcal{X} \cup \{v\}, [t_a, t_b])$ *is a* (Δ, γ)-*Clique.*
- $(\mathcal{X}, [t_a - 1, t_b])$ *is a* (Δ, γ)-*Clique. This applies only if* $t_a - 1 \geq t^{min}$.
- $(\mathcal{X}, [t_a, t_b + 1])$ *is a* (Δ, γ)-*Clique. This applies only if* $t_b + 1 \leq t^{max}$.

In this paper, we study the problem of enumerating the maximal (Δ, γ) -Cliques.

3   Proposed Methodology

The proposed methodology is broadly divided into two steps. Given all the links with time duration of the temporal network, initially, we find out the maximal cliques of cardinality two. Next, taking these duration wise maximal cliques, we add vertices into the clique without violating the definition of (Δ, γ)-clique.

3.1   Phase 1 (Initialization)

Algorithm 1 describes the initialization process. For a given temporal network $\mathcal{G}$, initially, we construct dictionary $\mathcal{D}_e$ with the static edges as the *keys* and correspondingly, the occurrence time stamps are the *values*. By the definition of (Δ, γ)-clique, if the end vertices of an edge is part of a clique, then the edge has to occur atleast γ times in the link stream. Hence, for each static edge (uv) of $\mathcal{G}$, if its frequency is at least γ, it is processed further. The occurrence time stamps of (uv) are fed into the list $\mathcal{T}_{(uv)}$. A temporary list, $Temp$, is created to store each current processing timestamps from $\mathcal{T}_{(uv)}$ with its previous occurrences, till it has maintained (Δ, γ)-clique property. Now, the for-loop from Line 8 to 32 computes all the (Δ, γ)-cliques with maximum duration where $\{u, v\}$ is the vertex set. During the processing of $\mathcal{T}_{(uv)}$, any of the following two cases can happen. In the first case, if the current length of $Temp$ is less than γ, the difference between the current timestamp from $\mathcal{T}_{(uv)}$ and the first entry of $Temp$ is checked (Line 10). Now, if the difference is less than or equal to Δ, current timestamp is appended in $Temp$. Otherwise, all the previous timestamps that have occurred within past Δ duration from the current timestamp are added in $Temp$ (Line 14). This process basically checks Δ timestamp backward from each occurrence times of the static edge (u, v). In the second case, when the current length of $Temp$ is greater than or equal to γ, it is checked whether the current processing time from $\mathcal{T}_{(uv)}$ falls within the interval of (last γ-th occurrence time + 1) to (last γ-th occurrence time + 1 + Δ). Now, if it is true, the current timestamp is appended in $Temp$. It can be easily observed that this appending is done iff the at least consecutive γ occurrences are within each Δ duration. Otherwise, the clique is added in $\mathcal{C}_T^I$ with the vertex set $\{u, v\}$ and time interval $[t_a, t_b]$ (Line 22), where t_a is the Δ ahead timestamp from the first γ-th entry in $Temp$ and t_b is the Δ on-wards timestamp from the last γ-th entry in $Temp$. Next, all the previous timestamps that have occurred within past Δ duration from the current timestamp are added in $Temp$ as before (Line 24). It allows to consider overlapping clique. Now, this may happen when we process the last occurrence from $\mathcal{T}_{(uv)}$, it is added in $Temp$. However, no clique can be added by the condition of 9 to 26 if the length of $Temp$ is greater than or equal to γ. This situation is handled by Line 27 to 31. This process is iterated for each key from the dictionary $\mathcal{D}_e$. We present few lemmas and they leads to the correctness of the methodology. Due to space limitation, we will not be able to give the proofs.

Lemma 1. *For a link (uv), if there exist any consecutive γ occurrences within Δ duration, then it has to be in 'Temp' at some stage, in Algorithm 1.*

Lemma 2. *In any arbitrary iteration of the 'for loop' at Line 8 in Algorithm 1, each consecutive γ occurrences of 'Temp' will be within Δ duration.*

Lemma 3. *Let, t^f and t^l be the first and last occurrence time of a link in $Temp$. In the interval $[t^f, t^l]$, $Temp$ contains at least γ links in each Δ duration.*

Lemma 4. *In Algorithm 1, the contents of $\mathcal{C}_T^I$ are (Δ, γ)-Cliques of size 2.*

Algorithm 1: First phase of Maximal (Δ, γ)-Clique enumeration

---

Data: The temporal network $\mathcal{G}(V, E, \mathcal{T})$, Δ, $\gamma \in \mathbb{Z}^+$.

Result: The initial clique set $\mathcal{C}_T^I$ of $\mathcal{G}$

1   Construct the Dictionary $\mathcal{D}_e$;

2   $\mathcal{C}_T^I = \phi$;

3   **for** *Every* $(uv) \in \mathcal{D}_e.keys()$ **do**

4   **if** $f_{(uv)} \geq \gamma$ **then**

5       $\mathcal{T}_{(uv)} = $ Time Stamps of (uv);

6       $Temp = [\,]$;

7       $Temp.append(\mathcal{T}_{(uv)}[1])$;

8   **for** $i = 2$ *to* $len(\mathcal{T}_{(uv)})$ **do**

9   **if** $len(Temp) < \gamma$ **then**

10   **if** $\mathcal{T}_{(uv)}[i] - Temp[1] \leq \Delta$ **then**

11           $Temp.append(\mathcal{T}_{(uv)}[i])$;

12   **else**

13           $Temp = [\,]$;

14           $Temp.append$(time stamps of the links occured in previous Δ *Duration*)

15   **end**

16   **else**

17   **if** $Temp[len(Temp) - \gamma + 1] + 1 + \Delta \geq \mathcal{T}_{(uv)}[i]$ **then**

18           $Temp.append(\mathcal{T}_{(uv)}[i])$;

19   **else**

20           $t_a = Temp[\gamma] - \Delta$;   // first γ-th occurrence of (u, v) in *Temp*

21           $t_b = Temp[len(Temp) - \gamma + 1] + \Delta$;   // last γ-th occurrence of (u, v) in *Temp*

22           $\mathcal{C}_T^I.addClique(\{u, v\}, [t_a, t_b])$;

23           $Temp = [\,]$;

24           $Temp.append$(time stamps of the links occured in previous Δ *Duration*)

25   **end**

26   **end**

27   **if** $i = len(\mathcal{T}_{(uv)})$ *and* $len(Temp) \geq \gamma$ **then**

28          $t_a = Temp[\gamma] - \Delta$;   // first γ-th occurrence of (u, v) in *Temp*

29          $t_b = Temp[len(Temp) - \gamma + 1] + \Delta$;   // last γ-th occurrence of (u, v) in *Temp*

30          $\mathcal{C}_T^I.addClique(\{u, v\}, [t_a, t_b])$;

31   **end**

32   **end**

33   **end**

34 **end**

Lemma 5. *All the cliques returned by Algorithm 1 and contained in $\mathcal{C}_T^I$ are duration wise maximal.*

Lemma 6. *All the duration wise maximal (Δ, γ)-cliques of size 2 are contained in $\mathcal{C}_T^I$.*

3.2   Phase 2 (Enumeration)

Algorithm 2 describes the enumeration strategy. For the given temporal network $\mathcal{G}$, we construct a static graph G where $V(G)$ is the vertex set of $\mathcal{G}$ and each link of $\mathcal{G}$ induces the corresponding edge in $E(G)$ without the time component, which we call as a static edge. Next, the dictionary $\mathcal{D}$ is built from the initial clique set C_T^I of Algorithm 1, where the vertex set of the clique is the key and corresponding occurrence time intervals are the values. This data structure is also updated in the intermediate steps of Algorithm 2. Now, two sets C^{T_1} and C^{T_2} are maintained during the enumeration process. At any i-th iteration of the while loop at Line 5, C^{T_1} maintains the current set of cliques which are yet to be processed for vertex addition and C^{T_2} stores the new cliques formed in that i-th iteration. At the beginning, all the initial cliques from C_T^I are copied into C^{T_1}. A clique $(\mathcal{X}, [t_a, t_b])$ is taken out from C^{T_1} which is duration wise maximal and the IS_MAX flag is set to true for indicating the current clique as maximal (Δ, γ)-clique. For vertex addition, it is trivial to see that only for the neighboring vertices of $\mathcal{X}$ ($v \in \mathcal{N}_G(\mathcal{X})$), there is a possibility of $(\mathcal{X} \cup \{v\}, [t_a', t_b'])$ to be a (Δ, γ)-clique. If the new vertex set $\mathcal{X} \cup \{v\}$ is found in $\mathcal{D}$ with one of its value as $[t_a, t_b]$, the IS_MAX flag is set to false, signifying that the processing clique $(\mathcal{X}, [t_a, t_b])$ is not maximal. Otherwise, if $\mathcal{X} \cup \{v\}$ is not present in $\mathcal{D}$, all the possible time intervals in which $\mathcal{X} \cup \{v\}$ can form a (Δ, γ)-clique are computed from Line 16 to 37. This process is iterated for all the neighboring vertices of $\mathcal{X}$ (Line 10 to 38). Now, we describe the statements from Line 17 to 36 in detail. As mentioned earlier, to form a (Δ, γ)-clique with the new vertex set $\mathcal{X} \cup \{v\}$ all the possible combinations from $\mathcal{X} \cup \{v\}$ of size $|\mathcal{X}|$, (represented as $\mathrm{C}(\mathcal{X} \cup \{v\}, \mathcal{X})$), has to be a (Δ, γ)-clique. Now, for all $z \in \mathrm{C}(\mathcal{X} \cup \{v\}, \mathcal{X}))$, if z is present in $\mathcal{D}.keys()$, it signifies the possibility of forming a new clique with the vertex set $\mathcal{X} \cup \{v\}$ (Line 17). Now, all the entries of these combinations are taken into a temporary data structure $\mathcal{D}_{Temp}$ from $\mathcal{D}$. For the clarity of presentation, we describe the operations from Line 19 to 35 for one vertex addition, i.e., $\mathcal{X} \cup \{v\}$ with the help of an example shown in Fig. 2. Now, let the entries of $\mathcal{D}_{Temp}$ are $z_1, z_2, \ldots z_n$, i.e., all $z_i \in \mathrm{C}(\mathcal{X} \cup \{v\}, \mathcal{X})$ and the length corresponding entries in $\mathcal{D}_{Temp}$ are $l_1, l_2, \ldots l_n$ respectively. So, one sample from $z_1 \otimes z_2 \otimes \cdots \otimes z_n$ is taken as $timeSet$ in Line 19 of Algorithm 2. One possible value of $timeSet$ is $[t_{11}, t_{21}, \ldots, t_{n1}]$. For this value, the resultant interval $[t_a', t_b']$ is computed as $t_{11} \cap t_{21} \cdots \cap t_{n11} = [max(t_{z_1}^{a^1}, t_{z_2}^{a^1}, \ldots, t_{z_n}^{a^1}),\ min(t_{z_1}^{b^1}, t_{z_2}^{b^1}, \ldots, t_{z_n}^{b^1})]$.

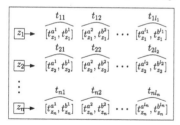

Fig. 2. The entries of $\mathcal{D}_{Temp}$

If the difference between t_b' and t_a' is more than or equal to Δ, then the newly formed (Δ, γ)-clique, $(\mathcal{X} \cup \{v\}, [t_a', t_b'])$, is added in C^{T_2} and $\mathcal{D}$. Also, if $[t_a', t_b']$ matches with the current interval of $\mathcal{X}$, then the flag IS_MAX is set to False, i.e., $(\mathcal{X}, [t_a, t_b])$ is not maximal. Now, this step is repeated for all the samples from $z_1 \otimes z_2 \otimes \cdots \otimes z_n$ from Line 19 to 35. This ensures that all the intervals in which $\mathcal{X} \cup \{v\}$ forms (Δ, γ)-clique are added in $\mathcal{D}$.

Now, if none of the vertices from $\mathcal{N}_G(\mathcal{X}) \setminus \mathcal{X}$ is possible to add in $\mathcal{X}$, $(\mathcal{X}, [t_a, t_b])$ becomes maximal (Δ, γ)-clique and added into final maximal clique set C_L at Line 40. Vertex addition checking is performed for all the cliques of C^{T_1} in the while loop from Line 7 to 42. When C^{T_1} is exhausted and C^{T_2} is not empty, the contents of C^{T_2} are copied back into C^{T_1} for further processing, signifying that all the maximal cliques have not been found yet. This is controlled using the flag $ALL\_MAXIMAL$ in the While loop at Line 5. If no clique is added into $\mathcal{C}^{T_2}$, the flag $ALL\_MAXIMAL$ is set to true so that in the next iteration the condition of the While loop at Line 5 will be false and finally Algorithm 2 terminates. At the end, for the temporal network $\mathcal{G}$, $\mathcal{C}_T$ contains all the maximal (Δ, γ)-cliques of it. An illustrative example of the enumeration Algorithm is given in Fig. 3.

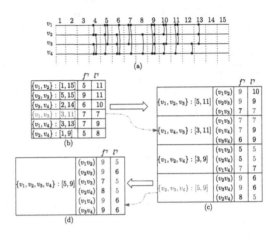

Fig. 3. Illustrative example of the proposed Maximal (Δ, γ)-Clique Enumeration Algorithm, (a)Input Temporal Graph with $\Delta = 4$ and $\gamma = 2$, (b)Output of the Algorithm 1 - First Phase, (c)-(d) The content of $\mathcal{C}^{T_1}$ at Different Iteration of Algorithm 2. The cliques in red are duration-wise maximal but not w.r.t. cardinality.

All the cliques are added in $\mathcal{C}_T$, only from $\mathcal{C}^{T_1}$ at Line 40 in Algorithm 2. Now, initially $\mathcal{C}^{T_1}$ contains the elements from $\mathcal{C}_{\mathcal{L}}^I$, which are (Δ, γ)-cliques from Lemma 4 and later it is updated with the entries of $\mathcal{C}^{T_2}$. So, if we show that the elements of $\mathcal{C}^{T_2}$ are (Δ, γ)-cliques, the statement will be proved. Now, all the cliques of $\mathcal{C}^{T_2}$ are of atleast Δ duration, from the condition at Line 28. Also, from the description of the Algorithm 2, it is easy to verify that in each iteration of vertex addition to a clique of $\mathcal{C}^{T_1}$ can only be made, if all the possible combinations of vertices form (Δ, γ)-cliques. This ensures that all the vertex pairs of the clique in $\mathcal{C}^{T_2}$ are linked atleast γ times in each Δ duration within the intersected time interval of all the combinations. Hence, the elements of $\mathcal{C}_T$ are (Δ, γ)-cliques.

Lemma 7. *In Algorithm 2, all the intermediate cliques are duration wise maximal.*

Lemma 8. *In Algorithm 2, at the begining of any i-th iteration, $\mathcal{C}^{T_1}$ holds all the duration wise maximal (Δ, γ)-cliques of size $i + 1$.*

Lemma 9. *All the (Δ, γ)-Cliques returned by Algorithm 2 and contained in $\mathcal{C}_T$ are maximal .*

Algorithm 2: Second phase of Maximal (Δ, γ)-Clique Enumeration

---

Data: A Temporal Network $\mathcal{G}$, Initial Clique Set $\mathcal{C}_T^I$, Δ, γ.
Result: Maximal (Δ, γ) Clique Set $\mathcal{C}_T$ of $\mathcal{G}$.

1   Construct the Static Graph G;
2   Prepare the dictionary $\mathcal{D}$ from $\mathcal{C}_{\mathcal{L}}^{\mathcal{I}}$ $\mathcal{C}^{T_1} \leftarrow \mathcal{C}_{\mathcal{L}}^{\mathcal{I}}$;
3   ALL_MAXIMAL $= False$;
4   **while** $\neg$ $ALL\_MAXIMAL$ **do**
5      $\mathcal{C}^{T_2} \leftarrow \phi$;
6   **while** $\mathcal{C}^{T_1} \neq \phi$ **do**
7   Take and remove a clique $(\mathcal{X}, [t_a, t_b])$;
8   IS_MAX $= True$;
9   **for** $Every\ v \in \mathcal{N}_G(\mathcal{X}) \setminus \mathcal{X}$ **do**
10            $\mathcal{X}_{new} = \mathcal{X} \cup \{v\}$;
11   **if** $\mathcal{X}_{new} \in \mathcal{D}$ **then**
12   **if** $[t_a, t_b] \in \mathcal{D}[\mathcal{X}_{new}]$ **then**
13   IS_MAX $= False$;
14   **end**
15   **else**
16   **if** $\forall z \in \{C(\mathcal{X}_{new}, \mathcal{X})\}\ and\ z \in \mathcal{D}$ **then**
17                 $\mathcal{D}_{Temp} \leftarrow$ Get the entries from $\mathcal{D}$ for $C(\mathcal{X}_{new}, \mathcal{X})$;
18   **foreach** $permutation\ of\ \mathcal{D}_{Temp}\ entries\ as\ timeSet$ **do**
19                     $max\_t_a = [\ ]$;
20                     $min\_t_b = [\ ]$;
21   **for** $t \in timeSet$ **do**
22                         $max\_t_a.append(t[1])$;
23                         $min\_t_b.append(t[2])$;
24   **end**
25                     $t_a^{'} = MAX(max\_t_a)$;
26                     $t_b^{'} = MIN(min\_t_b)$;
27   **if** $t_b^{'} - t_a^{'} \geq \Delta$ **then**
28                       $\mathcal{C}^{T_2}.add(\mathcal{X}_{new}, [t_a^{'}, t_b^{'}])$;
29                       $\mathcal{D}[\mathcal{X}_{new}].append([t_a^{'}, t_b^{'}])$;
30   **if** $t_a^{'} = t_a \wedge t_b^{'} = t_b$ **then**
31   IS_MAX $= False$;
32   **end**
33   **end**
34   **end**
35   **end**
36   **end**
37   **end**
38   **if** IS_MAX **then**
39            $\mathcal{C}_T.append(\mathcal{X}, [t_a, t_b])$;
40   **end**
41   **end**
42   **if** $len(\mathcal{C}^{T_2}) > 0$ **then**
43         $\mathcal{C}^{T_1} \leftarrow \mathcal{C}^{T_2}$;
44   **else**
45   ALL_MAXIMAL $= True$;
46   **end**
47   **end**

Theorem 1. *All the maximal (Δ, γ)-Cliques of $\mathcal{G}$ are contained in $\mathcal{C}_T$.*

Lemma 10. *Running time and space requirement of Algorithm 2 is of $\mathcal{O}(n^4.2^{2n}.f_{max} + n^5.2^{2n} + 2^n.f_{max}^n.n^3)$ and $\mathcal{O}(m + |\mathcal{C}_T^I|.f_{max} + n.f_{max} + n.2^n)$, respectively.*

Theorem 2. *The computational time and space requirement of the proposed methodology is of $\mathcal{O}(n^4.2^{2n}.f_{max} + n^5.2^{2n} + 2^n.f_{max}^n.n^3 + \gamma.m)$ and $\mathcal{O}(m + |\mathcal{C}_T^I|.f_{max} + n.2^n + n^2.f_{max})$, respectively.*

4   Experimental Results

Datasets and Experimental Setup. We use the following three **datasets**: (i) College Message Temporal Network *(College Message)* [17], (ii) Bitcoin OTC Trust Weighted Signed Network *(Bitcoin)*[1] [13], and (iii) Infectious SocioPatterns Dynamic Contact Network *(Infectious)* [12]. All the datasets contain set of tuples of the form (t, u, v), where u and v are the anonymous ids of the person who are in contact at time t. The basic statistics of the datasets are given in Table 1. The only **parameters** involved in our study are Δ and γ. For analyzing temporal network, one intuitive question becomes to find out the frequently connected groups for a given time duration which depends on the network lifetime. Due to the page limit, the Δ values are directly described in the result table. To select γ, we consider two points, (i)duration Δ, (ii)the ratio of the no. of links with the no. of static edges. For smaller value of Δ, we start with the γ value as 1, keep on increasing it by 1 till the maximal clique set becomes empty. For a larger value of Δ, the increment is performed in the order of 10 to 30. For, Bitcoin due to small ratio of link to static edge, we report the result only for $\gamma = 1$.

Table 1. Basic statistics of the datasets

Datasets	#Nodes	#Links	#Static Edges	Lifetime
College Msg	1899	59835	20296	193 Days
Bitcoin	5881	35592	21492	5.21 Years
Infectious	10972	415843	44516	80 Days

Experimental Goal. The goals of the experiments are to analyze, how the number of maximal cliques, maximum cardinality, maximum duration, computational time and space change with Δ and γ. We compare the performance with: (i) **Virad et al.'s Method** [20]: This is the first method proposed to enumerate maximal Δ-Clique of a temporal network. (ii) **Himmel et al.'s Method** [9]: This method incorporates the famous Born-Kerbosch Algorithm to improve the Virad et al.'s Method. (iii) **Banerjee and Pal's Method** [3]: This is the existing maximal (Δ, γ)-Clique enumeration algorithm. We obtain the source code of the first two methodologies as implemented by the respective authors. The proposed methodology is developed in Python 3.4 along with NetworkX 2.0. All the experiments have been carried out on a CPU server with 40 cores and 160 GB of RAM. Implementations of the algorithms are available at https://github.com/BITHIKA1992/Delta-Gamma-Clique.

---

[1] https://snap.stanford.edu/data/soc-sign-bitcoin-otc.html.

Results and Discussion. First, we focus on Δ-Clique, which is equivalent to (Δ, γ)-Clique with $\gamma = 1$. This result is shown in Table 2. In all the datasets, the maximal clique count decreases with the increment of Δ. This quantity can increase for a large Δ if there exist some user pairs who contact very frequently for long duration. This generates many maximal cliques with cardinality 2 and different $[t_a, t_b]$, like bitcoin and Infectious. The maximum cardinality(C) identifies if there is a large group and the maximum duration(D) signifies maximum how long the users contacted among them. Both C and D increase with the growth in Δ. It is observed that the proposed methodology is the fastest one compared to the existing methods. The computational time increases with Δ in [20], as the algorithm starts with the clique (link) with duration $=1$, and expands in both right and left by Δ and creates more intermediate cliques. Whereas, the processing time depends on the maximal clique count in both [9] and the proposed method. The computational space is mainly dependent on the size of intermediate clique set, and it get penalized more in Virad et al.'s method due to the same reason as discussed. The effect can be seen in Infectious for $\Delta = 6000$ and 12000. System's memory becomes insufficient to compute for these two Δ values. Comparing with the Himmel et al.'s method the proposed method the trade-off between space and time can be observed for the dense dataset Infectious.

Now, the results for $\gamma > 1$ are shown in Fig. 4. As the maximal clique set becomes null for $\gamma > 2$ in Bitcoin, we only show the plots for College Message and Infectious datasets. For a fixed Δ, the maximal clique count decreases with the increase in γ, which in turn reduce the computational time and space as

Table 2. Results for the Maximal Clique Count (N), Maximum Caridinality (C), Maximum Duration (D), Computational Time (in Secs.) / Space (in MB) for Maximal Δ-clique ((Δ, γ)-clique with $\gamma = 1$) Enumeration for different datasets

Dataset	Δ	N	C	D	Algorithm		
					Virad et al. [20]	Himmel et al. [9]	Proposed
College Message	3600	33933	4	21761	35.25/372	41/140	19.84/148
	43200	25635	5	403018	43.02/546	31.38/133	4.19/142
	88640	22701	5	896134	52.29/727	28.56/131	2.28/140
	259200	21019	5	2322612	84.05/1281	27.61/128	1.41/136
	604800	21658	6	6334253	133.53/2427	25.85/128	1.19/139
Bitcoin	3600	26577	7	10791	18.31/142.97	196.49/221	2.36/157
	43200	26091	8	129422	19.58/142.71	193.74/235	2.41/158
	88640	25970	8	265798	20.69/142.57	190.93/250	2.3/159
	259200	26290	8	777572	22.6/142.73	191.49/288	2.36/160
	604800	27149	8	1814344	29.69/143.39	193.52/367	2.34/162
Infectious	60	161066	6	3760	274.13/2591	1025.01/286	80.75/357
	120	138662	7	5180	405.84/3915	998.53/266	46.88/354
	600	128392	10	11200	2659.82/18117	1043.46/262	30.87/523
	1200	139684	13	12400	9824.58/72628	1062.69/278	84.13/1017
	6000	152121	16	22740	NA	1238.75/293	108.34/3076
	12000	152198	16	34740	NA	1266.8/293	108.03/3097

Clique Count   Max. Cardinality   Max. Duration   Time   Space

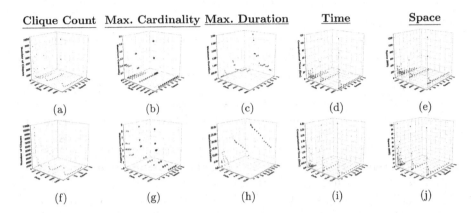

Fig. 4. Plots for the Clique Count, Maximum Caridinality, Maximum Duration, Computational time, Space with the change of Δ and γ; (a-e) for College Message, (f-j) for Infectious dataset; In computational time and space for the proposed algorithm is marked with blue and green respectively, and Banerjee et al.'s Method [3] is marked with red for the same (Color figure online)

well. Maximum cardinality and the duration also reduces with the increment in γ. For fixed γ, the same observation of $\gamma = 1$ is found. While comparing with the only existing method [3], it can be observed that the improvement is more significant for large value of Δ and the small value of γ (Refer Fig. 4 [d,e,i,j]). Lastly, we can conclude for both Δ and (Δ, γ)-Clique enumeration, the proposed methodology is better if the input dataset is sparse.

5   Conclusion

In this paper, we have studied the problem of enumerating maximal (Δ, γ)-Cliques and proposed a two-phase approach to solve this problem. We have shown that the proposed approach is correct. Experimental evaluation with real-world datasets shows that the proposed approach can be used effectively to study the contact patterns of temporal network datasets. Now, the proposed approach can be extended to study the temporal networks with probabilistic links.

References

1. Akkoyunlu, E.A.: The enumeration of maximal cliques of large graphs. SIAM J. Comput. **2**(1), 1–6 (1973)
2. Al-Naymat, G., Chawla, S., Arunasalam, B.: Enumeration of maximal clique for mining spatial co-location patterns. In: 2008 IEEE/ACS International Conference on Computer Systems and Applications (2007)
3. Banerjee, S., Pal, B.: On the enumeration of maximal (Δ, γ)-cliques of a temporal network. In: Proceedings of the ACM India Joint International Conference on Data Science and Management of Data (COMAD/CODS 2019), Kolkata, India, January 3–5 2019, pp. 112–120 (2019)

4. Bhowmick, S.S., Seah, B.S.: Clustering and summarizing protein-protein interaction networks: a survey. IEEE Trans. Knowl. Data Eng. **28**(3), 638–658 (2015)
5. Bron, C., Kerbosch, J.: Finding all cliques of an undirected graph (algorithm 457). Commun. ACM **16**(9), 575–576 (1973)
6. Eppstein, D., Löffler, M., Strash, D.: Listing all maximal cliques in large sparse real-world graphs. J. Exp, Algorith. (JEA) **18**, 1–3 (2013)
7. Garey, M.R., Johnson, D.S.: Computers and Intractability, vol. 29. WH freeman, New York (2002)
8. Himmel, A.S., Molter, H., Niedermeier, R., Sorge, M.: Enumerating maximal cliques in temporal graphs. In: IEEE/ACM International Conference on Advances in Social Networks Analysis and Mining (ASONAM 2016), pp. 337–344. IEEE (2016)
9. Himmel, A.S., Molter, H., Niedermeier, R., Sorge, M.: Adapting the bron-kerbosch algorithm for enumerating maximal cliques in temporal graphs. Soc. Netw. Anal. Mining **7**(1), 35 (2017)
10. Holme, P., Saramäki, J.: Temporal networks. Phys. Rep. **519**(3), 97–125 (2012)
11. Hulovatyy, Y., Chen, H., Milenković, T.: Exploring the structure and function of temporal networks with dynamic graphlets. Bioinformatics **31**(12), i171–i180 (2015)
12. Isella, L., Stehlé, J., Barrat, A., Cattuto, C., Pinton, J.F., Van den Broeck, W.: What's in a crowd? analysis of face-to-face behavioral networks. J. Theor, Biol. **271**(1), 166–180 (2011)
13. Kumar, S., Hooi, B., Makhija, D., Kumar, M., Faloutsos, C., Subrahmanian, V.: Rev2: Fraudulent user prediction in rating platforms. In: Proceedings of the Eleventh ACM International Conference on Web Search and Data Mining, pp. 333–341. ACM (2018)
14. Masuda, N., Holme, P.: Temporal Network Epidemiology. Springer, Singapore (2017)
15. Molter, H., Niedermeier, R., Renken, M.: Enumerating isolated cliques in temporal networks. In: International Conference on Complex Networks and Their Applications, pp. 519–531. Springer (2019)
16. Mukherjee, A.P., Xu, P., Tirthapura, S.: Enumeration of maximal cliques from an uncertain graph. IEEE Trans. Knowl. Data Eng. **29**(3), 543–555 (2016)
17. Panzarasa, P., Opsahl, T., Carley, K.M.: Patterns and dynamics of users' behavior and interaction: Network analysis of an online community. J. Assoc. Inf, Sci, Technol. **60**(5), 911–932 (2009)
18. Rossi, R.A., Gleich, D.F., Gebremedhin, A.H., Patwary, M.M.A.: Fast maximum clique algorithms for large graphs. In: Proceedings of the 23rd International Conference on World Wide Web, pp. 365–366. ACM (2014)
19. Viard, J., Latapy, M., Magnien, C.: Revealing contact patterns among high-school students using maximal cliques in link streams. In: Proceedings of the 2015 IEEE/ACM International Conference on Advances in Social Networks Analysis and Mining 2015, pp. 1517–1522. ACM (2015)
20. Viard, T., Latapy, M., Magnien, C.: Computing maximal cliques in link streams. Theor. Comput. Sci. **609**, 245–252 (2016)

Author Index

Printed in the United States
by Baker & Taylor Publisher Services